MATERIALS SCIENCE AND TECHNOLOGIES

COMPREHENSIVE GUIDE FOR MESOPOROUS MATERIALS

VOLUME 3

PROPERTIES AND DEVELOPMENT

MATERIALS SCIENCE AND TECHNOLOGIES

Additional books in this series can be found on Nova's website
under the Series tab.

Additional e-books in this series can be found on Nova's website
under the e-book tab.

MATERIALS SCIENCE AND TECHNOLOGIES

COMPREHENSIVE GUIDE FOR MESOPOROUS MATERIALS

VOLUME 3

PROPERTIES AND DEVELOPMENT

MAHMOOD ALIOFKHAZRAEI
EDITOR

New York

Copyright © 2015 by Nova Science Publishers, Inc.

All rights reserved. No part of this book may be reproduced, stored in a retrieval system or transmitted in any form or by any means: electronic, electrostatic, magnetic, tape, mechanical photocopying, recording or otherwise without the written permission of the Publisher.

We have partnered with Copyright Clearance Center to make it easy for you to obtain permissions to reuse content from this publication. Simply navigate to this publication's page on Nova's website and locate the "Get Permission" button below the title description. This button is linked directly to the title's permission page on copyright.com. Alternatively, you can visit copyright.com and search by title, ISBN, or ISSN.

For further questions about using the service on copyright.com, please contact:
Copyright Clearance Center
Phone: +1-(978) 750-8400   Fax: +1-(978) 750-4470   E-mail: info@copyright.com.

NOTICE TO THE READER

The Publisher has taken reasonable care in the preparation of this book, but makes no expressed or implied warranty of any kind and assumes no responsibility for any errors or omissions. No liability is assumed for incidental or consequential damages in connection with or arising out of information contained in this book. The Publisher shall not be liable for any special, consequential, or exemplary damages resulting, in whole or in part, from the readers' use of, or reliance upon, this material. Any parts of this book based on government reports are so indicated and copyright is claimed for those parts to the extent applicable to compilations of such works.

Independent verification should be sought for any data, advice or recommendations contained in this book. In addition, no responsibility is assumed by the publisher for any injury and/or damage to persons or property arising from any methods, products, instructions, ideas or otherwise contained in this publication.

This publication is designed to provide accurate and authoritative information with regard to the subject matter covered herein. It is sold with the clear understanding that the Publisher is not engaged in rendering legal or any other professional services. If legal or any other expert assistance is required, the services of a competent person should be sought. FROM A DECLARATION OF PARTICIPANTS JOINTLY ADOPTED BY A COMMITTEE OF THE AMERICAN BAR ASSOCIATION AND A COMMITTEE OF PUBLISHERS.

Additional color graphics may be available in the e-book version of this book.

Library of Congress Cataloging-in-Publication Data

ISBN: 978-1-63463-318-5

Published by Nova Science Publishers, Inc. † New York

CONTENTS

Preface		vii
Chapter 1	Ordered Mesoporous Silicates MCM-41 and SBA-15 As Matrices for Improving Dissolution Rate of Poorly Water Soluble Drugs *Valeria Ambrogi*	1
Chapter 2	Mesoporous Bioactive Glasses: Implants and Drug Delivery Systems for Bone Regenerative Therapies *Daniel Arcos and MaríaVallet-Regí*	19
Chapter 3	Synthesis of a Mesoporous Silica *Kulamani Parida, Suresh Kumar Dash and Dharitri Rath*	43
Chapter 4	Mesoporous Acid Catalysts for Renewable Raw-Material Conversion into Chemicals and Fuel *M. Caiado, J. Farinha and J. E. Castanheiro*	67
Chapter 5	Mesoporous TiO_2 as Efficient Photocatalysts and Solar Cells *Adel A. Ismail*	85
Chapter 6	Porous Si Structures for Gas, Vapor and Liquid Sensing *V. A. Skryshevsky*	123
Chapter 7	Hierarchical Zeolites: Preparation, Properties and Catalytic Applications *Ana P. Carvalho, Nelson Nunes and Angela Martins*	147
Chapter 8	Novel Mesoporous Materials for Electrochemical Energy Storage *J. Santos-Peña, J. Ortiz-Bustos, S. G. Real, A. Benítez de la Torre, C. Medel, J. Morales, M. Cruz and R. Trócoli*	213
Chapter 9	Optical Properties of Mesoporous Silicon/Silicon Oxide and the Light Propagation in these Materials *Joël Charrier and Parastesh Pirasteh*	241

Chapter 10	Behavior of Unmodified MCM-41 Towards Controlled Release of Pro-Drug Molecule, Cysteine *Anjali Patel and Varsha Brahmkhatri*	**271**
Chapter 11	Mesoporous Multivalent Transition Metal Oxides (V, Cr, Mn, Fe, and Co) in Catalysis *Altug S. Poyraz, Sourav Biswas, Eugene Kim, Yongtao Meng and Steven L. Suib*	**285**
Chapter 12	Nanocomposites Embedded by Mesoporous Materials *Chunfang Du, Yiguo Su and Zhiliang Liu*	**315**
Chapter 13	On the Conception and Assessment of Mesopore Networks: Development of Computer Algorithms *Fernando Rojas-González, Graciela Román-Alonso, Salomón Cordero-Sánchez, Miguel Alfonso Castro-García, Manuel Aguilar-Cornejo and Jorge Matadamas Hernández*	**343**
About the Editor		**371**
Index		**373**

PREFACE

This is the third volume of the four volume set of Comprehensive Guide for Mesoporous Materials, mainly discussing different properties and the development of mesoporous materials. Many applications for these materials have been recognized. Ultra precision methods also developed for commercialization of these materials. Like many nanostructured materials, mesoporous materials are also found in abundance in nature, such as in cell walls that are made of mesoporous membranes although they are much more complicated. During past decades in the oil industry, natural mesoporous materials known as zeolites are widely used. However, most of them are now produced artificially.

Chapter 1 - Ordered mesoporous silicates have been recently proposed as drug carriers because of their peculiar characteristics which enable them to host drugs inside their ordered mesopores and then to release them. Many different applications have been proposed such as drug controlled release, targeted delivery and poorly water-soluble drug dissolution rate improvement. Poor solubility of a drug is one of the most serious problems in the development of a pharmaceutical product and ordered mesoporous silicates have been proposed as agents to decrease/solve this problem. Among them, MCM-41 and SBA-15 are the most studied. This chapter reviews the research work in this area and presents results on different drug loading procedures, the influence of pore size and drug physical stability. A comparison with other porous inorganic matrices is discussed as well.

Chapter 2 - The development of bioceramics designed for bone tissue regeneration has experienced an important boost with the appearance of a new generation of materials: mesoporous bioactive glasses (MBG). MBGs are commonly prepared in the systems SiO_2-CaO-P_2O_5 or SiO_2-CaO, similarly to bioglasses prepared by conventional sol-gel processes. The incorporation of a structure directing agent to the synthesis results in sol-gel bioglasses with highly ordered mesoporous structures and much higher surface area. The consequence is that MBGs exhibit much higher solubility, better bioactive behavior and the capability of hosting and releasing drugs aimed to improve the regenerative healing process. In this chapter, the preparation methods, drug delivery applications and the development of scaffolds for in situ bone regeneration are reviewed and commented.

Chapter 3 - The substitution of various main group metals such as Al, Ga, In, Zn and Cd, transition metals such as Ti, V, Fe, Cu, Nb, Mo and Zr and rare earth elements like La and Ce into the frame-work of MCM-41 has received much attention to develop more efficient and stable materials for application in adsorption, catalysis, photocatalysis, separation and chemical sensing. The metal ions can be incorporated via ion-exchange, impregnation, co-

condensation and solid state vaporization. The incorporation of a metal into the frame-work of MCM-41 seems to be easy but the retention of mesoscopic order is often difficult and the stability depends on the amount of dopant material. In case of tetravalent metals viz. Ti, Zr, Ge and Sn the grafting could be done by isomorphous substitution of tetravalent Si, into the walls of MCM-41 during in-situ or Liquid Crystal Templating (LCT) mechanism where the charge on the walls is balanced and neutral. But the incorporation of trivalent metals (Al, B, Ga and Fe) introduces a negative charge and generates a cation exchange or Bronsted acid site on the surface of MCM-41. MCM-41 have been used to custom synthesize catalysts because of the controllable properties, such as pore size, active phase incorporation, crystal size, and morphology, among others. The materials prepared by hydrothermal methods provides a very high surface area and narrow uniform pore distribution in the mesopore region, and are highly thermally stable whereby the ions are not leached out after high temperature annealing, catalyst in electrochemical devices. The incorporation of platinum, ruthenium, and palladium onto Al-MCM-41 mesoporous silica by direct inclusion of various precursors indicate that the Al-MCM-41 mesoporous-ordered structure was not affected by metallic particle incorporation. Metallic nanoparticles dispersion on Al-MCM-41 was homogeneous, platinum and palladium samples have round shape particles and ruthenium sample exhibit a rod shape. In addition to this, the bimetal modified MCM-41 materials are equally important in various redox reactions. The incorporation of two different metals might create materials with different or new redox and acid properties. Supported bimetallic catalysts are very interesting materials in general terms since one metal can fine tune or modify the structural and electronic properties of the other. The bimetal systems like Cu/Zn, Cu/Ni, Co/(V, Nb, La), Ru/(Cr, Ni, Cu) modified MCM-41 can be used in some industrially important reactions such as oxidation, hydrogenation, hydro dehalogenation, H_2 production etc.

Chapter 4 - Mesoporous catalysts, such as silica (PMOs, MCM-41, SBA-15) and activated carbons, have been used in heterogeneous catalysis, due to a combination of high surface areas and controlled pore sizes. These mesoporous materials have been used as catalyst in a wide range of chemical reactions. Due to environmental pressure and a decrease in fossil fuel sources, alternative fuel sources, such as biomass or renewable feedstock sources, have become increasingly popular. Traditionally, the biomass conversion is carried out over homogeneous catalysts. However, homogeneous catalysts have some disadvantages, such as difficulty in separations and the production of toxic waste. Solid catalysts can replace the homogeneous ones in order to make the processes simpler and more environmentally benign. Heterogeneous catalysts have been used in different reactions biomass conversion. In this work, the use of mesoporous acid catalysts (silica and activated carbons) for the biomass conversion into chemicals and fuel will be reviewed.

Chapter 5 - Structuring of titania on a nanometer scale has attracted considerable attention in the past few years, since nanostructured titania shows outstanding properties and has widespread application potentials: e.g., in photovoltaics, photocatalysis, and gas sensing. Increasing attention has recently been focused on the simultaneous achievement of high bulk crystallinity and the formation of ordered mesoporous TiO_2 frameworks with high thermal stability. Mesoporous TiO_2 have continued to be highly active in photovoltaics and photocatalytic applications because it is beneficial for promoting the diffusion of reactants and products, enhancing photovoltaics and photocatalytic efficiency by facilitating access to the reactive sites on the surface of mesoporous TiO_2 films. This steady progress has demonstrated that mesoporous TiO_2 nanoparticles are playing and will continue to achieve an important role in the protections of the environment and in developing progress of dye-sensitized solar cells (DSSC). This chapter focuses on the preparation and characterisation of mesoporous titania as efficient photocatalysts and solar cells.

Chapter 6 - In the present chapter, the authors pay attention both to technical and physico-chemical aspects of porous silicon (PS) sensors operation. The transducers and sensors of resistivity, capacitance, Schottky barrier, MIS, FET, EIS, ISFET, LAPS and optical sensors with PS are considered. The set of gas sensor`s parameters for detection of humidity, CO, NO_2, different volatile organic compounds, alcohols, H_2S, H_2 and other gases is described.

Chapter 7 - Zeolites are crystalline aluminosilicates widely used in catalysis as well as in separation and purification fields. The unique combination of properties, such as, high surface area, well-defined microporosity, high thermal stability and intrinsic acidity underlie the successful performance of these materials for a great number of applications. However, when bulky reaction intermediate species are involved the purely microporous character of zeolites is a drawback, because it often imposes diffusion limitations due to restricted access to the active sites. This is the case of some applications in the petroleum, petrochemical and fine chemical industries.

This chapter aims to provide a comprehensive examination of the different approaches that can be adopted to enhance the accessibility and molecular transport in the zeolite framework, illustrated with examples from the literature.

Dealumination and, more recently, desilication are post-synthesis treatments that have been extensively used over numerous structures with academic and industrial interest. These treatments lead to Si/Al ratio changes and, simultaneously, hierarchical structures combining micro and mesopores are commonly obtained. As a consequence of the treatments, alterations of some important properties, such as, hydrothermal resistance and acidity are also observed. The catalytic performance is enhanced and generally the structures become less sensitive to deactivation due to coke deposition. There are also several examples pointing out the great advantage of consecutive treatments (e.g., dealumination by acid treatment followed by desilication, or vice-versa) to tune the samples catalytic properties.

In recent years, template based preparation strategies have been increasingly explored to obtain hierarchical structures with two or even three levels of porosity. This will result in a great increase in the accessibility and mass transport of the guest molecules towards the active sites within the zeolite framework.

In spite of the well documented benefits of hierarchical network for catalytic applications, the main industrial application is the steamed dealuminated Y zeolite in fluid catalytic cracking. The production of hierarchical porous materials, through the several methods before mentioned, are still at an exploratory phase, although some attempts to scale-up and supply them to industrial units were recently reported, which can be the start to a new era of hierarchical zeolites in industrial catalysis.

Chapter 8 - Since the discovery of the environmental issues associated to the use of fosil-fuel based energy, special attention has been paid to the development of electrochemical energy storage devices.In this context, lithium ion batteries and supercapacitors have been proposed for satisfying the future energy demands. However, the current state of the art for both devices indicates that advanced materials are required to a truly implantation of both technologies for stationary and mobile applications. Mesoporous materials can play an important role in these advances. They mainly show unique textural and morphological properties that can facilitate the electrochemical reactions involved in a lithium ion battery or can improve the necessary electrolyte/electrode surface interaction in a supercapacitor. After an introduction exposing the different mechanism of energy storage for lithium ion batteries and supercapacitors, the authors will focus on how the interesting properties of the mesoporous materials can influence the performance of themost common electrode materials. Finally, the authors will show their more recent research in the field, namely mesoporous LiFePO$_4$ and template mesoporous carbons for lithium ion battery and supercapacitor electrodes, respectively.

Chapter 9 - Mesoporous silicon owns a large range of refractive indices and is a material which offers many advantages for integrated photonic circuits. The light propagation will be presented for waveguides manufactured from porous silicon or oxidized porous silicon layers. Optical properties of these materials will also be discussed as a function of the oxidation degree, different functionalization steps of these porous layers. Some different types of waveguides will be described. Surface and volume scattering losses of these mesoporous materials will be modeled and discussed in order to determine the principal contributions to optical losses.

Chapter 10 - Mesoporous silica material, MCM-41 was synthesized and characterized by various physico-chemical techniques such as FT-IR, X-ray diffraction, N$_2$ adsorption-desorption, SEM and TEM. Cysteine was selected as a pro-drug molecule and the potential of MCM-41 as a drug delivery system was explored. The study on release profile was also carried out for cysteine loaded sample in stimulated body fluid (SBF) at room temperature. It was found that MCM-41 was able to release cysteine in a controlled manner and exhibit better effect than N-acetylcysteine in solution. 100% Cysteine release was observed in 12h.

Chapter 11 - Multivalent first row transition metal oxides (V, Cr, Mn, Fe, and Co) can form various oxide structures with unique catalytic properties. The catalytic performance of a mesoporous, high surface area TM oxide is known to be better than its nonporous counterpart. However, the direct synthesis of multivalent mesoporous TM oxide materials with desired crystal structure and structural properties is still a challenge to date. The multivalent nature, lack of proper sources, weak inorganic-organic interactions, and poor control of reaction rates are problems for the direct synthesis of mesoporous multivalent TM oxides. The authors summarize here some recent results of synthesized mesoporous hybrid materials with these TMs and their catalytic performances.

Chapter 12 - "Host-guest chemistry" is a potential subject, attracting a great deal of attention from various research areas, including large-molecule catalysis, separation techniques, adsorption procedures and drug delivery. Mesoporous materials with huge specific surface area, large pore volume and uniform pore size distribution could act as a kind of "micro-reactor" for synthesis of functional nanoparticles. The interaction of host species-mesoporous materials and guest nanoparitcles as well as the restriction effect of framework of host species endow the nanocomposites embedded by mesoporous materials unique and enhanced properties that facilitate future applications. Undoubtedly, the combination of various functional species and mesoporous matrices would open a realm of new possibilities for exploring of novel materials with unexpected functions.

Chapter 13 - There exist a great variety of phenomena taking place in porous solids that are strongly affected by the morphological and topological characteristics of these media, among these processes the authors can mention: (a) the immiscible displacement of a given fluid by another, (b) imbibition and drying processes, (c) separation of fluid mixtures, (d) heterogeneous catalysis, and (e) catalytic deactivation, etc. The characterization of mesoporous and macroporous materials, especially the issue regarding the determination of the pore size distribution of these substrates from experimental data, is a subject of great practical importance that involves the development of both theoretical and experimental methods. Some of the experimental techniques, as for instance NMR, SAXS, and SANS, require sophisticated instruments while some others such as Hg porosimetry and sorption of vapors require simple devices that are available to many laboratories. In order to understand the textural results provided by these methods, a crucial issue consists in the development of a theory that can appraise the topological properties of these media.

The present chapter thoroughly describes the computational techniques developed as well as the preliminary results that have been obtained for the simulation of porous networks subjected to geometrical restrictions. Also, the structure characterization of the constructed pore networks is updated within the framework of the fractal and percolation theory, something missing in previous publications. The chapter is organized as follows. Section 2 presents the theoretical background of the DSBM approach. Section 3 describes the incorporation of geometrical restrictions into the DSBM. Section 4 covers the topological characterization of pore networks. Section 5 accounts for the algorithms for the *in silico* construction of pore networks. Section 6 describes the parallel computational techniques for implementing the previous algorithms. Section 7 displays the simulated pore networks and the results obtained therefrom. Finally, in Section 8 the perspectives and the conclusion of this chapter are stated.

Chapter 1

ORDERED MESOPOROUS SILICATES MCM-41 AND SBA-15 AS MATRICES FOR IMPROVING DISSOLUTION RATE OF POORLY WATER SOLUBLE DRUGS

Valeria Ambrogi[*]

Dipartimento di Scienze Farmaceutiche,
Università degli Studi di Perugia, Perugia, Italy

ABSTRACT

Ordered mesoporous silicates have been recently proposed as drug carriers because of their peculiar characteristics which enable them to host drugs inside their ordered mesopores and then to release them. Many different applications have been proposed such as drug controlled release, targeted delivery and poorly water-soluble drug dissolution rate improvement. Poor solubility of a drug is one of the most serious problems in the development of a pharmaceutical product and ordered mesoporous silicates have been proposed as agents to decrease/solve this problem. Among them, MCM-41 and SBA-15 are the most studied. This chapter reviews the research work in this area and presents results on different drug loading procedures, the influence of pore size and drug physical stability. A comparison with other porous inorganic matrices is discussed as well.

---

[*] Tel/fax +39 0755855125, e-mail: valeria.ambrogi@unipg.it.

INTRODUCTION

Main statements	References
Bioavailability problems connected to poorly soluble drugs	[1], [2], [5]
Biopharmaceutics Classification System (BCS)	[3], [4]
Main features of mesoporous silicates MCM-41 and SBA-15	[6], [7], [8]
MCM-41 and SBA-15 as drug carriers for controlled drug release	[8], [9]
Mesoporous silicates as matrices for improving dissolution rate	[10], [11], [12]

Recently the ordered mesoporous silicates have been extensively studied for their application in pharmaceutical field and in particular as agents to improve dissolution rate of poorly water soluble drugs.

The poor water solubility is one of the most frequent cause of a new entity development failure and of reduction of marketed drug performance [1, 2]. In fact more than one third of the drugs listed in the US Pharmacopoeia and half of the new chemical entities or new active ingredients are poorly water soluble or insoluble. These negative characteristics obstacle drug development starting from the first steps such as in vitro studies which always required drug solubilisation, and then in clinical trials. Also when a drug is approved for market commercialization, poorly water solubility is a parameter that influences drug effectiveness and safety. In fact the first step for a solid dosage form administered through oral route is the drug dissolution and successively its absorption through the gastrointestinal mucosa until it reaches systemic circulation. Thus drug solubility and mucosa permeability are two relevant characteristics which determinate the drug bioavailability and consequently its efficacy and safety. A drug with poor water solubility often has an erratic or incomplete absorption, which can be strongly affected also by the presence of food, variability of gastric pH and gastrointestinal transit rate. In order to obtain the required efficacy it may be necessary to increase the dose but this often causes increased side-effects.

The importance of drug solubility is proved by the Biopharmaceutics Classification System (BCS) proposed by Amidon in 1995 [3] and provided by FDA [4] which classified drugs in four different classes according to their water solubility and permeability. BCS classified drug substances in:

Class I: high permeability, high solubility
Class II: high permeability, low solubility
Class III: low permeability, high solubility
Class IV: low permeability, low solubility

All poorly water soluble drugs are classified as BCS II or IV class. Whereas class I drugs do not present problems of formulation and bioavailability, class II and IV drugs present a big challenge for the formulator and are good candidates for reformulation using the new different technologies available for solving solubility problems. These new strategies offer both the opportunity to rescue promising developmental compounds which were abandoned because of solubility problems, and to improve marketed drug performance. In particular for well absorbed drugs (class II), the step limiting the bioavailability is the drug dissolution rate which according to the Noyes-Whitney equation is proportional to drug solubility. Thus poorly soluble drugs have a slow dissolution rate with the consequence of low absorption and low bioavailability and an increase of their dissolution rate could improve their performance [5].

Recently among the range of technologies currently available for improving dissolution rate of poorly water soluble chemical compounds, the use of mesoporous ordered silicates has been proposed.

The mesoporous silicates are inorganic materials synthesized in the presence of surfactants which act as templates for polycondensation of silicic species, a source of silica, a solvent and a catalyst. [6]. When a cationic surfactant, such as alkylamonium salt, is present as a template, the M41S materials are obtained [7]. The most famous representative of this group is MCM-41 which has a honeycomb structure that is the result of hexagonal packing of unidimensional hexagonal mesopores with pore size and wall thickness do not go beyond 4.0 and 2.0 nm respectively. When a non ionic surfactant such as Pluronic P123, a triblock copolymer of poly(ethylene glycol)-poly(propylene glycol)-poly(ethylene glycol), in acidic medium is used, materials called SBA (Santa Barbara University) are obtained. The most representative SBA material is SBA-15. It is characterized by hexagonal, unidirectional pores such as MCM-41, with thicken pore walls, larger pore size and interconnected micropores. Whereas SBA-15 material has a pore diameter of 6-8 nm and pore volume > 1.0 cm^3/g, MCM-41 has a pore diameter of 2-3 nm and a pore volume > 0.7 cm^3/g. Both materials have high surface area (more than 900 m^2) and the presence of superficial silanol groups (single, hydrogen bonded and geminal) which can interact with molecules that can be hosted inside the pores (Figure 1).

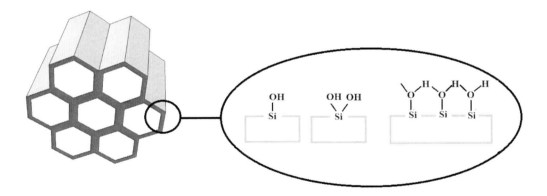

Figure 1. Schematic representation of MCM-41 and of the different kinds of superficial silanol groups (single, geminal and hydrogen bonded).

Originally the focus of mesoporous silica materials was on the development of slow release formulations after their compression in disks with the aim of obtaining bioactive biomaterials for drug controlled release [8, 9]. Then the attention of researchers was turned towards the possible use of these materials for obtaining fast drug release. First Charnay [10] evaluated the release property of MCM-41 as a powder and ibuprofen was selected as a model molecule since it is a well documented and used anti-inflammatory drug. Furthermore, it has a lipophilic character and its molecular size is suitable for inclusion within the mesopores of MCM-41. After this paper publication, the effects of mesoporous silicate materials such as MCM-41 and SBA-15 on dissolution rate of poorly water soluble drugs has been largely evaluated.

Ordered mesoporous silicates such as MCM-41 and SBA-15 result proper materials for improving drug dissolution rate because (i) their high surface area allows a wide contact between solid particles and biological fluids and this favors dissolution rate according to Noyes-Whitney equation, (ii) the unidirectional and size uniform pores shun any tortuosity and narrowing that could slow the diffusion of the adsorbed drug, (iii) the light interactions between silicate silanols and the adsorbed molecules break easily in the presence of water with consequent rapid guest release in molecular form, (iv) once adsorbed, the drug molecules are not organized in crystalline form as they are confined in the pore space which is only a few times larger than the drug molecule and thus the physicochemical properties differ dramatically from those of the corresponding bulk material; the lack of an ordered form facilitates the drug dissolution especially when there is a high lattice energy, v) the improvement of wettability properties favors the contact between the fluid and the powder surface [11,12].

Figure 2 illustrates the release of adsorbed drug molecules from MCM-41 or SBA-15 in the presence of water.

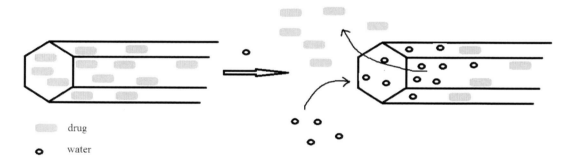

Figure 2. Schematic representation of the release of loaded drug molecules from the mesoporous silicates (MCM-41 or SBA-15) in the presence of water.

The drug loading procedure can influence the dissolution rate as it plays an important role in determining the physical state of the drug inside the pores. The most common drug loading procedures in mesoporous silicates are achieved by: (i) adsorption from an almost saturated drug solution in an organic solvent, usually ethanol but also others such as hexane, dichloromethane and dimethylsulfoxide. After equilibration the loaded silicate is recovered by filtration [10, 11, 13]; (ii) the melt method in which the physical blend of drug and matrix is heated until drug melts and enters into the pores [14, 15]; (ii) the incipient wetness impregnation method in which a controlled amount of concentrated drug solution, generally in an organic solvent, is added to the inorganic particles, giving a moist powder, and allowed to diffuse by capillarity into the pores, followed by solvent evaporation [10, 16]; (iv) the solvent evaporation method in which the silicate is dispersed in a drug solution (e.g., ethanol, dichloromethane, etc.), sometimes sonicated and finally dried, upon solvent removal, by evaporation [14, 17]; (v) co-spray drying [18]; and (vi) near-critical (liquid) [19] and supercritical [20, 21] CO_2 methods which avoid the use of organic solvents and the steps for their elimination.

Influence of drug loading procedure on drug dissolution rate

Most common drug loading procedures	Drugs	References
Adsorption and filtration	Ibuprofen Carbamazepine Piroxicam Methotrexate	[10] [11] [13] [22]
Melting	Itraconazole, ibuprofen Econazole	[14] [15]
Incipient wetness impregnation	Ibuprofen, Carbamazepine, indomethacin, ketonazole, griseofulvin, diazepam, phenylbutazone, cinarrizine, fenofibrate, danazol	[10] [16]
Solvent evaporation	Itraconazole, ibuprofen Furosemide Carbamazepine	[14] [17] [29]
Co-spray drying	Ibuprofen	[18]
Near-critical (liquid) and supercritical CO_2	Ibuprofen Fenofibrate Carbamazepine	[19], [20], [21], [26]

In Figure 3 the chemical structures of some poorly soluble drugs loaded into MCM-41 and SBA-15 mesopores are reported.

Melaerts [14] evaluated the influence of the loading procedure on drug dissolution rate. Ibuprofen and itraconazole, two drugs belonging to Class II of the BCS, were loaded into SBA-15 using three different loading procedures, such as solvent evaporation, incipient wetness impregnation and melting method. The incipient wetness impregnation resulted the most efficient procedure for dispersing itraconazole in SBA-15, probably due to the high solubility of itraconazole in the employed solvent (dichloromethane) and to the low viscosity of the solvent itself. Itraconazole molecules were located both on the mesopore walls and inside the micropores of the mesopore walls. When the solvent evaporation procedure was used, because of the fast solvent evaporation, the itraconazole molecules ended up in the mesopores that they plugged locally. At last the melting procedure gave glassy itraconazole particles deposited externally on the SBA-15 particles, maybe due to the high viscosity of the molten itraconazole (Figure 4).

These three different locations influenced the drug release. In fact, even if for all samples the release of itraconazole occurred fast and after 5 min was almost complete, itraconazole loaded SBA-15 samples obtained by solvent evaporation and incipient wetness impregnation methods exhibited faster release kinetics than the corresponding melt inclusion compound. This means that a better dispersion of drug molecules gives a faster release.

Ibuprofen was the second drug loaded by these three different procedures. The drug location into the mesopores was the same in the samples obtained by the different methods and no differences could be detected among the three different samples. In this case the melt method too resulted successful because the very low viscosity of molten ibuprofen enabled the drug diffusion into the pores and its dispersion over the SBA-15 surfaces. Thus the effectiveness of the loading method was found to be strongly compound dependent (Figure 4).

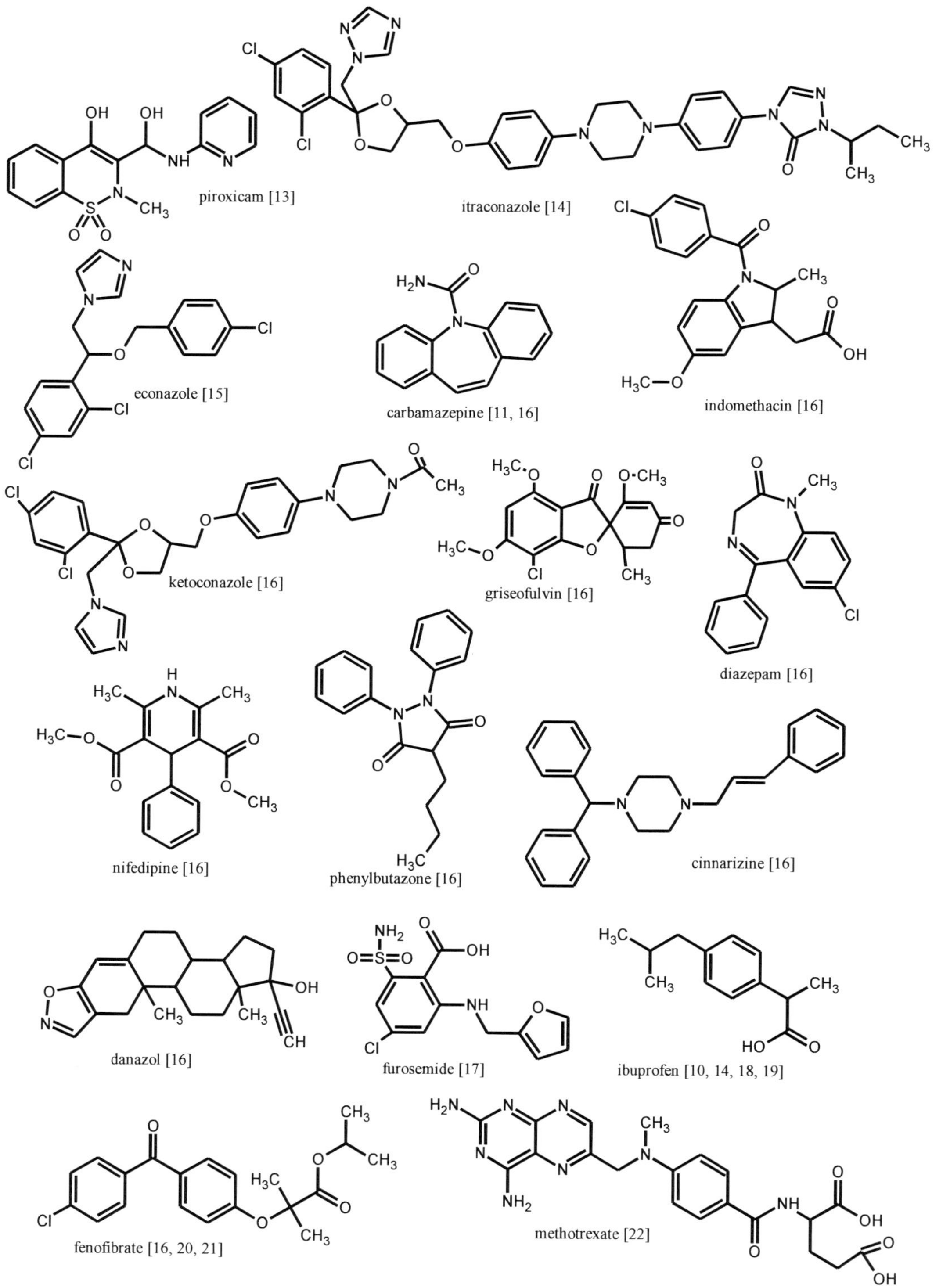

Figure 3. Structures of compounds used in studies where MCM-41 and SBA-15 were employed to increase dissolution rate.

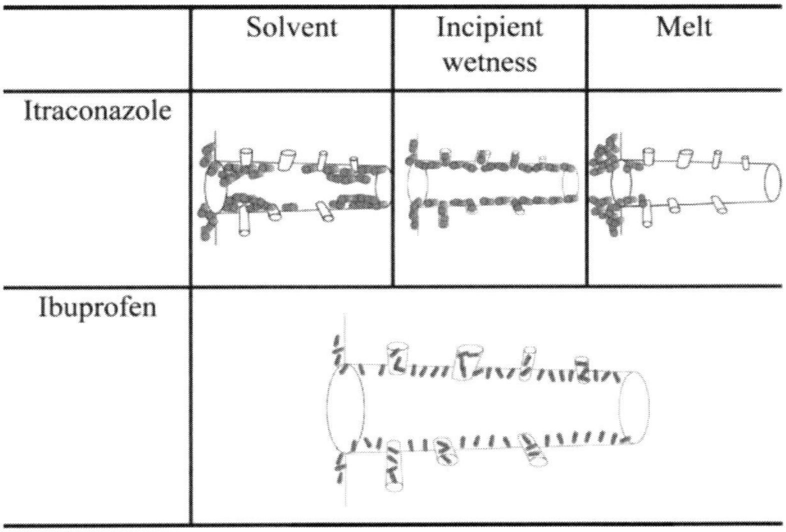

	Solvent	Incipient wetness	Melt
Itraconazole			
Ibuprofen			

Figure 4. Physical state of itraconazole and ibuprofen upon association with SBA-15 loaded according to the solvent method, incipient wetness method and melt method "Reprinted with permission from [14] Copyright 2008 American Chemical Society".

The release of ibuprofen from SBA-15 in simulated gastric fluid was very fast and more than 80% of the drug was released within 5 min. The sample prepared with the solvent method gave the fastest release. This confirms that a good drug molecule dispersion is the major factor influencing the drug release.

Carbamazepine is another BCS class II drug largely studied for evaluating the effects of the mesoporous silicates on drug dissolution. The peculiarity of this drug is not only its low solubility but also its high dose (200 mg pro dose) with the consequence of a slow and irregular absorption and delay of the plasmatic peak (4-8h) [11]. Polymorphism is an additional characteristic of carbamazepine. In fact it exists in different polymorphic and solvate forms [23, 24] which are characterized by different thermodynamic stability and different solubility. Thermodynamically polymorphic and amorphous metastable forms show improved solubility and faster dissolution rate and thus they could be useful tools for solving drug solubility problems, but unfortunately these forms with better solubility cannot find pharmaceutical applications as polymorphic conversion occurs both during manufacturing process and storage in the presence of humidity. Besides carbamazepine amorphous form is physically unstable too and transforms in other polymorphic forms depending on thermal and mechanical preparative techniques and storage conditions [25].

Carbamazepine loading into MCM-41 was performed by different authors following different methods such as adsorption/equilibration of an ethanol/dichloromethane (7/3 v/v) mixture [11] and a supercritical carbon dioxide process combined with various organic solvents such as ethanol, acetone and methanol [26]. The percentage of loaded drug was almost 15% when the first method was used and 27–33% in descending order EtOH>Acetone>MeOH in the presence of supercritical CO_2.

Whereas the adsorption/equilibration method afforded carbamazepine loaded MCM-41 with no presence of crystalline drug, when supercritical CO_2 procedure was used, the presence of different polymorphic forms of carbamazepine, depending on the employed solvent, could be detected. In this case the real inclusion into mesopores was not accurately verified and thus polymorphic carbamazepine could be present on the external surface of the silicate.

The dissolution profiles of carbamazepine from loaded MCM-41 showed a fast drug release from both samples obtained by the two different methods, but it was faster from the sample obtained by adsorption/equilibration method. In fact, after 15 min, the release from drug loaded MCM-41 reached 83% in pH 1.2 fluid, 97% in deionized water and 90% in phosphate buffer, while from the crystalline form the drug dissolution was only 35%, 60% and 65%, respectively (Figure 5).

Samples obtained by supercritical carbon dioxide process demonstrated increased carbamazepine dissolution profiles with more than 65% of drug released after 60 min. In conclusion with this method an increased drug loading and the formation of different polymorphic forms were obtained, but no information on their physical stability were reported. The sample obtained by adsorption/equilibration method showed a higher and faster dissolution rate and a good physical stability in two different storage conditions [11]. In fact when it was stored at 40°C in presence of $CaCl_2$ for 60 days and at 40°C and 75% of relative humidity (RH) for 60 days no traces of crystalline drug could be detected by scanning differential calorimetry (DSC) and X-ray powder diffraction.

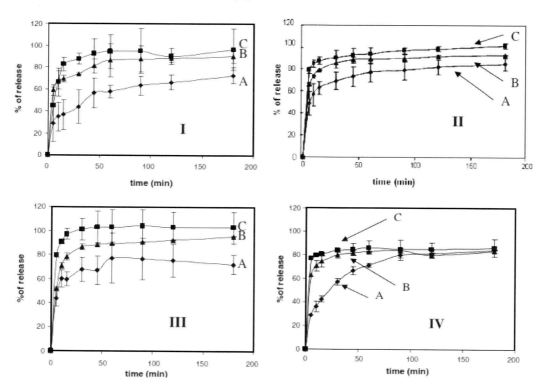

Figure 5. Release profile of carbamazepine from crystalline drug (A), physical mixture (B) and carbamazepine loaded MCM-41 (C) in gastric fluid at pH 1.2 (I), in intestinal fluid at pH 6.8 (II), in deionized water (III) and in aqueous solution of sodium lauryl sulfate 1% (IV) "Reprinted with permission from [11] Copyright 2008 Elsevier".

This improved stability is probably due to two different factors: (i) the interactions between MCM-41 silanols and the drug molecules which stabilize the system; (ii) the small pore size (an average diameter of 3.29 nm) of the silicate if compared to carbamazepine molecule size (0.90x0.62x0.43 nm). In fact the pore sizes are only ca. 3.9 times bigger that the longest drug dimension and thus the drug molecules are confined in a nanospace which prevents their crystallization [27].

However, the limit of this product was the low drug loading (ca. 15 wt.%). This aspect represents a limit in the use of mesoporous silicates for carbamazepine oral administration. In fact oral administered dose of carbamazepine is 200 mg/pro dose and thus a high amount of the inclusion product (1.33 g) should be required in order to administer the effective dose. The causes of the low loading could be associated both to the loading procedure (adsorption process), which consists on the formation, on the particle surface, of a monolayer of both solvent and drug molecules and to the silicate pore volume. In fact this last characteristic is reported to affect the drug loading capacity of porous silica carriers [28].

In order to obtain an increased drug loading, the use of the SBA-15 characterized by higher pore volume was proposed [17, 29]. Carbamazepine inclusion into SBA-15 was reported by Van Speybroeck et al. [16] by the incipient wetness procedure, using dichloromethane as a solvent for drug solubilization reaching a final loading of 20 wt.%. Successively the properly modified solvent evaporation method was proposed by Ambrogi [29]. In order to remove air from matrix pore and to make easy the drug entrance, the vacuum was applied at the dispersion until air was removed from the sample. Afterwards, before removing the solvent, the mixture was kept under stirring for 30 minutes. Following this procedure a drug loading up to 40% was obtained, with no traces of the drug crystalline form.

Carbamazepine *in vitro* release from this sample reached 70% after 10 min in comparison to ~50% and ~20% from the physical mixture and crystalline carbamazepine respectively (Figure 6).

Interesting results were obtained from the physical mixture between carbamazepine and SBA-15 as well. In fact, an improvement of the drug release was observed from the physical mixture too. This effect could be due to the partial spontaneous carbamazepine amorphization demonstrated by DSC thermal profile [29], but also to the improved wettability of the drug. Wang et al. evaluated the effect of SBA-15 on drug wettability by measuring the time required for water powder permeation and the contact angle [30]. Pure carbamazepine had a contact angle of 139.2±1.00 degrees, but when the drug was adsorbed on SBA-15 or was physically mixed with it, water permeation was so quick that it could not be possible to measure the contact angle accurately.

Unfortunately the use of the physical mixture, which involves numerous advantage such as reproducibility, scalability, time saving and environmental respect because of lack of solvent, does not guarantee the same improvement of dissolution rate and the same physical stability as the sample in which the drug is included into the silicate mesopores.

Shen et al. [18] formulated ibuprofen with SBA-15 to achieve a solid dispersion by co-spray drying.

The ratio of ibuprofen and SBA-15 which guaranteed the lack of crystalline form was 50:50 w:w. This drug amount was ca. the maximum theoretical loading of ibuprofen inside the pore channel. The amorphous state obtained resulted physically stable after storage under conditions of 40°C and 75% of RH for 1 year. Dissolution test performed in 0.1N HCl solution showed a dissolution rate much faster from drug loaded SBA-15 than from crystalline ibuprofen. Same results were obtained when the solid dispersion was compressed with cornstarch into tablets. Within the first 15 min the percentage of dissolved ibuprofen from the spray dried solid dispersion reached 95%, whereas only 16% of original crystals were dissolved. The complete release was obtained after 50 min. The co-spray-dried solid dispersion exhibited excellent dissolution performance both in powder and in tablet form. Physical stability of ibuprofen co-spray dried with MCM-41 and SBA-15 was verified as well [31].

The techniques traditionally employed to load drugs into mesoporous silicates often require the employment of organic solvents which have to be removed with complex processes and high cost.

In order to avoid these problems, Ahern et al. proposed the use of supercritical fluids such as carbon dioxide [21] without the employment of organic solvents. The model drug was fenofibrate, owing to BCS Class II. The loaded drug was found to be in an amorphous state and this state was maintained after storage under 75% RH and 40°C for 12 months. Fenofibrate release from all samples (1:5 drug : silica) was rapid with over 50% released by 5 min. After ca. 20 min, fenofibrate release reached a plateau at ca. 80% and the total drug release was not obtained, maybe because of the presence of drug molecules in depth in the mesopores or micropores which are inaccessible to the dissolution media due to surface tension effects.

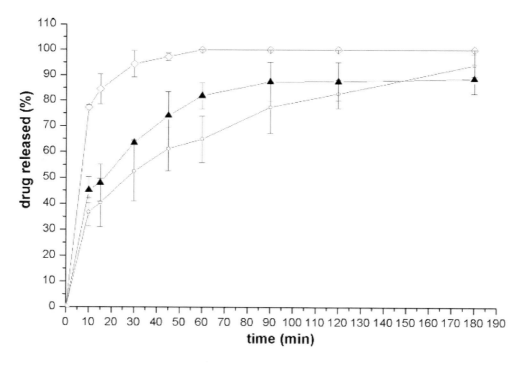

Figure 6. In vitro release profiles at pH 1.2, 37.0°C ± 0.5 of crystalline carbamazepine (o), SBA-15/carbamazepine physical mixture (▲) and carbamazepine loaded SBA-15 (◊), n=3, mean ± SD "Reprinted with permission from [29] Copyright 2012 Elsevier".

Influence of pore size on drug dissolution rate

Mesoporous silicates	Drugs	Results	References
MCM-41 and SBA-15	ibuprofen	higher drug loading and dissolution rate improvement for SBA-15 than for MCM-41	[28]
MCM-41	furosemide	low drug loading and good dissolution rate improvement	[17]
SBA-15	furosemide	good drug loading and good dissolution rate improvement	[32]

Heikkilä et al. [28] compared the effect of MCM-41 and SBA-15 on dissolution rate of ibuprofen. They loaded the drug by the same loading procedure by contact of the mesoporous material with an ibuprofen solution (carrier:drug ratio 1:14 w/w), filtration and washing in order to eliminate the excess drug. MCM-41 and SBA-15 calculated BET specific surface area were 1067 m^2/g and 625 m^2/g respectively, pore volumes were 0.717 cm^3/g and 1.067 cm^3/g and pore diameters were 2.6 nm and 6.9 nm respectively. The authors found that the drug load capacity was dependent on its total mesopore volume and the highest load was obtained for SBA-15 (drug load ca. 1:1 in weight in relation to the carrier mass). In addition the larger pore diameter allowed the drug diffusion deep down the pores without formation of blockages in the pore channel.

Ibuprofen release was performed at different pH values (5.5, 6.8 and 7.4) to mimic the conditions between duodenum and jejunum of the small intestine. The dissolution rate of crystalline ibuprofen was strongly affected by the pH of the dissolution medium as it displays low aqueous solubility at acidic pH values below and close to its pKa of 4.42. At all tested pHs the release of ibuprofen from the mesoporous carriers was faster compared to the dissolution rate of crystalline drug, but the release from SBA-15 was higher than that from MCM-41. The more narrow mesopores of MCM-41 sterically hindered the diffusion of the drug molecules into the dissolution media, slowing down the drug release in comparison to SBA-15.

It is noteworthy that the presence of the carrier diminished the pH dependency of ibuprofen dissolution and this yielded great improvements in the dissolution of the drug at low pH conditions where the drug solubility is very low.

The effect of pore diameter on dissolution rate was confirmed by Ambrogi et al. [17, 32] who loaded furosemide into MCM-41 and SBA-15 by the solvent evaporation method. Because of its weak acidity (pKa = 3.8), furosemide shows the lowest solubility at the low pH values of the stomach which in turn is its preferential absorption site. Thus drug release was performed at acidic pH (ca. 3.0). After 5 min the amount of drug released from the SBA-15 inclusion product was 65% vs 11% (of crystalline drug) and 50% (from furosemide loaded MCM-41) reaching, after 2 h, ca. 80% vs 68% (of crystalline furosemide) and 95% (from drug loaded MCM-41). It can be hypothesized that the main mechanism governing the drug release from SBA-15 loaded product was the rapid matrix imbibitions that promoted faster drug dissolution in comparison to MCM-41 that is characterized by more narrow pore size and lower concentrations of silanol groups. As previously reported [21], the drug release from

SBA-15 was not complete, maybe because of the presence of drug molecules in the micropores hardly accessible to the aqueous medium.

The fast drug release at acidic pH values allows the presence of high drug concentration in the stomach where the drug has the lowest solubility but the best absorption.

STUDIES ON EFFECT ON ORAL DRUG BIOAVAILABILITY

Whereas many studies proved the positive effect of MCM-41 and SBA-15 on dissolution rate of poorly water soluble drugs, only a few papers evaluated the biopharmaceutical performance of drug loaded mesoporous silicates. Mellaerts et al. [33] determined the bioavailability of itraconazole from silicate formulations in rabbits and dogs and compared them with those of the marketed product Sporanox® and crystalline itraconazole. Sporanox ® is a capsule formulation containing 100 mg of amorphous itraconazole coated on sugar spheres. Ordered mesoporous silicate had a pore diameter of ca. 7.3 nm and the drug was loaded by solvent evaporation method (drug loading ca. 21%). The in vitro release profile of itraconazole from the silicate in simulated gastric fluid was fast with 76.4% of drug released after 5 min whereas drug dissolution of crystalline itraconazole was only 19.0% within the same time. Capsules were prepared mixing the crystalline drug and the drug loaded silicate with proper excipients, such as croscarmellose, lactose and sodium lauryl sulfate, in order to ensure rapid capsule disintegration. Bioavailability curves were obtained after capsule oral administration. The area under the curve (AUC_{0-24}), determined after 24 hours, from crystalline itraconazole resulted 521±159 nM h and T_{max} resulted 9.8±1.8 h. After itraconazole loaded silicate administration, AUC_{0-24} increased up to 1069±278 nM h with a T_{max} of 4.2±1.8 h. Similar bioavailability was obtained after Sporanox® administration. Great differences in biopharmaceutical aspects resulted after capsule administration in dogs too (AUC_{0-8} 681±566 from capsules containing drug loaded silicate, AUC_{0-8} 760±364 from Sporanox® and 0±0 from crystalline itraconazole capsules).

Carbamazepine bioavailability from SBA-15 formulated as pellets was evaluated by Wang et al. [30]. They loaded carbamazepine into mesoporous SBA-15 via the wetness impregnation method and then the loaded SBA-15 was processed into pellets by extrusion/spheronization after mixing with microcrystalline cellulose, sorbitol and sodium carboxymethyl starch. After oral administration in beagle dogs plasma concentration-time profiles were determined and compared with those of carbamazepine from commercial tablets. The main pharmacokinetic parameters, AUC, C_{max} and T_{max} resulted 72580±25283 ng h/mL, 528.83±106.85 ng/mL and 65.00±15.49 min respectively from commercial tablet and 113709±17150 ng h/mL, 803.7±296.78 ng/mL and 65.00±12.25 min respectively from carbamazepine-SBA-15 pellets. By Student's *t*-test, the results showed that there was significant difference in AUC ($P<0.05$) but no significant difference in C_{max} ($P>0.05$). The relative bioavailability (AUC from pellets/AUC from commercial tablet) resulted 156.67%, that means a higher bioavailability than that from the commercial formulation.

COMPARISON WITH OTHER MESOPOROUS SILICATES

A comparison among different mesoporous materials was evaluated by Heikkilä et al. [28]. They studied the effect of mesoporous materials with different porous characteristics on the ibuprofen release profile at different pH values. The chosen materials were: i) thermally carbonized porous silicon (TCPSi) characterized by hydrophilic mesopore surface, wide pore size distribution (2-30 nm), 2D mesopores akin to a spruce tree and a superficial area of ca. 300 m^2/g; ii) Technische Universiteit Delft (TUD-1), a silica-base mesoporous with highly accessible pore networks. This material is sponge-like reservoir for the drug with a pore volume <0.5 cm^3/g, pore diameter varying between 2.5-20 nm, a surface area ca. 400-500 m^2/g and pores randomly connected in three dimensions. These materials were compared to MCM-41 and SBA-15. The maximal drug load capacity depended mainly on the total pore volume of the mesoporous solid and thus was obtained with SBA-15 (ca. 1:1 w/w drug/SBA-15) whereas the lowest drug load was obtained for TUD-1, despite its most accessible mesopore network.

All mesoporous materials positively influenced ibuprofen release which was remarkably faster from the mesopore materials than from the bulk drug. The amount of dissolved ibuprofen in the phosphate buffer at pH 5.5 was 25% after 45 min whereas ibuprofen released from the mesoporous carriers was 2.5-3.5 fold higher (64-86%).

The fastest ibuprofen release was observed for TUD-1 carrier, probably due to its highly accessible random 3D mesopore network. It was followed by SBA-15 with larger pores and then MCM-41. In conclusion, SBA-15 resulted the most versatile carrier among the tested ones, because of its higher drug loading, which allows to formulate high dosage drugs too, and its fast drug release.

The effect of mesoporous silica MCM-41 on dissolution rate of indometacin was compared to that of the non ordered silica gel Syloid 244 FP EU [34]. Syloid is an amorphous commercial silica gel-type material with randomly oriented pores, conforming to the specifications for "silica, colloidal hydrated" in the Ph. Eur. Its particle size is less than 5 μm and it has wide pore size distribution with average pore diameter more than five times larger (ca. 19.0 nm) than of the ordered mesoporous material MCM-41 (3.4 nm). The size of one indomethacin molecule is 4.27 × 8.97 × 12.66 Å and thus inside the pores of Syloid there may be enough space for indomethacin to crystallize. Three techniques were used to load indomethacin into the mesoporous silica particles: the solvent evaporation method, the fluid bed, typically used in granulating or drying of pharmaceutical masses, and the adsorption method. At last the loaded mesoporous silica particles were formulated and compressed into tablets to evaluate the pharmaceutical functionality. The release of indomethacin was faster from the non ordered silica (Syloid 244 FP EU) than from the ordered mesoporous silica (MCM-41), presumably due to the larger pore size and smaller particle size. The silica materials were also compressed into tablets and no major alterations in the porous structures of the silicates were detected as a result of the compression forces used in tableting. MCM-41 particles retained the ability of improving release of poorly soluble drugs even after compression into tablets and this property was further optimized through the formulation. In fact the dissolution of indomethacin was enhanced by adding excipients and further improved after compression into a tablet formulation. On the other hand, Syloid tablets had a different behavior and their disintegration was much slower than that of MCM-41 tablets. Therefore, the release of indomethacin was slower from the Syloid tablets than from Syloid particles. Notwithstanding the good properties of Syloid in improving drug dissolution rate, there are some doubts on physical stability of drug inside the pores, as the amorphous drug inside the wide pores of Syloid would start to crystallize during storage and thus would influence the drug release rate. Takeuchi et al. [35] studied the stability of indomethacin loaded in porous silica gel Sylysia 350 with slightly larger pore size than the Syloid (21 vs. 19 nm, respectively). The particles were kept at 40°C and 75% RH for 2 months, and while the reference bulk amorphous indomethacin crystallized during this time, drug crystallization was not observed inside of the pores. However the tendency of a drug to crystallize inside the pores of Syloid and its possible effect on the dissolution rate should, however, be studied further.

Another comparative study among different silicates has been performed with indomethacin [36].

EFFECT OF THE TEMPLATE OCCLUDED SBA-15 ON DRUG DISSOLUTION

Fu et al. [37] proposed the employment of template occluded SBA-15. They evaluated the effect of this matrix on the dissolution rate of piroxicam. SBA-15 was selected because of its template, poly(ethylene oxide)–poly(propylene oxide) triblock copolymer (Pluronic P123), has been widely used in pharmaceutics as a safe drug delivery agent to improve hydro-solubility, stability and bioavailability of drugs for a long time [37]. Piroxicam was introduced into as-prepared SBA-15 occluded with Pluronic P123.

Dissolution tests were performed at pH 1.2 and pH 7.5. At pH 1.2 nearly 10 wt% of the drug was released from the crystalline powder in the first 5 min, while the released percentage from piroxicam loaded SBA-15 was 25% and from occluded SBA-15 62 wt%. These differences should be ascribed to the surfactant inside SBA-15 piroxicam pores, which disperses the drug more efficiently and provides more efficient area to contact water. For comparison piroxicam release from a dispersion of the drug and P123 was evaluated as well, but not great improvement of drug release was obtained. This could be explained by the condensed paste state of Pluronic P123 which encapsulates the drug and restricts its dissolution. While in the template occluded SBA-15 the surfactant molecules were dispersed within the pores of SBA-15 like the spokes of the wheel, the nanospaces between the micelle and the wall, and the nanopores among the poly(propylene oxide) blocks, both provide efficient channels for the diffusion of water and piroxicam molecules. This phenomenon suggests that the special distribution of micelles inside the channel of the mesoporous samples is the crucial factor for the high dissolution rate.

The same improvement of drug dissolution was obtained with the poorly soluble drug silybin [38].

CONCLUSION

The results presented in this chapter clearly indicate that the mesoporous silicates MCM-41 and SBA-15 are promising carriers for poorly water soluble drugs. The main conclusions that can be drawn are: i) both silicates greatly improve drug dissolution rate in comparison to that from the crystalline drug and the physical mixture constituted by the drug and the mesoporous silicate; ii) the physical characteristics of the loaded drug are maintained after storage under defined humidity and temperature conditions (usually 75% RH and 40°C); iii) the drug loading procedure effects the drug disposition inside the pores and thus effects the dissolution rate. The adsorption procedure, even if guarantees a homogenous drug distribution and a great improvement of dissolution rate, is not a convenient loading procedure as it gives a low drug loading and does not offer possibility of reproducibility and scalability. Among traditional drug loading methods the solvent evaporation procedure is the most suitable. Other more recently proposed methods, such as spray drying and fluid bed and the methods based on the use of supercritical fluids seem promising but need further studies; iv) the use of SBA-15 in comparison to MCM-41 offers both the possibility of increased drug loading and of a higher improvement of dissolution rate because of its larger pore volume and easier access to the entry of water molecules and release of drug molecules; v) SBA-15 and MCM-41 result suitable for proper formulations; vi) at last the improvement of bioavailability reported after oral administration encourages the continuation of in vivo drug delivery studies with MCM-41 and SBA-15.

In conclusion the use of both MCM-41 and SBA-15 appears to be a promising strategy for the development of rapidly dissolving and improved absorbed oral formulations for poorly soluble drugs.

REFERENCES

[1] Stegemann, S.; Leveiller, F.; Franchi, D.; de Jong, H.; Lindén, H. (2007). When poor solubility becomes an issue: from early stage to proof of concept. *Eur. J. Pharm. Sci, 31*, 249-261.

[2] Sugano, K.; Okazaki, A.; Sugimoto, S.; Tavornvipas, S.; Omura, A.; Mano, T. (2007) Solubility and Dissolution Profile Assessment in Drug Discovery. *Drug Metab. Pharmacokinet., 22*, 225–254.

[3] Amidon, G.L.; Lennernäs, H.; Shah, V.P.; Crison, J.R. (1995). A theoretical basis for a biopharmaceutic drug classification: the correlation of in vitro drug product dissolution and in vivo bioavailability. *Pharm. Res., 12*, 413–420.

[4] Food and Drug Administration, Guidance for Industry-Waiver of in vivo Bioavailability and Bioequivalence Studies for Immediate-release Solid Oral Dosage Forms based on a Biopharmaceutics Classification Systems, August 2000.

[5] Kawabata, Y.; Wada, K.; Nakatani, M.; Yamada, S.; Onoue, S. (2011). Formulation design for poorly water-soluble drugs based on biopharmaceutics classification system: basic approaches and practical applications. *Int. J. Pharm., 420*, 1– 10.

[6] Øye, G.; Sjöblom, J.; Stöcker, M. (2001). Synthesis, characterization and potential applications of new materials in the mesoporous range. *Adv. Coll. Interf. Sci., 89-90*, 439-466.

[7] Beck, J.S.; Vartuli, J.C.; Roth, W.J.; Leonowicz, M.E.; Kresge, C.T.; Schmitt, K.D.; Chu, C.T-W.; Olson, D.H.; Sheppard, E.W.; McCullen, S.B.; Higgins, J.B.; Schlenker, J.L. (1992). A new family of mesoporous molecular sieves prepared with liquid crystal templates. *J. Am. Chem. Soc., 114*, 10834–10843.

[8] Vallet-Regi, M.; Rámila, A.; del Real R.O.; Pérez-Pariente, J. (2001). A new property of MCM-41: drug delivery release. *J. Chem. Mater.,13*, 308-311.

[9] Vallet-Regi, M.; Balas, F.; Arcos, D. (2007). Mesoporous Materials for Drug Delivery. *Angew. Chem. Int. Ed., 46*, 7548 – 7558.

[10] Charnay, C.; Bégu, S.; Tourné-Péteilh, C.; Nicole, L.; Lerner, D.A.; Devoisselle, J.M. (2004). Inclusion of ibuprofen in mesoporous templated silica: drug loading and release property. *Eur. J. Pharm. Biopharm., 57*, 2004, 533–540.

[11] Ambrogi, V., Perioli, L., Marmottini, F., Accorsi, O., Pagano, C., Ricci, M., Rossi, C. (2008). Role of mesoporous silicates on carbamazepine dissolution rate enhancement. *Micropor. Mesopor. Mater. , 113*, 445-452.

[12] Hirvonen, J., Laaksonen, T., Peltonen, L., Santos, H., Lehto, V. P., Heikkilä, T., Riikonen, J., Mäkilä, E., Salonen, J. (2008). Feasibility of silicon-based mesoporous materials for oral drug delivery applications. *Dosis, 24*, 129-149.

[13] Ambrogi, V., Perioli, L., Marmottini, F., Giovagnoli, S., Esposito, M., Rossi, C. (2007). Improvement of dissolution rate of piroxicam by inclusion into MCM-41 mesoporous silicate. *Eur. J. Pharm. Sci., 32*, 216–222.

[14] Mellaerts, R., Jammaer, J.A.G., Van Speybroeck, M., Chen, H., Van Humbeeck, J., Augustijns, P., Van den Mooter, G., Martens, J.A. (2008). Physical state of poorly water soluble therapeutic molecules loaded into SBA-15 ordered mesoporous silica carriers: a case study with itraconazole and ibuprofen. *Langmuir, 24*, 8651–8659.

[15] Ambrogi, V., Perioli, L., Pagano, C., Marmottini, F., Moretti, M., Mizzi, F., Rossi, C. (2010). Econazole Nitrate-Loaded MCM-41 for an Antifungal Topical Powder Formulation. *J. Pharm. Sci. 99*, 4738-4745.

[16] Van Speybroeck, M., Barillaro, V., Thi, T. D., Mellaerts, R., Vermant, J., Van Humbeeck, J., Annaert, P., Van Den Mooter, G., Martens, J., Augustijns, P. (2009). Ordered Mesoporous Silica Material SBA-15: A Broad-Spectrum Formulation Platform for Poorly Soluble Drugs. *J. Pharm. Sci., 98*, 2648-2658.

[17] Ambrogi, V., Perioli, L., Pagano, C., Latterini, L., Marmottini, F., Ricci, M., Rossi, C. (2012). MCM-41 for furosemide dissolution improvement. *Micropor. Mesopor. Mater., 147*, 343–349.

[18] Shen S.-C., Kiong Ng, W., Chia, L., Dong, Y.C., Tan, R. B. H. (2010). Stabilized amorphous state of ibuprofen by co-spray drying with mesoporous SBA-15 to enhance dissolution properties. *J. Pharm. Sci., 99*, 1997-2007.

[19] Hillerström, A., Stamb, J.V., Andersson, M. (2009). Ibuprofen loading into mesostructured Silica using liquid carbon dioxide as a solvent. *Green Chem., 11*, 662–667.

[20] Sanganwar, G.P., Gupta, R.B. (2008). Dissolution-rate enhancement of fenofibrate by adsorption onto silica using supercritical carbon dioxide. *Int. J. Pharm., 1-2*, 213–218.

[21] Ahern, R. J., Crean A. M., Ryan K. B. (2012). The influence of supercritical carbon dioxide (SC-CO$_2$) processing conditions on drug loading and physicochemical properties. *Int. J. Pharm., 439,* 92– 99.

[22] Vadia, N., Rajput, S. (2012). Study on formulation variables of methotrexate loaded mesoporous MCM-41 nanoparticles for dissolution enhancement. *Eur. J. Pharm. Sci. 45,* 8–18.

[23] Grzesiak, A. L., Lang, M., Kim, K., Matzger, A. J. (2003). Comparison of the four anhydrous polymorphs of carbamazepine and the crystal structure of form I. *J. Pharm. Sci. 92,* 2260-71.

[24] Harris, R.K., Ghi, P.Y., Puschmann, H., Apperley, D.C., Griesser, U.J., Hammond, R.B., Ma, C., Roberts, K.J., Pearce, G. J., Yates, J. R., Pickard, C. J. (2005). Structural Studies of the Polymorphs of Carbamazepine, Its Dihydrate, and Two Solvates. *Org. Process Res. Dev., 9,* 902-910.

[25] Patterson, J. E., James, M. B., Forster, A. H., Lancaster, R. W., Butler, J. M., Rades, T. (2005). The influence of thermal and mechanical preparative techniques on the amorphous state of four poorly soluble compounds. *J. Pharm. Sci., 94,* 1998-2012.

[26] Patil, A., Chirmade, U. N., Trivedi, V., Lamprou, D. A., Urquhart, A., Douroumis, D. (2011). Encapsulation of Water Insoluble Drugs in Mesoporous Silica Nanoparticles using Supercritical Carbon Dioxide. *J. Nanomed. Nanotechnol., 2:111.* doi: 10.4172/2157-7439.1000111.

[27] Sliwinska-Bartkowiak, M., Dudziak, G., Gras, R., Sikorski, R., Radhakrishnan, R., Gubbins, K. E. (2001). Freezing behavior in porous glasses and MCM-41, *Colloids Surf. A: Physicochem. Eng. Aspects, 187–188,* 523–529.

[28] Heikkilä, T., Salonen, J., Tuura, J., Kumar, N., Salmi, T., Murzin, D. Y., Hamdy, M. S., Mul, G., Laitinen, L., Kaukonen, A. M., Hirvonen, J., Lehto, V.-P. (2007). Evaluation of mesoporous TCPSi, MCM-41, SBA-15, and TUD-1 materials as API carriers for oral drug delivery. *Drug Delivery, 14,* 337-347.

[29] Ambrogi, V., Marmottini, F., Pagano, C. (2013). Amorphous carbamazepine stabilization by the mesoporous silicate SBA-15. *Micropr. Mesopor. Mater., 177,* 1-7.

[30] Wang, Z., Chen, B., Quan, G., Li, F., Wu, Q., Dian, L., Dong, Y., Li, G., Wu, C. (2012). Increasing the oral bioavailability of poorly water soluble carbamazepine using immediate-release pellets supported on SBA-15 mesoporous silica, *Int. J. Nanomed., 7,* 5807-5818.

[31] Shen, S.-C.; Ng, W. K.; Chia, L.; Hu, J.; Tan, R.B.H. (2011). Physical state and dissolution of ibuprofen formulated by co-spray drying with mesoporous silica: Effect of pore and particle size. *Int. J. Pharm., 410,* 188–195.

[32] Ambrogi, V., Perioli, L., Pagano, C., Marmottini, F., Sagnella, A., Rossi, C., Ricci, M. (2012). Use of SBA-15 for furosemide oral delivery enhancement. *Eur. J. Pharm. Sci., 46,* 43–48.

[33] Mellaerts, R., Mols, R., Jammaer, A.G. J., Aerts, C. A., Annaert, P., Van Humbeeck, J., Van der Mooter, G., Augustijins, P., Martens L. A. (2008). Increasing the oral bioavailability of the poorly water soluble drug itraconazole with ordered mesoporous silica. *Eur. J.Pharm. Biopharm. 69,* 2008, 223-230.

[34] Limnell,T., Kumar, N., Santos, H. E. A., Laaksonen, T., Akil, E. M., Teemu Heikkil, A., Peltonen, L., Salonen, J., Hirvonen, J., Murzin, D.Y. (2011). Drug Delivery

Formulations of Ordered and Nonordered Mesoporous Silica: Comparison of Three Drug Loading Methods. *J. Pharm. Sci., 100,* 3294-3306.

[35] Takeuchi H., Nagira S., Yamamoto H., Kawashima Y. (2005). Solid dispersion particles of amorphous indomethacin with fine porous silica particles by using spray-drying method. *Int. J. Pharm., 293,* 155–164.

[36] Wang, Y.; Zhao, Q.; Hu, Y.; Sun, L.; Bai, L.; Jiang, T.; Wang, S. (2013). Ordered nanoporous silica as carriers for improved delivery of water insoluble drugs: a comparative study between three dimensional and two dimensional macroporous silica. *Int. J. Nanomed., 8,* 4015-4031.

[37] Fu T., Guo, L., Le, K., Wang, T., Lu, J. (2010). Template occluded SBA-15: An effective dissolution enhancer for poorly water-soluble drug. *Appl. Surface Sci. 256,* 6963–6968.

[38] Chiappetta, D.A., Sosnik, A. (2007). Poly(ethylene oxide)-poly(propylene oxide) block copolymer micelles as drug delivery agents: improved hydrosolubility, stability and bioavailability of drugs. *Eur. J. Pharm. Biopharm., 66,* 303–317.

[39] Fu T., Lu, J., Guo, L., Zhang, L., Cai, X., Zhu, H. (2012). Improving bioavailability of silybin by inclusion into SBA-15 mesoporous silica materials. *J. Nanosci. Nanotechnol., 12,* 1–10.

In: Comprehensive Guide for Mesoporous Materials. Volume 3   ISBN: 978-1-63463-318-5
Editor: Mahmood Aliofkhazraei   © 2015 Nova Science Publishers, Inc.

Chapter 2

MESOPOROUS BIOACTIVE GLASSES: IMPLANTS AND DRUG DELIVERY SYSTEMS FOR BONE REGENERATIVE THERAPIES

Daniel Arcos[1,2] *and María Vallet-Regí*[1,2]
[1]Networking Research Center on Bioengineering, Biomaterials and Nanomedicine, CIBER-BBN, Spain
[2]Department of Inorganic and Bioinorganic Chemistry, Faculty of Pharmacy, UCM, Instituto de Investigación Sanitaria Hospital, Madrid, Spain

ABSTRACT

The development of bioceramics designed for bone tissue regeneration has experienced an important boost with the appearance of a new generation of materials: mesoporous bioactive glasses (MBG). MBGs are commonly prepared in the systems SiO_2-CaO-P_2O_5 or SiO_2-CaO, similarly to bioglasses prepared by conventional sol-gel processes. The incorporation of a structure directing agent to the synthesis results in sol-gel bioglasses with highly ordered mesoporous structures and much higher surface area. The consequence is that MBGs exhibit much higher solubility, better bioactive behavior and the capability of hosting and releasing drugs aimed to improve the regenerative healing process. In this chapter, the preparation methods, drug delivery applications and the development of scaffolds for in situ bone regeneration are reviewed and commented.

1. INTRODUCTION

In the last decades, the improvements in public health have resulted in a difficult challenge for our societies. The enhancement of life expectancy has brought a new scenario, which comprises an ever-ageing population with millions of people between 60 and 100 years demanding a good quality of life. Considering that living tissues begin a progressive path of deterioration from 30 years until the individual death, achieving this goal is not an easy task.

In the case of bone tissue related degenerative pathologies, the challenge is even greater. The musculo-skeletal disorders due to fractures (especially those of osteoporotic origin) or tumor extirpation, result in more than 2.2 million patients every year requiring bone prostheses and implants to restore the structure and function of the affected bone [1]. The successful substitution of aged, diseased or impaired bone tissue is a routine activity thanks to the improvements of surgical techniques, new biomaterials and rehabilitation strategies. However, the limited lifetime of implants and prostheses is still a very serious drawback, whose significance increases insofar the patient life expectancy also increases. It is estimated that the prosthesis failure rate is between 15 to 50%, for those implanted after 15 to 30 years [2]. This situation has prompted a change in the goals of biomaterials scientists. The improvements introduced in bone implants are not enough to resolve the limited useful lifetime of the prostheses. For this reason, many efforts are currently addressed towards the synthesis of biomaterials for regenerative purposes, instead of substitution of the natural tissues with permanent artificial devices [3]. The new goal is developing biomaterials able to stimulate the osteogenic (bone forming) response in the patient, by means of gathering and stimulating the self-repairing mechanism of natural bone. This shift of concept is expressed in the development of the so named 3rd generation biomaterials, which are designed, synthesized and processed to stimulate the bone tissue regeneration [4]. The most representative members of 3rd generation bioceramics are bioactive glasses, since they are intended and tailored for regenerative purposes in small bone defects (see table 1).

At the end of the 1960s, L.L Hench discovered the capability of certain SiO_2 based glasses to bond to the bone through a chemically stable and mechanically strong linkage [5]. Since then, the development and clinical applications of bioactive glasses have growth during the last 40 years. Compounds such Bioglass 45S5, commercialized under different trade marks (Novamin Tech., Perioglass^{TM}, Novabone^{TM}, etc.) are currently used in orthopedic and periodontal surgery, as well as dentine regeneration agents included in tooth pastes (Sensodine^{TM}).

Table 1. The different generations of bioceramics

	1st generation bioceramics	2nd generation bioceramics	3rd generation bioceramics
Purpose	Bone replacement	Bone replacement	Bone regeneration
Biological response	Almost inert	Bone bonding/dissolution	Bone bonding and regeneration
Examples	Alumina Zirconia Carbon Silicon nitride Silicon carbide	Synthetic Hydroxyapatite Natural origin hydroxyapatite (animal, coralline and algae) A/W glass-ceramics α-TCP, β-TCP, BCPs	Bioactive glasses (melt derived, sol-gel glasses, mesoporous bioactive glasses) Nanocrystalline HA associated to osteo-stimulatory agents. Nanocrystalline HA manufactured as porous scaffolds

This glass has a composition based on SiO_2-P_2O_5-Na_2O-CaO, where SiO_2 and P_2O_5 are network formers, whereas CaO and Na_2O are network modifiers and play a fundamental role in the bioactive behavior of this material. In 1991 the incorporation of the sol-gel chemistry to the preparation of bioceramics resulted into a new generation of bioactive glasses, which exhibited a very high potential to develop better implants with osteogenic capabilities [6]. This potential relayed on the enhanced textural properties inherent to the sol-gel method, i.e., surface area and porosity, respect to the conventional melt derived bioglasses (like Bioglass 45S5). Since then, the advances in this research field have been numerous and significant, as can be deduced from the amount of scientific works published so far [7-11]. However, more than twenty years after the transition *from the bench to the bedside* has not occurred. This fact can be somehow explained because the added value of sol-gel glasses does not compensate the high costs associated to the incorporation a new product to the clinical market.

Very recently, a new generation of mesoporous bioactive glasses (MBGs) has been developed. These new materials exhibit textural and bioactive properties not observed in previous generations of glasses. MBGs result from the incorporation of structure directing agents to the sol-gel synthesis of bioglasses. The result is the preparation of highly ordered mesoporous materials with surface and porosity five times higher than those obtained for the conventional sol-gel bioglasses. MBGs exhibit the fastest *in vitro* bioactivity observed up to date. However the real clinical significance is still unknown, as the ordered mesoporous structure could allow incorporating osteogenic agents, osteoclasts inhibitors, antitumoral drugs, etc. thus providing an excellent potential for the treatment of bone diseases.

In this chapter, we will study the synthesis methods and properties of MBGs, paying special attention to the chemical compositions and mesoporous structures developed up to date. The advances in surface functionalization of MBGs and the incorporation of therapeutic agents for local drug delivery are also discussed. Finally, this chapter emphasizes on the preparation of three dimensional macroporous MBGs scaffolds by rapid prototyping techniques. The preparation of implants based on these materials is called to play a fundamental role in the new tissue engineering techniques and bone regenerative therapies.

2. SILICA MESOPOROUS MATERIALS FOR BONE GRAFTING PURPOSES

Between 1992 and 1995, several studies suggested that SiO_2 glasses synthesized by the sol-gel method could exhibit bioactive behavior [12-14]. The bioactivity would be mainly due to the higher surface area and porosity exhibited by the sol-gel glasses compared with the melt-derived ones. Based on these results, three mesoporous materials - MCM-41, MCM-48 and SBA-15- were submitted to *in vitro* bioactivity tests in simulated body fluid (SBF) [15, 16]. SBF is an inorganic solution containing the ionic components of human plasma in an almost equal concentration. This simulated body fluid has been widely accepted by the scientific community as a single and cheap test to predict the *in vivo* bioactive behavior of biomaterials for bone grafting purposes. The composition of SBF can be seen in Table 2.

When a SiO_2 based bioactive glass is soaked into SBF at 37°C, a cascade of chemical reactions is initiated. Figure 1 shows the sequence of reaction at the interface between a SiO_2-P_2O_5-Na_2O-CaO glass surface and SBF.

Table 2. Ionic concentrations of SBF (mM) and human plasma

Ion	SBF	Human plasma
Na^+	142.0	142.0
K^+	5.0	5.0
Mg^{2+}	1.5	1.5
Ca^{2+}	2.5	2.5
Cl^-	148.8	103.0
HCO_3^-	4.2	27.0
HPO_4^{2-}	1.0	1.0
SO_4^{2-}	0.5	0.5

These reactions result in the nucleation and growth of a nanocrystalline carbonate-hydroxyapatite (CHA) phase – very similar to the mineral component of the mammal's bone - onto the materials surface. It is established that any material able to reproduce this CHA when soaked in SBF will be bioactive under *in vivo* conditions [17], i.e., *the materials will bond to the bone ensuring a good integration with living tissues.* MCM-41, MCM-48 and SBA-15 mesoporous materials exhibit a slight *in vitro* bioactivity in contact with SBF [16]. The high surface area of these materials, together with the ionic oversaturation of SBF, leads to the nucleation of an apatite like phase onto their surfaces. However, this process was limited to the formation of a small fraction of hydroxyapatite HA after 30 days or more soaked in SBF. The very slow bioactive kinetic of pure SiO_2 mesoporous materials indicates that these materials are not likely to osseointegrate with living bone. Such a slow reactivity necessarily would result in the formation of a fibrous tissue around the implant after implantation, instead of the desired bone bonding. These results evidenced the necessity of changes in the chemical composition of these materials in order to obtain bioactive behaviors. In this way, the synthesis of mesoporous materials with chemical compositions similar to conventional bioactive glasses became a priority research line in this field.

3. SYNTHESIS OF MESOPOROUS BIOACTIVE GLASSES

In 2004 Zhao et al. synthesized the first MBGs [18]. This group combined the sol-gel chemistry of the multicomponent system SiO_2-P_2O_5-CaO, with the benefits of adding a structure directing agent. As a result, they obtained mesoporous materials with chemical compositions analogous to the conventional SiO_2-P_2O_5-CaO sol-gel glasses [19-21], but exhibiting porosities and surface areas very similar to MCM-41, SBA-15, etcmesoporous materials.

MBGs cannot be prepared following the conventional hydrothermal synthesis, used for most of silica mesoporous materials. MBGs commonly contain CaO and P_2O_5, in addition to SiO_2. The multicomponent nature of MBGs requires a synthesis method that led to more robust mesoporous structures.

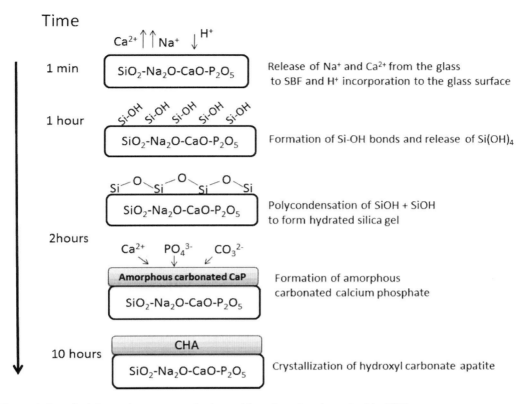

Figure 1. Interfacial reactions occurred when a bioactive glass is soaked in SBF.

Otherwise, the presence of network modifiers such as Ca^{2+} cations interferes in the surfactant-silica interaction leading to very defective mesostructures. For this reason, Zhao et al. turned to the evaporation induced self-assembly (EISA) route, previously developed by J.C. Brinker [22]. This method is schemed in figure 2 and is based on the preparation of diluted solutions containing the precursors and the surfactant in a volatile solvent, for instance ethanol. Insofar the solvent slowly evaporates, the critical micelles concentration is reached resulting in the self-assembly of the micelles into ordered phases. Another important topic is the kind of structure directing agent. In the presence of Ca^{2+} cations, non-ionic surfactants (P123, F127, etc.) work better compared with ionic ones (CTAB, etc).

The mesoporous structures for the firsts MBGs were limited to 2D-hexagonal *p6m* ordering. These MBGs were prepared by using Pluronic 123 (P123) as structure directing agent. P123 is an amphiphilic triblock polymer with composition $EO_{20}PO_{70}EO_{20}$, where EO is poly(ethylene oxide) and PO is poly(propylene oxide). Tetraethyl orthosilicate (TEOS), triethyl phosphate (TEP) and calcium nitrate tetrahydrate ($Ca(NO_3)_2 \cdot 4H_2O$) are commonly used as SiO_2, P_2O_5 and CaO precursors, respectively. In a typical synthesis P123 is dissolved in ethanol/acid solution at room temperature. Afterward the appropriate amounts of TEOS, TEP and/or $Ca(NO_3)_2 \cdot 4H_2O$ are added under continuous stirring. The resulting sols are stirred at room temperature for 24 hours and then they must be transferred into dishes with high surface contact with atmosphere. The EISA process takes place at room temperature for about 7 days, until homogeneous and flexible membranes of surfactant-inorganic components composite are obtained.

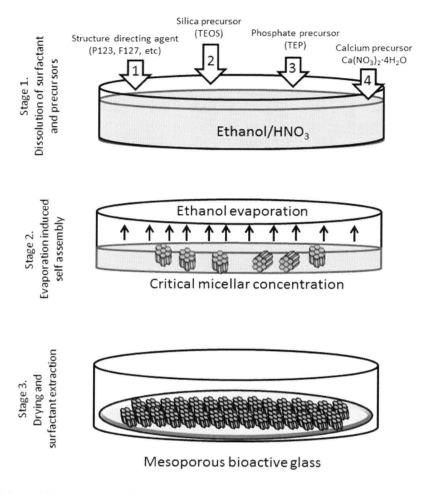

Figure 2. Scheme of the synthesis of a MBG by the EISA process.

Finally, the membranes are heating in air at 700°C for 6 hours, thus obtaining ordered mesoporous powders of SiO_2-P_2O_5-CaO composition.

4. MESOPOROUS BIOACTIVE GLASSES. STRUCTURES AND COMPOSITIONS

The mesoporous structure and textural properties are strongly dependent of several factors such as:

- Type of surfactant and concentration
- Chemical composition of the MBG
- Temperature of evaporation during the EISA process, etc.

Table 3 shows the most important MBGs prepared so far, considering the factors indicated above.

Table 3. Structural and textural properties of MBG, as a function of the type of surfactant, chemical composition and evaporation temperature

Composition (% mol)	Surfactant	Temp (°C)	Symmetry	SBET (m²/g)	Vp (cm³/g)	D_P (nm)	Ref.
58SiO$_2$-37CaO-5P$_2$O$_5$	P123	40	2D-Hexagonal (p6mm)/wormlike	195	0.46	9.45	[24]
75SiO$_2$-20CaO-5P$_2$O$_5$	P123	40	2D-Hexagonal (p6mm)/ Orthorombic (p2mm)	393	0.59	6.0	[24]
85SiO$_2$-10CaO-5P$_2$O$_5$	P123	40	3D-Cubic (Ia-3d)	427	0.61	5.73	[24]
85SiO$_2$-10CaO-5P$_2$O$_5$	P123	20	2D-Hexagonal (p6mm)	473	0.63	5.37	[25]
90SiO$_2$-10CaO	P123	40	2D-Hexagonal (p6mm)	468	0.63	5.37	[25]
75SiO$_2$-21CaO-4P$_2$O$_5$	F127	40	Cubic (Im-3m)	400-506	0.25-0.53	3.2-5.4	[27]
90SiO$_2$-5CaO-5P$_2$O$_5$	P123	r.t	2D-Hexagonal (p6mm)	338	0.46	5.5	[49]
80SiO$_2$-15CaO-5P$_2$O$_5$	P123	r.t	2D-Hexagonal (p6mm)	229	0.31	5.2	[49]
80SiO$_2$-15CaO-5P$_2$O$_5$	P123	r.t	2D-Hexagonal (p6mm)	351	0.49	4.6	[18]
70SiO$_2$-25CaO-5P$_2$O$_5$	P123	r.t	2D-Hexagonal (p6mm)	319	0.49	4.6	[18]
60SiO$_2$-35CaO-5P$_2$O$_5$	P123	r.t	2D-Hexagonal (p6mm)	310	0.43	4.3	[18]
70SiO$_2$-25CaO-5P$_2$O$_5$	F127	r.t	wormlike	300	0.36	5.0	[18]
60SiO$_2$-35CaO-5P$_2$O$_5$	B50-6600	r.t	wormlike	228	0.42	7.1	[18]
80SiO$_2$-15CaO-5P$_2$O$_5$	PN-430 + Brij 70	r.t	2D-Hexagonal (p6mm)	485	0.27	<2.0	[23]
80SiO$_2$-15CaO-5P$_2$O$_5$	P85	r.t	2D-Hexagonal (p6mm)	328	0.36	3.4	[23]
80SiO$_2$-15CaO-5P$_2$O$_5$	P123	r.t	2D-Hexagonal (p6mm)	325	0.40	5.0	[23]
80SiO$_2$-15CaO-5P$_2$O$_5$	B50-6600	r.t	2D-Hexagonal (p6mm)	301	0.41	6.4	[23]

In 2006, the research team of Prof. Vallet-Regí prepared SiO$_2$-P$_2$O$_5$-CaO based MBGs with different mesoporous structures, using the same structure directing agent (P123) in all the cases [24]. Keeping constant the molar ratio between the network formers (SiO$_2$ and P$_2$O$_5$) and the surfactant (P123), the dependence of the mesoporous structure on the CaO content was demonstrated. For high CaO contents (about 36% mol), the already known 2D hexagonal *p6m* phase was obtained. However, for lower CaO content, the formation of a cubic 3D bicontinuous *Ia-3d* phase takes place. This cubic phase is related with more hydrophobic systems, as it is formed from micelles with less curvature. In these cases, the evaporation temperature during EISA process is extremely important. For instance the material MBG-85 with composition 85SiO$_2$– 5 P$_2$O$_5$– 10 CaO (% mol) can present *p6m* or *Ia-3d* structures when the EISA process is carried out at 20°C or 40°C, respectively [25]. These observations indicate that high CaO contents and low temperatures favor the formation of hydrophilic mesoporous systems, derived from micelles with high curvature. The micelles exhibit cylindrical rod shapes resulting in 2D hexagonal structures (*p6m*) after the self-assembly stage. On the contrary, low CaO contents and higher evaporation temperatures favors the formation of less hydrophilic systems, derived from micelles with lower curvature. The micelles exhibit 3D bicontinuous structures resulting in 3D cubic *Ia-3d* after the self-assembly process. The incorporation of CaO favors the formation of hydrophilic systems, because the volume ratio between the inorganic precursors (SiO_2, P_2O_5, CaO) and organic phase (surfactant) increases during the formation of the inorganic-surfactant micelle composite. CaO is a network modifier that increases the volume of the inorganic network as impedes the formation of covalent Si-O-Si bonds. Insofar the volume of the hydrophilic inorganic component increases, 2D hexagonal phases are favored.

Temperature strongly affects the hydrogen bonding between the non-ionic triblock copolymers, used as structure directing agents, and the solvent (ethanol/water). High evaporation temperature results in the weakening of hydrogen interactions, thus interfering in the mesophase formation.

In the case of nonionic triblock copolymers such as P123, the micelle size is strongly dependent on the hydrogen-bond interactions with the solvent, which becomes greater when hydrogen interactions are reduced. Consequently, the hydrophilic/hydrophobic ratio is reduced, favoring hydrophobic mesostructures such as cubic *Ia-3d*, as previously indicated by Zhao and co-workers for pure silica mesoporous materials obtained via the hydrothermal method [26]. This fact explains that MBG 85S, with low CaO content, exhibits a 2D hexagonal *p6m* (hydrophilic) phase and 3D cubic *Ia-3d* (hydrophobic) phase when EISA is carried out at 20°C and 40°C, respectively.

Yun et al. obtained new bioactive mesoporous structures by adding a surfactant with higher hydrophilic character [27, 28]. By using F127, which contains longer chains of ethylene oxide (hydrophilic head) compared with P123, MBGs with cubic *Im3m* cage-type structures were prepared [29]. This structure derives from spherical micelles characteristic of highly hydrophilic systems, that is, when the volume of the hydrophilic part (inorganic components and polar chains of surfactant) clearly predominates over the hydrophobic part (in this case polypropylene oxide chain of F127). In this way, this group prepared for the first time MBGs with spherical pores instead of channels–like structures prepared so far. Figure 3 shows the reconstruction of *Im3m*, *p6m* and *Ia-3d* structures exhibited for the different MBGs prepared so far.

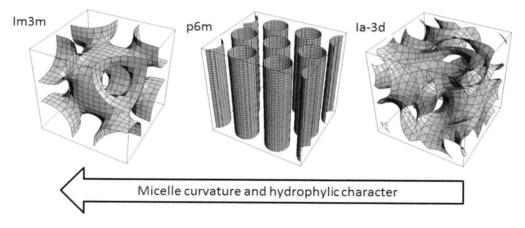

Figure 3. Reconstruction of Im3m, p6m and Ia-3d structures and their relative hydrophilic character.

The Local Environment in the Walls of MBGs

The structure of MBGs can be described as an ordered arrangement of mesopores with sizes between 5 to 9 nm. The wall thickness has been calculated to be around 3 nm, which means very thin walls composed of glassy SiO_2-P_2O_5-CaO material. Such a small volume of the walls initially led to think that the three components would be homogeneously distributed [18], that is, without calcium phosphate clustering as was observed in conventional sol-gel glasses of identical compositions. However, the local environment of MBGs is very heterogeneous even in such narrow walls. Solid state NMR studies can help to understand the local environment of Si, Ca and P within the MBGs walls. Figure 4 shows the solid state ^{31}P single-pulse excitation and $^1H \rightarrow ^{31}P$ CP MAS NMR spectra for MBG-85 with composition 85 SiO_2- 10 CaO- 5 P_2O_5 (% mol). MBG-85 shows a mean maximum of ~2 ppm assigned at the q^0 environment, which is typical of an amorphous orthophosphate. This result indicates that most of phosphate is forming amorphous calcium phosphate (ACP) clusters instead being homogeneously distributed into the SiO_2 matrix. Only a second weak signal sited around -7 ppm appears for these samples. This resonance falls in the range of q^1 tetrahedra [30] and can be assigned to P-O-P or P-O-Si environments [31]. Both single-pulse and $^1H \rightarrow ^{31}P$ CP spectra show very similar results. This fact indicates that the CaO presence in this system leads to the nucleation of acalcium phosphate phase [32], consuming all of the P_2O_5 and thus avoiding polyphosphate formation.

^{29}Si NMR spectroscopy can be used to evaluate the network connectivity of MBGs as a function of chemical composition. The single-pulse ^{29}Si NMR spectrum for MBG-85 is shown in Figure 5 and shows resonances corresponding to Q^4, Q^3, and Q^2_H, but not for Q^2_{Ca}. The absence of non-bonding oxygen associated to the presence of Ca^{2+} is due to the small amount of free CaO (not entrapped as ACP clusters) in this composition. However, $^1H \rightarrow ^{29}Si$ CP spectrum shows a small signal at -85.4 ppm assigned to the Q^2_{Ca} environment, highlighting the presence of this species (although very small) is close to the protons sited at the material surface. The MBG-85 sample evidences a disrupted network at the surface (16.0% of Q^4 units for the MBG-85).

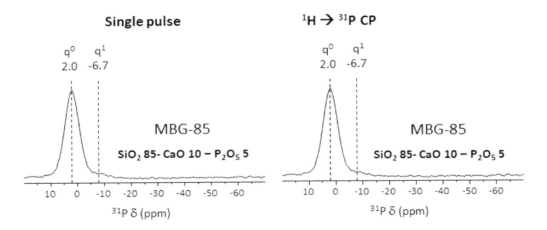

Figure 4. Solid-state ^{31}P single-pulse (left) and cross polarization (right) MAS NMR spectra (with their respective q^n phosphorous environments shown at the top) of MBG-85.

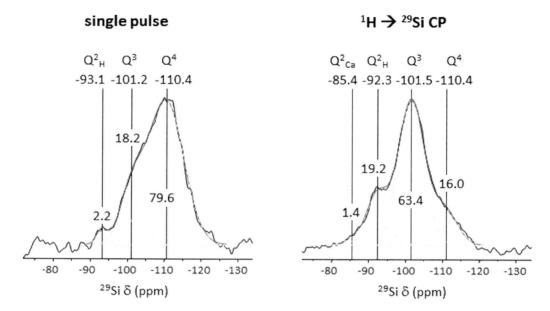

Figure 5. Solid-state ^{29}Si single-pulse (left) and cross polarization (right) MAS NMR spectra (with their respective Q^n silicon environments shown at the top) of MBG-85.

This is clearly indicative of the joint presence of Ca^{2+} and PO_4^{3-}, resulting in calcium phosphate clusters located at the wall surface and, consequently, very accessible to the surrounding fluids when implanted.

A similar scenario has been observed for MBGs with high CaO contents, for instance MBG-58S with 37% of CaO (% mol). Calcium phosphate clusters are also formed in these compositions, but an important amount of Ca^{2+} is not entrapped by phosphate and remains within the silica network acting as network modifier. Samples with high calcium content also evidence the higher presence of Ca^{2+} cations near de MBG surface, whereas the inner part of the pore walls are enriched in more connected SiO_2, as can be observed from the emphasized Q^3_{Ca}, Q^2_{Ca} and Q^1_{Ca} signals respect to Q^4 in the $^1H \rightarrow {}^{29}Si$ CP spectrum (figure 6a). Figure 6.b represents a model of the local environment in this MBG.

5. THE BIOACTIVE BEHAVIOR OF MBGS

S.F. Hulbert et al. defined a *bioactive material* as "one that elicits a specific biological response at the interface of the material which results in the formation of a bond between the tissues and the material" [33]. In the case of bioceramics, the bioactive bond takes place through the formation of an apatite-like phase at the implant-bone interface, without the interposition of a fibrous tissue between them.

Besides the attractive idea of preparing different structures with multicomponent systems, these achievements involve important consequences for their biomedical applications. The bioactive process comprises a set of reactions with the living tissues, which start at the interface between the implant surface and the bone (see figure 1). The amount of material exchanged between the MBGs and the bone determines the kinetic and the type of reaction that will take place between both surfaces.

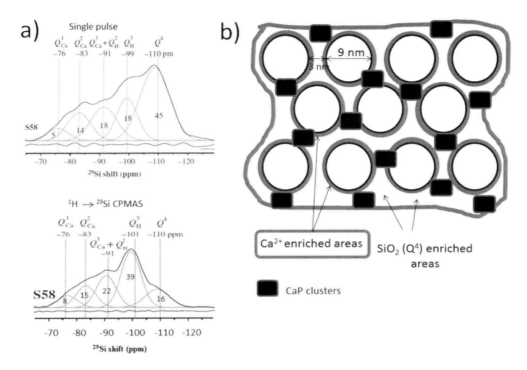

Figure 6. (a) Solid-state ^{29}Si single-pulse (up) and cross polarization (down) MAS NMR spectra of MBG-58 material. The areas for the Q^n units were calculated by a Gaussian line-shape deconvolutions and were displayed by grey dotted lines with their relative populations in %. (b) Scheme of the local environment in MBG-58 with calcium phosphate clusters and Ca^{2+} cations close to the MBG surface.

The material diffusion is higher in those structures with 3D interconnected pores, which extend all through the MBG volume, compared with bidimensional hexagonal structures with non-connected channels [34].

Conventional non mesoporous sol-gel glasses exhibit better bioactive behaviors insofar the CaO content increases. In this sense, the bioactivity of conventional sol-gel glasses is ruled by the CaO amounts. However, the bioactive behavior of MBGs seems to be ruled by the mesoporous ordering and the textural properties. Low CaO contents (about 10% mol) lead to bicontinuous 3D cubic structures with the fastest bioactive response observed so far. This fact is explained because its structure exhibits higher connectivity, surface area and pore volume than that presented by MBGs with more CaO contents (about 36% in mol). These compositions result into *p6m* bidimensional hexagonal structures. Anyway, a minimum CaO amount is required to observe acceptable bioactive behaviors independently of the mesoporous structure.

In vitro test in SBF demonstrates that MBG-85 (85SiO$_2$-10CaO-5P$_2$O$_5$,% mol) material develops an amorphous calcium phosphate after a few minutes of soaking, which evolves to HA after 30 minutes. Although this composition comprises a low CaO amount (10% in mole), the outstanding textural properties of the mesoporous glass greatly accelerates the surface reactions, such as ionic exchange and calcium phosphate nucleation and growth. Besides, MBGs with high calcium content like MBG-58 (58SiO$_2$-37CaO-5P$_2$O$_5$, % mol) together with the good textural properties lead to a new mechanism of bioactivity that mimics the natural mineral maturation of bone in mammals. Figure 7 shows the TEM images obtained from the MBG-58 surface after soaking in SBF at different times. After 1 hour, the surface is coated by a thick layer of amorphous calcium phosphate (ACP). This ACP does not straightly maturate to HA, but evolves forming octacalcium phosphate (OCP) that transform into HA a few hours after. The ACP-OCP-HA mineral maturation pathway is equivalent to that proposed for *in vivo* in mammals. No other bioactive bioceramic is able to reproduce this mineral maturation.

The unique characteristics of MBG with high CaO content explain this fact. MBG-58 has a high amount of Ca^{2+} cations very close to the material surface and accessible to the surrounding fluids. When MBG-58 get in contact with SBF, an ionic Ca^{2+}-H^+ exchange occurs, increasing the Ca^{2+} concentration in SBF while incorporating H^+ to the MBG surface. This exchange is so intense that the MBG surface is transiently acidified, thus falling in the pH range stability for OCP nucleation. Thereafter, the presence of pre-formed calcium phosphate clusters facilitates the fast nucleation of nanocrystalline HA.

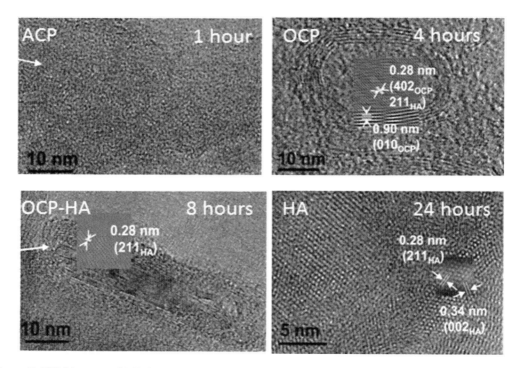

Figure 7. TEM images of MBG-58 after being soaked in SBF at different times. 1 hour in SBF: A large amount of ACP is observed next to MBG grain; 4 hours in SBF: nanometrical OCP oval nuclei are observed within the ACP matrix; 8 hours in SBF. Needle shaped apatite crystallizes within the ACP matrix; 24 hours: evidence of the microstructural evolution of OCP nuclei to apatite crystallites.

6. CONTROLLED DRUG DELIVERY FROM MBGS

In 2001 the SiO_2 based mesoporous material MCM-41was firstly proposed as a potential drug delivery system (DDS) [35]. MCM-41 belongs to the M41S materials family that resulted from the incorporation of a structure directing agent, cetylammonium bromide CTAB, to the SiO_2 synthesis via sol-gel method [36].

The CTAB molecules self-organize into micelles that co-assemble with the soluble SiO_2 precursors resulting into ordered SiO_2-CTAB structures. After CTAB removing by calcinations or acid extraction, a SiO_2 mesoporous material is obtained exhibiting excellent textural properties for drug delivery purposes [37] such as:

- Ordered mesoporous network with homogeneous pore size, which allows a close control on the kinetic release of the drugs incorporated.
- High surface are and pore volume to adsorb the required drug dosage.
- High density of silanol (Si-OH) groups at the surface, which can be functionalized for a better control of drug load and release.

The fast development of DDSs based on pure SiO_2 mesoporous materials (such as MCM-41 or SBA-15) facilitated the incorporation of this function to MBGs. Since MBGs keep the excellent surface and porosity properties, as well as the capability to be functionalized, several research groups understood the potential of MBGs a matrix for local drug delivery in bone tissue [38]. Thus, their excellent bioactive behavior could be added to the local delivery of antibiotics, antitumoral, antiosteoporotic and osteogenic agents as can be seen in table 4.

One of most common and complicated problems in orthopedic surgery is the implant infection. The incidence of osteomyelitis caused by *S. aureous* and *S. epidermidis* is relatively high. The problem is more complicated when these microorganisms develop a biofilm over the implant surface.

In this scenario, the systemic administration of antibiotics is not efficient and the only safe solution is the implant withdrawal to avoid septicemia. For this reason, the controlled and local drug delivery from the implant is considered as an excellent alternative treatment against infection. Xia and Chang proposed for the first time the incorporation of antibiotics within the porous network of MBGs [47, 48].

Table 4. List of therapeutic agents incorporated into MBGs

Therapeutic Purpose	Drug/growth factor	Ref.
Osteogenic drugs and growth factors	Dexamethasone	[39]
	Bone morphogenetic proteins (BMP)	[40]
	β-Fibroblast growth factor (β-FGF)	[41]
Angiogenic agents	Vascular endothelial growth factor (VEGF)	[42]
	Dimethyloxallyl glycine	[43]
Anti-inflammatory	Ibuprofen	[44]
Antibiotic/antiseptic	Gentamicin	[47]
	Ampicilin	[45]
	Triclosan	[55]
Antitumoral	Doxorubicin	[46]

These authors applied a conventional drug impregnation strategy by soaking the 58S-MBGs into a highly concentrated solution of gentamicin sulphate. They could observe that MBGs could incorporate 3 times more drug than a conventional sol-gel glass with the same composition. Besides, the kinetic of the gentamicin delivery profile can be explained in terms of the different sites of the pores that gentamicin molecules can occupy and schemed in figure 8:

- Outside window of the pore (location 1)
- Linked to the inner pore wall by hydrogen bonding (location 2)
- Inner part of the pore linked to another gentamicin molecule (location 3)
- External surface of MBG (location 4)

The drug loading capability and the kinetic release no only depends on the porosity, but also on the chemical composition, as Zhao et al. demonstrated [49]. For instance, the amount of tetracycline absorbed in MBGs is also function of the CaO content. CaO acts as chelating agent in the pore wall. Consequently, the antibiotic release is slower insofar the CaO content increase in the systems SiO_2-CaO-P_2O_5, as the affinity of tetracyclines to Ca^{2+} would control the release to the surrounding fluids. On the other hand, several studies have been carried out to test the drug loading and release as a function of the structure directing agent used, for instance P123 and F127. Some differences could be observed, which were explained in terms of the different textural properties for each MBG [50].

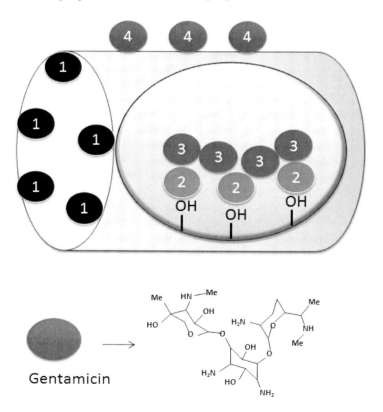

Figure 8. Different locations of gentamicin molecule respect to the mesopores in MBGs.

Even considering the importance of the textural properties for the control of drug release, the previous experience with pure SiO$_2$ mesoporous materials had demonstrated that only the surface functionalization could provide a satisfactory control on drug release [37]. Otherwise, the burst effect and the impossibility of achieving zero order kinetics constrained the use of MBGs as control drug delivery matrixes. However, the functionalization of MBGs can result in the bioactivity loss as the surface is modified [51]. Recently, a systematic study about how functionalization agents control the drug release without significantly affecting the MBGs bioactivity [52]. By means of functionalization with thiol, amino, hydroxyl, and phenyl groups zero order kinetics were obtained. The kinetic constants could be controlled by tailoring the hydrogen bonding strength between the drug and the different functional groups.

7. BIOACTIVE MESOPOROUS MICROSPHERES

In addition to the mesoporous structure and chemical composition, the morphology and size of the particles is one the most important features of MBGs when are intended for DDS. A system formed by small size spheres offers much better possibilities for reproducibility and control release compared with irregular bulky particles with different sizes. The research team of Prof. Stucky proposed the preparation of bioactive mesoporous microspheres and tested their hemostatic properties [53]. These materials have been proposed as first aid agents in those hemorrhages with massive blood loosening [54]. Moreover, these hemostatic materials do not show the thermal effect commonly shown by other systems thus avoiding the necrosis of surrounding healthy tissues. These spheres have size diameters ranging between 100 nm and 1 micrometer and can be prepared in the system SiO$_2$-CaO-P$_2$O$_5$ using the surfactant P$_{123}$ as structure directing agent.

There are several synthesis methods for the preparation of bioactive mesoporous spheres. Among them, aerosol assisted methods and basic precipitation methods must be highlighted. Aerosol assisted methods combine the EISA method with the formation of very small aerosol droplets containing all the precursors that, subsequently, will constitute the bioactive mesoporous microspheres. The ordered phase is formed within these droplets when the solvent is partially evaporated. Thereafter, the particles are pyrolised to form systems like SiO$_2$-CaO-P$_2$O$_5$ or similar. Some of these microspheres have been proposed for bone grafting and DDS [55]. These particles are intended for bone grafting and augmentation in small bone defects, for instance in periodontal surgery before the implantation of endosseous titanium abutments. In this sense, we can obtain a therapeutic synergy by combining the regenerative capability of MBGs and the pharmacologic effect when the particles are loaded with antiseptic agents.

A second strategy to prepare mesoporous bioactive microspheres is the precipitation from diluted solutions of precursors and surfactant in alkaline medium. This method produces smaller and monodisperse spheres than those prepared by aerosol. However the agglomeration of the particles is very likely to occur, [56] and the Ca^{2+} precipitation as calcium hydroxide often takes place under basic pH. Obtaining monodisperse nanoparticles would allow the incorporation into the blood system. The hydrodynamic stability in the blood torrent is reached when the particle size is kept between 50 and 300 nm. Larger particles are retained in lungs and liver, whereas those with smaller sizes can cross the vascular endothelium and being non-specifically distributed all over the body. Finally, the combination of MBGs with biocompatible polymers must be highlighted. These systems keep the most of the bioactive properties while ensuring a controlled drug delivery kinetic [57]. In these composites the MBGs does not form the microsphere itself but takes part as a discrete phase into the continuous polymeric phase. In these microspheres the drug can be incorporated into the mesoporous structure of MBGs, within the polymeric matrix and in both.

Ultrafine hollow fibers have been prepared con MBGs [58] by means of electrospinning technique. Through the control of the water/ethanol ratio in the precursors solution and the addition of a phase separation agent, the spinoidal decomposition of the solution is reached during the fibers fabrication. Consequently, the formation of hollow fibers with mesoporous structures in the walls takes place. The drug loading and controlled release capabilities are closely related with length of the fibers while exhibiting excellent bioactive behavior.

Stimuli-responsive or smart systems are very interesting drug delivery devices to achieve highly controlled and specific drug releases. These systems have been successfully applied to pure SiO_2 mesoporous materials [37]. Stimuli responsive mesoporous systems consist on the design of open/close gates at the entrance of the pores, which are controlled by external stimuli. In this sense, smart devices have been tailored based on molecular gates, nanoparticles and nanomachines. In the field of MBGs, there are a few works comprising stimuli-responsive systems. Recently, Lin et al. have prepared a molecular gate system based on the photodimerization of the cumarine molecule [59]. This molecular gate is activated by means of UV light, in such a way that irradiation with light of 310 nm or higher leads to cumarine dimerization thus closing the gate. On the contrary, when the system is irradiated with UV light of wavelengths of 250 nm or lower, the monomer is regenerated thus opening the gate and releasing the drug entrapped within the mesopores. Another stimuli-responsive DDS are those combining magnetic properties and controlled drug delivery. The incorporation of magnetic nanoparticles within mesoporous systems supplies new capabilities to MBGs. Through the action of an external magnetic field, the MBG nanoparticle can be vectored toward localized regions. Moreover, the capability of superparamagnetic nanoparticles of heating under an AC field opens the possibility of using them as thermoseeds for the treatment of tumor by hyperthermia. These systems have been studied for pure SiO_2 mesoporous systems [60-63] as well as for SiO_2-CaO-P_2O_5 MBGs [64].

8. MBGs Scaffolds for Regenerative Bone Therapies

The regeneration of critical bone defects is one of the most challenging and difficult issue to be tackled by biomaterials science. In the case of small bone defects, for instance defects in periodontal locations, the material can be implanted as granules or mixed with blood or physiological serum to form a moldable paste. However, the bone regeneration in a critical defect requires pieces that must be fitted to the defect size and morphlogy. Very often, the implantation site bears high mechanical loads, thus constraining the use of ceramic and porous materials. Finally, the implants must have pores large enough to let the bone cells colonization and blood vessels formation [65, 66]. Considering all together, we can envision the magnitude of the quest tackled by MBGs in bone regenerative therapies.

The first attempt to shape MBGs as pieces was carried out by Stucky et al. [67]. These authors prepared an injectable paste by mixing a SiO_2-CaO-P_2O_5 MBG with a buffered solution of ammonium phosphate. This paste settled in a similar way to calcium phosphate cements, allowing the formation of solid pieces with certain mechanical strength and high bioactive behavior. However, these materials do not exhibit the porous architecture required for scaffolding purposes in regenerative treatments. For such purpose, interconnected macropores with sizes about 400 μm are needed. The pioneering team shaping MBGs as macroporous scaffolds was the research group of Prof. Yun [68]. This group incorporated the rapid prototyping methods to the preparation of MBGs macroporous scaffolds with ordered macroporous architecture, as well as to the free form fabrication of MBGs implants through computer assisted designs (Figure 9). Thee scaffolds show three different porous systems in the nanometer, micrometric and macroscopic levels. For this purpose, the following combination is required:

- A structure directing agent to achieve an ordered mesoporous arrangement at the nanometer level
- A polymer, such as methylcellulose, to leave pores of several microns after calcination.
- A macroporous architecture tailored by rapid prototyping.

The main drawback of the scaffolds so prepared is their fragility, which limits their clinical application in the orthopedic field. Although the formation of the new apatite phase during the bioactive behavior reinforces the mechanical strength of MBG based scaffolds, this enhancement is not enough for satisfying the mechanical requirements in load bearing locations [69]. The combinations of MBGs with biocompatible polymers, such as polycaprolactone (PCL), have been proposed [70]. These composite scaffolds have shown excellent results regarding the mechanical improvement, while keeping their bioactive properties under in vitro conditions. In vitro biocompatibility tests using osteoblastic cells showed a better cell adhesion and osteoblasts proliferation compared to those scaffolds made of only PCL. The excellent cell biocompatibility has been subsequently confirmed by the same group and other authors with different cell cultures [71-76].

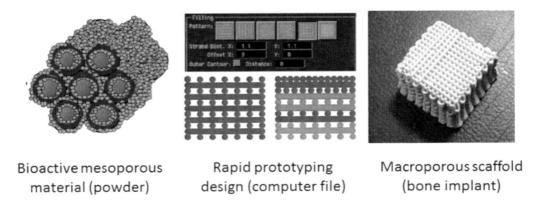

Figure 9. The three main stages in the fabrication of MBG macroporous implants for bone tissue regeneration.

The macroporous architecture of the scaffolds can be also designed from polymers that are easily removable by calcination. This methodology has been also applied using polyurethane sponges [77]. After being removed by calcination pieces with highly interconnected macropores are obtained, thus mimicking the morphology of the cancellous bone. Based on this kind of structures and considering the antiosteoporotic activity of strontium renalate, Wu et al. prepared mesoporous scaffolds in the SrO-SiO$_2$ system [78]. This composition does not have bioactive properties, but the capability to release strontium comprises antiosteoporotic action and facilitates the scaffold degradation. In this sense, the development of macroporous scaffolds manufactured with MBGs, and combined with antiosteoporotic agents, opens very promising perspectives for the treatment of fractures in osteoporotic patients [79].

CONCLUSION AND PERSPECTIVES

The regenerative therapies in the field of bone pathologies require the synergy of materials and pharmaceutical sciences. In the last decade, the strategy in biomaterials design has shift towards the tissue regeneration instead of substitution. The current trend comprises to supply to the bone the appropriated scaffolding to stimulate its self-regeneration properties. In this sense, mesoporous bioactive glasses are called to play an important role as scaffolds for bone tissue regeneration and drug delivery systems.

Mesoporous bioactive materials can supply unique features in terms of surface and porosity that contributes to better regeneration properties and provides excellent matrixes for drug delivery. Moreover, the incorporation of rapid prototyping techniques to the implants fabrication allows the materials preparation, by controlling the different organization levels of the materials, i.e., from the nanometer scale up to the macroscopic size and morphology.

REFERENCES

[1] Gentleman, E., and Polak J. M. (2006) Historic and current strategies in bone tissue engineering: Do we have a hope in Hench? *J. Mater. Sci. Mater. Med.*, 17, 1029-1035.

[2] Hench, L. L. Genetic design of bioactive glass. (2009). *J. Eur. Ceram. Soc.*, 29, 1257-1265.

[3] Vallet-Regí, M. (2006). M. *Revisiting ceramics for medical application Dalton Trans.*, 44, 5211-5220.

[4] Hench, L. L., Polak, J. M. (2002). Third-generation biomedical materials. *Science*, 295, 1014-1017.

[5] Hench, L. L., Splinter, R. J., Greenlee, T. K., and Allen, W. C. (1971). Bonding mechanisms at the interface of ceramic prosthetic materials. *J. Biomed. Mater. Res.*, 2, 117-141.

[6] Li, R., Clark, A. E., and Hench, L. L. (1991). An Investigation of Bioactive Glass Powders by Sol-Gel Processing. *J. Appl. Biomater.*, 2, 231-239.

[7] Arcos, D., Vallet-Regí, M. (2010). Sol-gel silica based biomaterials and bone tissue regeneration. *Acta. Biomaterialia*, 6, 2874-2888.

[8] Mahony, O., and Jones, J. R. (2008). Porous bioactive nanostructured scaffolds for bone regeneration: a sol-gel solution. *Nanomedicine*, 3, 233-245.

[9] Coradin, T., Boissière, M., and Livage, J. (2006). Sol-gel Chemistry in Medicinal Science *Curr. Med. Chem.*, 13, 99-108.

[10] Vallet-Regí, M. (2001). Ceramics for medical applications. *J. Chem. Soc. Dalton Trans.*, 97-108.

[11] Vallet-Regí, M., Ragel, C. V., and Salinas, A. J. (2003). Glasses with medical applications. *Eur. J. Inorg. Chem.*, 1029-1042.

[12] Li, P., Ohtsuki, C., Kokubo, T., Nankanishi, K., Soga, N., Nakamura, T., and Yamamuro, T. (1992). Apatite formation induced by silica gel in a simulated body fluid. *J. Am. Ceram. Soc.*, 75, 2091-2097.

[13] Li, P., Ohtsuki, C., Kokubo, T., Nankanishi, K., Soga, N., Nakamura, T., and Yamamuro, T. (1993). Effects of ions in aqueous media on hydroxyapatite induction by silica gel and its relevance to bioactivity of bioactive glass and glass-ceramics. *J. Appl. Biomat.*, 4, 221-229.

[14] Pereira, M. M., Clark, A.E., and Hench. L. L. (1995). Effect of texture on the rate of hydroxyapatite formation on silica gel surface. *J. Am. Ceram. Soc.*, 78, 2463-2468.

[15] Horcajada, P., Rámila, A., Boulahya, K., González-Calbet, J., andVallet-Regí, M. (2004). Bioactivity in ordered mesoporous silica materials. *Solid State Sci.*, 6, 1295-1300.

[16] Izquierdo-Barba, I., Ruiz-González, L., Doadrio, J. C., González-Calbet, J. M., andVallet-Regí, M. (2005).Tissue regeneration: a new property of mesoporous materials. *Solid State Sci.*, 7, 983-989.

[17] Kokubo, T., Kushiani, H., Sakkta, S., Kitsugi, T., Yamamuro, T. (1990). Solutions able to reproduce in vivo surface-structure changes in bioactive glass-ceramic A-W. *J. biomed. Mater. Res.*, 24, 721-734.

[18] Yan, X. X., Yu, C. Z., Zhou, X. F., Tang, J. W., and Zhao, D. Y. (2004). Highly ordered mesoporous bioactive glasses with superior in vitro bone forming bioactivities. *Angew. Chem. Int. Ed.*, 43, 5980-5984.

[19] Rámila, A., Balas, F., andVallet-Regí, M. (2002). Synthesis routes for bioactive sol-gel glasses: alkoxides vs. nitrates. *Chem. Mater.*, 14, 542-548.

[20] Pereira, M. M., Clark, A. E., and Hench, L. L. (1994). Calcium phosphate formation on sol-gel derived bioactive glasses in vitro. *J. Biomed. Mater. Res.*, 28, 693-698.

[21] Vallet-Regí, M., Arcos, D., and Pérez-Pariente, J. (2000). Evolution of porosity during in vitro hydroxycarbonate apatite growth in sol-gel glasses. *J. Biomed. Mater. Res.*, 51, 23-28.

[22] Brinker, C. J., Lu, Y. F., Sellinger, A., and Fan, H. Y. (1999).Evaporation-induced self-assembly: nanostructures made easy. *Adv. Mater.*, 11, 579.

[23] Yan, X., Wei, G., Zhao, L., Yi, J., Deng, H., Wang, L., Lu, G. and Yu, C. (2010). Synthesis and in vitro bioactivity of ordered mesosructured bioactive glasses with adjustable pore sizes. *Microporous Mesoporous Mater.*, 132, 282-289.

[24] López-Noriega, A., Arcos, D., Izquierdo-Barba, I., Sakamoto, Y., Terasaki, O., and Vallet-Regí, M. (2006). Ordered mesoporous bioactive glasses for bone tissue regeneration. *Chem. Mater.*, 18, 3137-3144.

[25] García, A., Cicuéndez, M., Izquierdo-Barba, I., Arcos, D., and Vallet-Regí, M. (2009). Essential role of calcium phosphate heterogeneities in 2D-hexagonal and 3D-cubic SiO2-CaO-P2O5 mesoporous bioactive glasses. *Chem. Mater.,* 21, 5474-5484.

[26] Li, Z., Chen, D. H., Tu, B., Zhao, D. Y. (2007). Synthesis and phase behaviors of bicontinuous cubic mesoporous silica from triblock copolymer mixed anionic surfactant. *Microporous Mesoporous Mater.,* 105, 34–40.

[27] Yun, H. S., Kim, S. E., andHyeon, Y. T. (2007). Highly ordered mesoporous bioactive glasses with Im3m symmetry. *Matter Lett.,* 61, 4569-4572.

[28] Yun, H. S., Kim, S. E., andHyeon, Y. T. (2008). Preparation of 3D cubic ordered mesoporous bioactive glasses. *Solid State Sci.,* 10, 1083-1092.

[29] Yun, H., Kim, S., Hyeon, Y. (2007) Highly ordered mesoporous bioactive glasses with Im3m symmetry. *Mater. Lett.,* 61, 4569–4572.

[30] MacKenzie, K. J. D.; Smith, M. E. Multinuclear Solid-State NMR of Inorganic Materials; Pergamon Press: Amsterdam, 2002.

[31] Leonova, E., Izquierdo-Barba, I., Arcos, D., López-Noriega, A., Hedin, N., Vallet-Regí, M., Eden, M. J. (2008) Multinuclear Solid-State NMR Studies of Ordered Mesoporous Bioactive Glasses. *Phys. Chem. C,* 112, 5552–5562.

[32] Mathew, R., Turdean-Ionescu, C., Stevensson, B., Izquierdo-Barba, I., García, A., Arcos, D., Vallet-Regí, M., Edén, M. (2013). Direct Probing of the Phosphate-Ion Distribution in Bioactive Silicate Glasses by Solid-State NMR: Evidence for Transitions between Random/Clustering Scenarios. *Chem. Mater.,* 25, 1877-1885.

[33] Hulbert, S. F., Hench, L. L., Forbers, D., Bowman, L. S. (1982) History of Bioceramics. *Ceram. Int.,* 8, 131 – 140.

[34] Izquierdo-Barba, I., Arcos, D., Sakamoto, Y., Terasaki, O., López-Noriega, A., and Vallet-Regí, M. (2008). High performance mesoporous bioceramics mimicking bone mineralization. *Chem. Mater.,* 20, 3191-3198.

[35] Vallet-Regí, M., Rámila, A., del Real, R. P., and Pérez-Pariente, J. (2001). A new property of MCM- 41: Drug delivery system. *Chem. Mater.,* 13, 308-311.

[36] Kresge, C. T., Leonowicz, M. E., Roth, W. J., Vartuli, J. C., and Beck, J. S. (1992). Ordered mesoporous molecular sieves synthesized by a liquid-crystal template mechanism. *Nature,* 359, 710-712.

[37] Vallet-Regí, M., Balas, F., and Arcos, D. (2007). Mesoporous materials for drug delivery. *Angew. Chem. Int. Ed.,* 46(40), 7548–7558.

[38] Wu, C., and Chang, J. Multifunctional mesoporous bioactive glasses for effective delivery of therapeutic ions and drug/growth factors. *J. Control. Rel.,* (2014), http://dx.doi.org/10.1016/j.jconrel.2014.04.026

[39] Wu, C., Miron, R., Sculeaaan, a., Kaskel, S., Doert, tT., Schullze, R., and Zhang, Y. (2011). Proliferation, differentiation and gene expression of osteoblasts in boron-containing associated with dexamethasone deliver from mesoporous bioactive glass scaffolds. *Biomaterials,* 32, 7068-7078.

[40] Dai, C., Guo, H., Lu, J., Shi, J., Wei, J. and Liu, C. (2011) Osteogenic evaluation of calcium/magnesium-doped mesoporous silica scaffold with incorporation of rhBMP-2 by synchrotron radiation-based muCT. *Bioaterials,* 32, 8506-8517.

[41] Perez, R. A., El-Fiqi, A., Park J. H., Kim, T. H., Kim, J. H. and Kim, H. W. (2013). Therapeutic bioactive microcarriers: co-delivery of growth factors and stem cells for bone tissue engineering. *Acta. Biomater.,* 10, 520-530.

[42] Wu, C., Fan, J., Chang, Y. and Xiao, Y. (2013) Mesoporous bioactive glass scaffolds for efficient delivery of vascular endothelial growth factor. *J. Biomater. Appl.*, 28, 367-374.

[43] Wu, C., Zhou, Y., Chang, J. and Xiao, Y. (2013). Delivery of diethyloxallyl glycine in mesoporous bioactive glass scaffolds to improve angiogenesis and osteogenesis of human bone marrow stromal cells. *Acta. Biomater.*, 9, 9159-9168.

[44] Lin, J., Fan., Y., Yang, P. P., Huang, S. S., Jiang, J. H., and Lian, H. Z. (2009). Luminescent and mesoporous europium-doped bioactive glasses (MBG) as a drug carrier. *J. Phys. Chem.*, C 113, 7826-7830.

[45] Wu, C., Zhou, Y., Fan, W. Han, P., Chang, J., Yuen, J., Zhang, M. and Xiao, Y. Hyoxia-mimicking mesoorous bioactive glass scaffolds with controlable cobalt ion reléase for bone tissue engineering. (2012). *Biomaterials*, 33, 2076-2085.

[46] Wu, C. T., Fan, W., and Chang, J. (2013). Functional mesoporous bioactive glass nanospheres: synthesis, high loading efficiency, controllable delivery of doxorubicin and inhibitory effect on bone cancer cells. *J. Mater. Chem. B*, 1, 2710-2718.

[47] Xia, W., and Chang, J. (2006). Well-ordered mesoporous bioactive glasses (MBG): A promising bioactive drug delivery system. *J. Control Rel.*, 110, 522-530.

[48] Xia, W., and Chang, J. (2008). Preparation, in vitro bioactivity and drug release property of well-ordered mesoporous 58S bioactive glass. *J. Non-Cryst. Solids*, 354, 1338-1341.

[49] Zhao, L. Z., Yan, X. X., Zhou, X. F., Zhou, L., Wang, H. N., Tang, J. W., and Yu, C. Z. (2008). Mesoporous bioactive glasses for controlled drug release. *Microporous Mesoporous Mater.*, 109, 210-215.

[50] Zhao, Y. F., Loo, S. C. J., Chen, Y. Z., Boey, F. Y. C., and Ma, J. (2008).In situ SAXRD study of sol–gel induced well-ordered mesoporous bioglasses for drug delivery. *J. Biomed. Mater. Res.*, 85A, 1032-1042.

[51] Sun, J., Li, Y. S., Li, L., Zhao, W. R., Li, L., Gao, J. H., Ruan, M. L., and Shi, J. L. (2008). Functionalization and bioactivity in vitro of mesoporous bioactive glasses. *J. Non-Cryst. Solids*, 354, 3799-3805.

[52] López-Noriega, A., Arcos, D., andVallet-Regí, M. (2010). Functionalizing Mesoporous Bioglasses for Long-Term Anti-Osteoporotic Drug Delivery. *Chem. Eur. J.*, 16, 10879-10886.

[53] Ostomel, T. A., Shi, Q. H., Tsung, C. K., Liang, H. J., andStucky, G. D. (2006). Spherical Bioactive Glass with Enhanced Rates of Hydroxyapatite Deposition and Hemostatic Activity. *Small*, 2, 1261-1265.

[54] Ostomel, T. A., Shi, Q., and Stucky, G. D. (2006). Oxide hemostatic activity. *J. Am. Chem. Soc.*, 128, 8384-8385.

[55] Arcos, D., López-Noriega, A., Ruiz-Hernández, E., Terasaki, O., and Vallet-Regí, M. (2009). Ordered mesoporous microspheres for bone grafting and drug delivery. *Chem. Mater.*, 21, 1000-1009.

[56] Zhao, S., Li, Y. B., and Li, D. X. (2010). Synthesis and in vitro bioactivity of CaO–SiO2–P2O5 mesoporous microspheres. *Microporous Mesoporous Mater.*, 135, 67-73.

[57] Li, X., Wang, X. P., Zhang, L. X., Chen, H. R., and Shi, J. L. (2009). MBG/PLGA Composite Microspheres with Prolonged Drug Release. *J. Biomed. Mater. Res. Appl. Biomater.*, 89B, 148-154.

[58] Hong, Y. L., Chen, X. S., Jing, X. B., Fan, H. S., Gu, Z. W., and Zhang, X. D. (2010). Fabrication and Drug Delivery of Ultrathin Mesoporous Bioactive Glass Hollow Fibers. *Adv. Funct. Mater.,* 20, 1503-1510.

[59] Lin, H.-M., Wang, W.-K., Hsiung, P.-A., andShyu, S.-G. (2010). Light-sensitive intelligent drug delivery systems of coumarin-modified mesoporous bioactive glass. *Acta. Biomaterialia,* 6, 3265-3263.

[60] Vivero-Escoto, J. L., Slowing, I. I., Trewyn, B. G., and Lin, V. S.-Y. (2010). Mesoporous Silica Nanoparticles for Intracellular Controlled Drug Delivery. *Small,* 6, 1952-1967.

[61] Martín-Saavedra, F., Ruíz-Hernández, E., Boré, A., Arcos, D., Vallet-Regí, M., and Vilaboa, N. (2010). Magnetic mesoporous silica spheres for hyperthermia therapy. *Acta. Biomaterialia,* 6, 4522-4561.

[62] Julian-López, B., Boissiere, C., Chaneac, C., Grosso, D., Vasseur, S., Miraux, S., Duguet, E., and Sanchez, C. (2007). Mesoporous maghemite–organosilica microspheres: a promising route towards multifunctional platforms for smart diagnosis and therapy. *J. Mater. Chem.,* 17, 1563-1569.

[63] Ruiz-Hernández, E., López-Noriega, A., Arcos, D., Izquierdo-Barba, I., Terasaki, O., andVallet-Regí, M. (2007). Aerosol-assisted synthesis of magnetic mesoporous silica spheres for drug targeting. *Chem. Mater.,* 19, 3455-3463.

[64] Li, X., Wang, X. P., Hua, Z., and Shi, J. L. (2008). One-pot synthesis of magnetic and mesoporous bioactive glass composites and their sustained drug release property. *Acta. Materialia,* 56, 3260-3265.

[65] Hutmacher, D. W. (2000). Polymeric Scaffolds in Tissue Engineering Bone and Cartilage. *Biomaterials,* 21, 2529-2543.

[66] Stevens, M. M., and George, J. (2005). Exploring and Engineering the Cell Surface Interface. *Science,* 310, 1135-1138.

[67] Shi, Q. H., Wang, J. F., Zhang, J. P., Fan, J., and Stucky, G. D. (2006). Rapid-setting, mesoporous, bioactive glass cements that induce accelerated in vitro apatite formation. *Adv. Mater.,* 18, 1038-1042.

[68] Yun, H.-S., Kim, S.-E., and Hyeon, Y.-T. (2007). Design and preparation of bioactive glasses with hierarchical pore networks. *Chem. Comm.,* 2139-2141.

[69] Arcos, D., Vila, M., López-Noriega, A., Rossignol, F., Champion, E., Oliveira, F. J., and Vallet-Regí, M. Mesoporous bioactive glasses: mechanical reinforcement by means of a biomimetic process. (2011). *Acta. Biomaterialia,* 7, 2952-2959.

[70] Yun, H.-S., Kim, S.-E, Hyun, Y.-T, Heo, S.-J, and Shin, J.-W. (2007). Three-Dimensional Mesoporous−Giant porous Inorganic/Organic Composite Scaffolds for Tissue Engineering. *Chem. Mater.,* 19, 6363-6366.

[71] Yun, H.-S., Kim, S.-E, Hyun, Y.-T, Heo, S.-J, Shin, J.-W. (2008). Hierarchically mesoporous-macroporous bioactive glasses scaffolds for bone tissue regeneration. *J. Biomed. Mater. Res. Part B: Appl. Biomater.,* 87B, 374-380.

[72] Wang, X. P., Li, X., Onuma, K., Ito, A., Sogo, Y., Kosuge, K., and Oyane, A. (2010). Mesoporous bioactive glass coatings on stainless steel for enhanced cell activity, cytoskeletal organization and AsMg immobilization. *J. Mater. Chem.,* 20, 6437-6445.

[73] Alcaide, M., Portolés, P., López-Noriega, A., Arcos, D., Vallet-Regí, M., and Portolés, M. T. (2010). Interaction of an ordered mesoporous bioactive glass with osteoblasts,

fibroblasts and lymphocytes demonstrates its biocompatibility as a potential bone graft material. *Acta. Biomaterialia*, 6, 892-899.

[74] Zhu, Y. F., Wu, C. T., Ramaswamy, Y., Kockrick, E., Simon, P., Kaskel, S., and Zreiqat, H. (2008). Preparation, characterization and in vitro bioactivity of mesoporous bioactive glasses (MBGs) scaffolds for bone tissue engineering. *Microporous Mesoporous Mater.*, 112, 494-503.

[75] Shih, C. J., Chen, H. T., Huang, L. F., Lu, P. S., Chang, H. F., and Chang, I. L. (2010). Synthesis andin vitro bioactivity of mesoporous bioactive glass scaffolds. *Mater. Sci. Eng. C,* 30, 657-663.

[76] Wei, G. F., Yan, X. X., Yi, J., Zhao, L.Z., Zhou, L., Wang, Y. H., and Yu, C. Z. (2011). Synthesis and in-vitro bioactivity of mesoporous bioactive glasses with tunable macropores. *Microporous Mesoporous Mater.*, 143, 157-165.

[77] Zhu, Y. F., and Kaskel, S. (2009). Comparison of the in vitro bioactivity and drug release property of mesoporous bioactive glasses (MBGs) and bioactive glasses (BGs) scaffolds. *Microporous Mesoporous Mater.*, 118, 176-182.

[78] Wu, C. T., Fan, W., Gelinsky, M., Xiao, Y., Simon, P., Schulze, R., Doert, T., Kuo, Y. X., and Cuniberti, G. (2011). Bioactive SrO–SiO2 glass with well-ordered mesopores: Characterization, physiochemistry and biological properties. *Acta. Biomaterialia*, 7, 1797-1806.

[79] Arcos, D., Boccaccini, A. R., Bohner, M., Díez-Pérez, A., Epple, M., Gómez-Barrena, E., Herrera, A., Planell, J. A, Rodríguez-Mañas, L., and Vallet-Regí, M. (2014). The relevance of biomaterials to the prevention and treatment of osteoporosis. *Acta. Biomaterialia,* 10, 1793-1805.

Chapter 3

SYNTHESIS OF A MESOPOROUS SILICA

Kulamani Parida[1], Suresh Kumar Dash[2] and Dharitri Rath[2]

[1]Centre for Nanoscience and Nanotechnology, Institute of Technical Education and Research, Siksha 'O'Anusandhan University, Bhubaneswar, Odisha, India
[2]Department of Chemistry, Institute of Technical Education and Research, Siksha 'O'Anusandhan University, Bhubaneswar, Odisha, India

ABSTRACT

The substitution of various main group metals such as Al, Ga, In, Zn and Cd, transition metals such as Ti, V, Fe, Cu, Nb, Mo and Zr and rare earth elements like La and Ce into the frame-work of MCM-41 has received much attention to develop more efficient and stable materials for application in adsorption, catalysis, photocatalysis, separation and chemical sensing. The metal ions can be incorporated via ion-exchange, impregnation, co-condensation and solid state vaporization. The incorporation of a metal into the frame-work of MCM-41 seems to be easy but the retention of mesoscopic order is often difficult and the stability depends on the amount of dopant material. In case of tetravalent metals viz. Ti, Zr, Ge and Sn the grafting could be done by isomorphous substitution of tetravalent Si, into the walls of MCM-41 during in-situ or Liquid Crystal Templating (LCT) mechanism where the charge on the walls is balanced and neutral. But the incorporation of trivalent metals (Al, B, Ga and Fe) introduces a negative charge and generates a cation exchange or Bronsted acid site on the surface of MCM-41. MCM-41 have been used to custom synthesize catalysts because of the controllable properties, such as pore size, active phase incorporation, crystal size, and morphology, among others. The materials prepared by hydrothermal methods provides a very high surface area and narrow uniform pore distribution in the mesopore region, and are highly thermally stable whereby the ions are not leached out after high temperature annealing, catalyst in electrochemical devices. The incorporation of platinum, ruthenium, and palladium onto Al-MCM-41 mesoporous silica by direct inclusion of various precursors indicate that the Al-MCM-41 mesoporous-ordered structure was not affected by metallic particle incorporation. Metallic nanoparticles dispersion on Al-MCM-41 was homogeneous, platinum and palladium samples have round shape particles and ruthenium sample exhibit a rod shape. In addition to this, the bimetal modified MCM-41 materials are equally important in various redox reactions. The incorporation of two different metals might create materials with different or new redox and acid properties. Supported bimetallic catalysts are very interesting materials in general terms since one metal can fine tune or modify the structural and electronic properties of the other. The bimetal systems like Cu/Zn, Cu/Ni, Co/(V, Nb, La), Ru/(Cr, Ni, Cu) modified MCM-41 can be used in some industrially important reactions such as oxidation, hydrogenation, hydro dehalogenation, H_2 production etc.

1. INTRODUCTION

The self-assembled mesoporous silicate materials formed by the co-condensation of silica and surfactant are of great interest because of their potential uses as adsorbents, catalysts, hosts for inclusion compounds and molecular sieves [1, 2]. The mesoporous materials having a highly ordered structure over a long range have some characteristics of a crystal that can be studied by diffraction and other structural analytical methods [3].The mesopores can be tailored by sophisticated choice of templates (surfactants), inclusion of auxiliary chemicals and reaction conditions (temperature and composition). This highly ordered pore structure comprising of hexagonally packed cylindrical channels can accommodate metals, large organic molecules, metal oxides and various metal complexes [4].

1.1. Mechanistic Pathway for Synthesis of Mesoporous Materials

The synthesis of mesoporous materials are generally controlled by a liquid crystal templating (LCT) mechanism, which proceeds via co-precipitation of organic templates with silica source [5]. The template used is not a single, solvated organic molecule or metal ion but rather a self-assembled molecular array called micelles. The three different mesoporous materials of this family are hexagonal MCM-41, cubic MCM-48 with a three dimensional pore system and lamellar MCM-50[6].

In a true liquid crystal template mechanism there must exists a strong interaction between the surfactant and inorganic precursor. During synthesis, the concentration of SDA (surface directing agent) is taken in excess and the lyotropic interactions facilitate the templating ions (positive, negative or neutral) to take different geometry in the absence of the precursor (usually tetraethylorthosilicate, TEOS). Various pathways are suggested for LCT in different media of solution viz. basic, acidic and neutral [7]. According to Huo et al., these pathways are as follows; a) If the reaction takes place under basic conditions (whereby the silica species are present as anions) and cationic quaternary ammonium surfactants are used as the SDA, the synthetic pathway is termed S^+I^- (Figure 1a; S: surfactant; I: inorganic species). The preparation can also take place under acidic conditions (below the iso-electric point of the Si–OH⁻ bearing inorganic species; pH–2), whereby the silica species are positively charged. To produce an interaction with the cationic surfactant, it is necessary to add a mediator ion X^- (usually a halide) [$S^+X^-I^+$; pathway (b)]. Conversely, when negatively charged surfactants (e.g., long-chain alkyl phosphates) are used as the SDA, it is possible to work in basic media, whereby again a mediator ion M^+ must be added to ensure interaction between the equally negatively charged silica species [$S^-M^+I^-$; pathway (c)]; a mediator ion is not required in acidic media [S^-I^+; pathway (d)]. Thus, the dominating interactions in pathways (a–d) are of an electrostatic nature. Moreover, it is still possible for the attractive interactions to be mediated through hydrogen bonds. This is the case when nonionic surfactants are used (e.g., S^0: a long-chained amine; N^0: polyethylene oxide), whereby uncharged silica species [S^0I^0; pathway (e)] or ion pairs [$S^0(XI)^0$; pathway (f)] can be present[8].

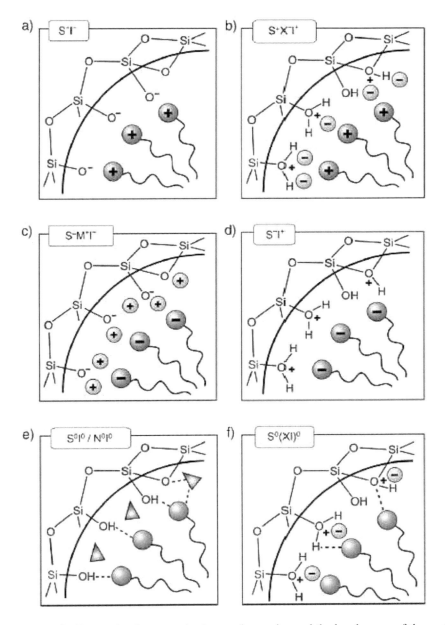

Figure 1. Pathways for Interaction between the inorganic species and the head group of the surfactant under basic, acidic and neutral medium. For positive and negative surfactants the interaction is electrostatic and hydrogen-bonding for neutral surfactants.

1.2. MCM-41

MCM-41, one of the members of this family of M41S materials, contains a regular hexagonal array of mesoporous (resembling to honey bee-hive) with a pore diameter of 2-10 nm and large surface area of 1000 m^2/g depending on the synthesis conditions [9]. The worldwide resurgence of MCM-41 was because of its applications as adsorbent, catalysis, pharmaceuticals, separation technique, sensors, petrochemical processes, gas storage, environmental pollution control and applications in synthesis of fine chemicals.

1.3. Synthesis of MCM-41

The synthesis of MCM-41 requires four gradients, a silica source, a template (surface directing agent), mineralizing agent (to dissolve silica) and a solvent. The usual pathway is the LCT (liquid crystal templating) one. The silica sources are either silica glass or organic silicon alkoxides that are soluble in water, sodium hydroxide or concentrated ammonia. The templates are the ammonium salt of long chain alkyl (cetyl = hexadecyl) halides e.g., cetyl trimethylammonium bromide (CTAB) having hydrophobicity nature. During synthesis the CTA$^+$ ions condense to spherical micelles to avoid repulsion [10].

As the spherical shape is energetically unstable due to surface reasons, the spherical micelles prefer to form a rod like shapes, which further aggregate to closely-packed hexagonals (Figure 2). The negatively charged silicate ions get attracted towards the positive surfaces on liquid crystalline phase and form a hexagonal array to attain stability, thus giving a geometrical shape. The amorphous nature of MCM-41 is concealed because of this long-range order arrangement and characteristic peaks are obtained when subjected to X-ray diffraction. To get a complete mesoporous MCM-41, the templates are removed by calcination at 550^{0}C.

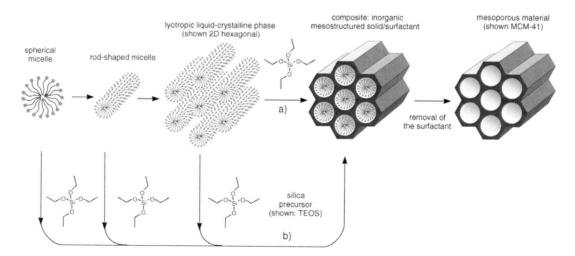

Figure 2. Formation of mesoporous materials by structure-directing agents: a) true liquid-crystal template mechanism, b) cooperative liquid crystal template mechanism.

Table 1. Various methods for the synthesis of MCM-41

Sl. No.	Method of synthesis	Conditions used	Reference
1.	Sol-gel	TEOS as silica source, CTAB as template and varying the CTAB/TEOS, EtOH/H_2O ratio and calcination time.	[11]
2.	Sol-gel	TEOS as silica source, Cetyl Trimethyl Phosphonium Bromide as template	[12]
3.	Hydrothermal	Sodium metasilicate as silica source, CTAB as template in H_2SO_4 medium	[13]
4	Hydrothermal	Sodium silicate as silica source CTAB as template, pH=11 in the autoclave at 120^0C for 96 h.	[14]
5.	Hydrothermal	TEOS as silica source, CTAB as template, Ph=12.5 in the autoclave 120^0C for 2 h	[15]

1.4. Characterization of MCM-41

The scanning Electron Microscopy (SEM) indicated the 2D hexagonal long range mesoscopic morphology of MCM-41 (Figure 3). The uniform distribution of spherical particles representing the outer surface shows the typical siliceous material. The Transmission Electron Microscopy (TEM) exhibits (Figure 4) the arrangement of the pore symmetry resembling to the bee-hive indicating the ordered arrangement of hexagonal pore systems similar to a crystalline state [16]. The spherical appearance of the TEM is due to the surface constraints by the silanol groups residing on the walls of MCM-41, having a tendency to acquire minimum surface area.

Figure 3. SEM image of mesoporous MCM-41.

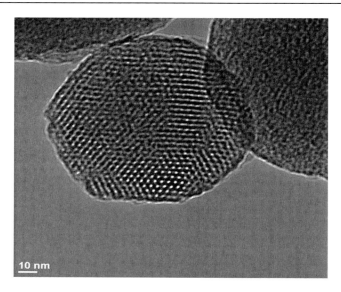

Figure 4. TEM image of mesoporous MCM-41.

As shown in Figure 5, the X-ray diffraction of MCM-41 depicts the highly ordered mesopores over a long-range in the amorphous material. Three peaks corresponding to 2θ value 2.4^0, 4.2^0 and 5.0^0 are obtained for diffraction planes (100), (110) and (200) respectively.

As shown in Figure 6, the nitrogen physisorption method helps in determining the surface area, pore size and pore volume of MCM-41. The BET (Brunauer, Emmett and Teller) method is usually followed to find out the surface area. The BET surface area of MCM-41 is very large and about 1000 m^2 g^{-1}. The different regions of isotherms shown explains I) the monolayer adsorption inside the pores and on the surface, II) the multilayer adsorption in the pores, III) capillary condensation into the pores, IV) multilayer adsorption on surface and V) subsequent desorption of nitrogen [17].

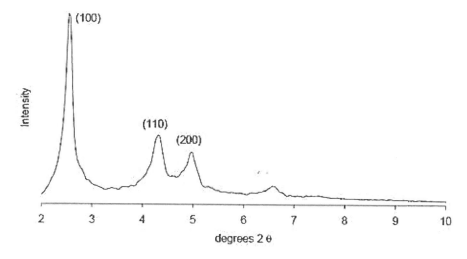

Figure 5. X-ray diffractogram of MCM-41 showing (100), (110) and (200) reflections assigned to the hexagonal lattice of the mesoporous material.

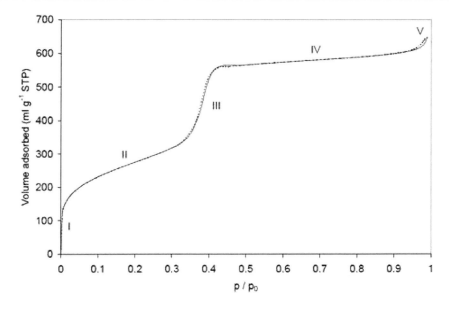

Figure 6. Nitrogen adsorption-desorption isotherm of MCM-41. The straight line is for adsorption and dotted line for desorption.

The BJH (Barrett, Joyner and Halenda) method is adopted to determine pore size distribution. A very narrow pore size distribution centered on 2.7 nm is obtained for MCM-41 (Figure 7). The absolute micropore volume can be obtained by extrapolating the nitrogen sorption curve for MCM-41 was found to be 0.97 mL g^{-1}. It has been reported that MCM-41 is acidic in nature. It possesses high thermal and mechanical stability, but the hydrothermal stability is poor. The regular hexagonal structure of MCM-41 is destroyed by hydrothermal treatment, reaction with strong alkali and hydrofluoric acid.

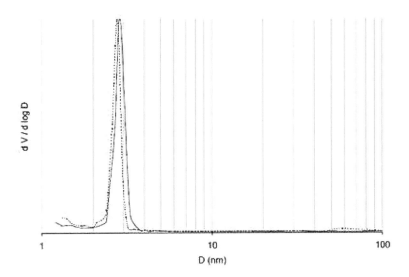

Figure 7. Pore size distribution plots of MCM-41 calculated from nitrogen adsorption-desorption.

NMR studies revealed that in MCM-41, estimated 8-27% silicon atoms are linked with pendant –OH groups = Si-OH after the template molecules are removed [18]. The well-defined pore shapes, either hexagonal (lattice parameter = 44.5 A^0) or cylindrical (lattice parameter = 30.0 A^0), narrow distribution of pore sizes; restricted pore blocking and high degree of pore-ordering are prominent characters of MCM-41[14]. The fine-tuning of pore diameter (1.5-20 nm), large pore volume (› 0.6 cm^3g^{-1}) , exceptional sorption capacity (64% of benzene at 50 torr, 298 K), high surface area to the tune of 1000 m^2/g and ease modification of surface properties have recognized MCM-41 as an excellent catalyst for organic as well as light-induced photochemical reactions.

1.5. FTIR Investigations

At room temperature, MCM-41 exhibits a sharp absorption peak at 3738 cm^{-1} ascribed to free silanol (SiOH) groups. A broad band at 3222 cm^{-1} assigned to hydrogen bonded SiOH groups perturbed by physically adsorbed water.All the silanol groups should be pendant to the internal surface of MCM-41. The band obtained at 3738 cm^{-1} is also assigned to the geminal silanol groups coming from external surfaces or internal lattice defects. The band at 1640 cm^{-1} ascribed to Si-OH vibration, a broad band at 1090 cm^{-1} for symmetric vibrations are also given by IR measurements. The pendant hydroxyl groups of Si-OH play an important role during functionalization and modification of various precursors onto MCM-41.

1.6. Perspective

The various special characteristics of MCM-41 such as (1) well-defined pore structure with apertures in the range of 15-100 Å which can be controlled by careful choice of surfactants, auxiliary chemicals, and reaction parameters, (2) high thermal stability, (3) mild acidity, (4) large BET surface area and pore volume, and (5) hydrophobic/hydrophilic property have made the material a potential adsorbent, catalyst, host for guest encapsulation and a support material for modifications.

2. METAL INCORPORATED MCM-41

The substitution of various main group metals such as Al, Ga, In, Zn and Cd, transition metals such as Ti, V, Fe, Cu, Nb, Mo and Zr and rare earth elements like La and Ce into the frame-work of the MCM-41 has received much attention to develop more efficient and stable materials for application in adsorption, catalysis, photodegradation, separation and chemical sensing [19-26]. The metal ions can be incorporated via ion-exchange, impregnation, co-condensation and solid state vaporization [27-29].

The incorporation of a metal into the frame-work of MCM-41 seems to be easy but the retention of mesoscopic order is often difficult and the stability depended on the amount of dopant material [42]. In case of tetravalent metals viz. Ti, Zr, Ge and Sn the grafting could be done by isomorphous substitution of tetra-valent Si, into the walls of MCM-41 during in-situ or LCT mechanism where the charge on the walls in balanced and neutral. But the incorporation of trivalent metals (Al, B, Ga and Fe) introduces a negative charge and generates a cation exchange or Bronsted acid site on the surface of MCM-41[43]. In addition to acidic sites, the multivalent cations of the frame-work can also create isolated redox sites which can act as good heterogeneous oxidation catalysts in presence of a mild catalyst [44]. The amount of metal loaded into the silica matrix determines the acid strength, catalytic activity and ion exchange capacity [45].

The doping of transition metals such as Cu [46], Fe [47], V [48] and Nb [49] onto MCM-41 were carried out by various synthesis methods like direct insertion, chemical vapor deposition (CVD) and hydrothermal route. The most of the cases the metal incorporated MCM-41 retained its regular hexagonal order and mesoporosity. Ying et al. mentioned that the cation size and its co-ordination state played an important role in the possibility of incorporation [50]. Introduction of a metal (Me) into the silicate structure results in generation of tension of various degrees, This is due to the different bond lengths and angles between Si-O-Si and Si-O-Me and similarly the difference between the ionic radii of Si^{4+} and Me^{2+} and Me^{3+}. The closer the ionic radii to each other the less is the tension generated in the silicate structure [51].

Table 2. Various methods used for preparing metal/bimetal incorporated MCM-41

Fe-MCM-41, Co-MCM-41 and Ni-MCM-41	metal incorporated into the framework	[30]
Bimetallic Ru-(Co, Nr, Cu) and La-(Co or Mn) modified-MCM-41	by direct synthesis using surfactant templating process	[31]
(Cr, V, Fe, Cu, Mn, Co, Ni, Mo, and La) incorporated MCM-41	hydrothermal method	[32]
V, Fe, and Cr incorporated into MCM-41	by hydrothermal methods and were loaded with TiO_2 utilizing a sol–gel technique	[33]
Tungsten (0–30%)-promoted University of Connecticut mesoporous materials (UCT-X, X = 55 (Ti), 50 (Zr), and 56 (Hf))	one step approach relies on inverse micelle formation and unique NO_x chemistry to control the sol–gel chemistry of inorganic components	[34]
Cu–Al-MCM-41, Fe–Al-MCM-41 and Zn–Al-MCM-4	metal incorporated in the framework	[35]
V-MCM-41	by direct hydrothermal and grafting	[36]
Ce–MCM-41, Al–MCM-41 and Ce–Al–MCM-41	hydrothermally by refluxing the gel with magnetic stirring under atmospheric pressure for 24–36 h	[37]
Fe-MCM-41	alkaline hydrothermal conditions	[38]
Zr–Mn-MCM-41 and Mn-MCM-41	hydrothermal conditions	[39]
Ti, V, Cr-MCM-41	surfactant-assisted direct hydrothermal (DHT) methods	[40]
M–MCM-41 (M = Ti, V, Cr, Mn, Co,Ni, Zr, Cu, Nb, Ce, Sn, and Mo)	hydrothermal method	[41]

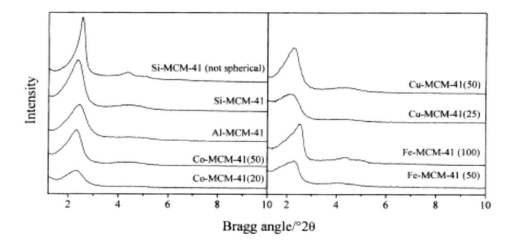

Figure 8. XRD Pattern of different transition metal containing MCM-41.

Table 3. Surface properties of some transition metal and Al- incorporated MCM-41

Sample	S_{BET}^a (m^2/g)	a_0^b (A^0)	Pore dia.(A^0)	Ref.
Cr- MCM-41	838	43.6	27.8	[23]
Mn- MCM-41	870	44.5	28.4	[23]
Ti-MCM-41	959	44.4	29.2	[24]
V- MCM-41	905	49.3	28.2	[30]
Cu- MCM-41	1403	40.3	30.9	[31]
Ce-MCM-41	840	26.9	17.4	[32]
Zr-MCM-41	598·1	32.3	23.5	[33]
Mo-MCM-41	982	44.5	30.2	[34]
Fe-MCM-41	1016	46.5	30.0	[35]
Al-MCM-41	830	44.3	27.5	[36]

[a]Surface Area (BET), [b]Unit cell Parameter

3. PREPARATION METHODS OF METAL/METAL OXIDE INCORPORATED MCM-41 CATALYSTS

3.1. Impregnation

The most frequently applied preparation method for heterogeneous catalysts is impregnation. The major advantage of this method is its simplicity. During impregnation a suitable support material is contacted with a solution containing a precursor of the active phase. Upon drying of the support material after impregnation solvent (usually water) is evaporated and as a result the precursor of the active phase adheres to the surface of the support [52]. Prior to impregnation support materials are frequently shaped into robust bodies, which facilitate their handling. Because of the shaping process voids are generated between the primary particles of support material and upon impregnation these voids (also denoted as "pores") are filled with precursor solution. Generally, there are two different methods of impregnation. In "wet" impregnation the amount of precursor solution added to the support material exceeds the pore volume. Although this is the simplest impregnation method it can result in the deposition of a vast amount of precursor material at the exterior parts of the support bodies during drying and the resulting heterogeneous catalysts display an egg-shell distribution of the active component. Nevertheless such a distribution might be beneficial from an application point of view, since it alleviates the need of reactant penetration deep inside the catalyst bodies, thus improving the catalytic process. Since MCM-41 offers a relatively large pore volume, impregnation appears to be a suitable method for the application of precursors of active phase(s) inside the mesopores. However, care should be taken to avoid redistribution of precursors during drying after impregnation.

3.2. Precipitation

3.2.1. Homogeneous Deposition Precipitation (HDP)

Another method of catalyst preparation involves the precipitation of a precursor of the active phase onto the surface of a support material. The most illustrious method probably is Homogeneous Deposition Precipitation(HDP), also referred to as precipitation from homogeneous solution [53]. For HDP a suitable, powderedsupport material is suspended into a solution containing precursor ions of the catalytically active phase. During the HDP process the pH of the thus generated suspension is homogeneously raised, resulting in the precipitation of the precursor ions. Key feature of this process is that precipitation of the precursor ions does not occur in the bulk of the solution, but on the surfaces of the suspended powdered support. i.e., the powdered support material acts as a source of nuclei onto which the precursor ions precipitate once the pH has reached a critical value.

3.2.2. Co-precipitation

The other method for the preparation of heterogeneous catalysts *via* precipitation involves co-precipitationof both a support material and a precursor of the active phase. Co-precipitation is sometimes used for the preparation of MCM-41 supported heterogeneous catalysts. In this case the process is usually referred to as, in-situ incorporation, since metal ions are incorporated into the silica pore walls during the synthesis of MCM-41.

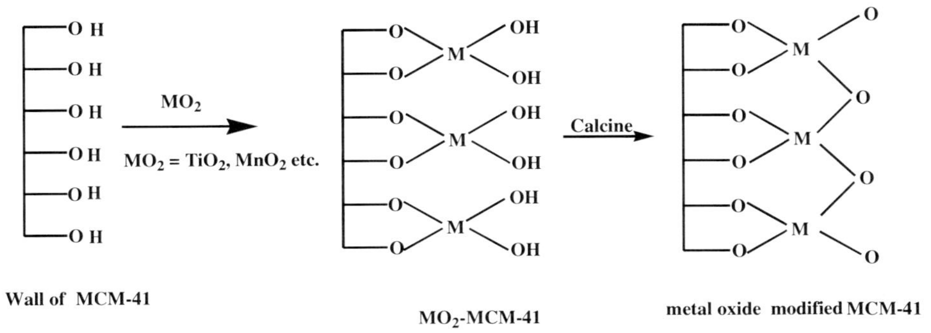

Scheme 1. Proposed mechanism for modification of MCM-41 by metal oxides.

This process is basically the same as for the incorporation of aluminium or titanium into the pore walls of theMCM-41 structure, as described above in the section on the synthesis ofMCM-41. Although this method is viable for the incorporation of aluminiumand titanium the results for a number of other elements are ratherless appreciated. An explanation for this finding can be that the presence ofmetal ions during synthesis of MCM-41 changes the chemistry of the synthesis gel in such a way that the precipitation of an ordered mesoporousmaterial is no longer kinetically favored [54]. Another explanation is that incorporation of hetero-elements inside the pore walls decreases the stability of these walls to such an extent that structural collapse of the framework occurs upon removal of template. Nevertheless, whatever the exact cause of the instability, this method is not at all useful for the application of nickel inside the mesopores of MCM-41 [55]. For the application of molybdenum some satisfying results have been obtained, albeit at rather low Mo loadings.

Other Preparation Methods

A very simple method for catalyst preparation relates to the spreading behavior of certain metal oxides, *viz.* MoO_3 and V_2O_5, over the surface of a support material when a physical mixture of metal oxide and support material is annealed at elevated temperatures for a sufficient long period of time.Unfortunately the interactions of these metal oxides with silica support materials are rather weak, resulting in either low dispersions or low loadings of thus prepared catalysts. The very large surface area of MCM-41 alleviates these constraints to some extent.

3.3. Chemical Vapor Deposition (CVD)

This method relies on the adsorption of a volatile metal precursor compound onto the surface of a support material. Upon interaction the precursor compound usually decomposes, resulting in the deposition of the selected element onto the support surface. CVD has been applied for the deposition of molybdenum onto MCM-41 *via* $Mo(CO)_6$, but the Mo loadings thus obtained are inadequate.

3.4. Ion-Exchange Method

Another, very common procedure for the application of metal ions onto a support material is ion-exchange. This method implies that a support material has an excess negative framework charge in order to adsorb the metal cations. Unfortunately, in order to generate sufficient excess negative charge on the framework of MCM-41 it is necessary to incorporate aluminium ions inside the pore walls during synthesis, resulting in a decreased stability of the support material (*vide supra*). During ion-exchange simple charge compensating cations, such as alkali metal ions, ammonium ions or protons are replaced by the metal ions of interest, a process which is generally entropy-driven. However, it should be noted that ion-exchange is generally diffusion-limited and as a result rather long equilibration times are required. Moreover, metal loadings are generally limited. Ion-exchange has been used to apply nickel inside the mesopores of MCM-41.

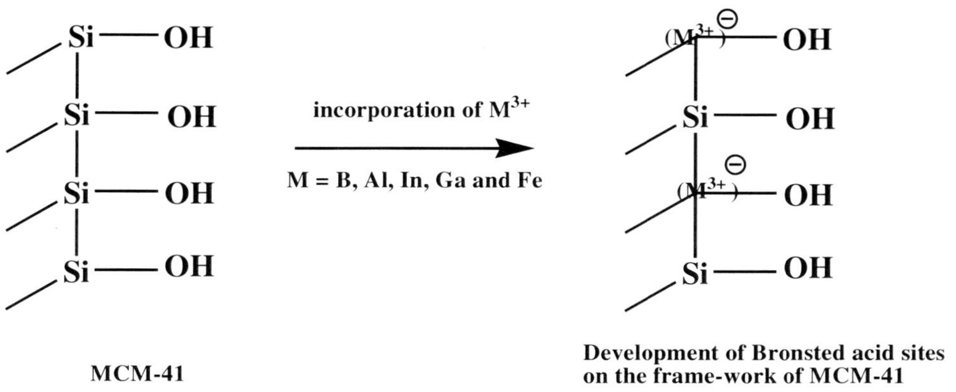

Scheme 2. Proposed mechanism for incorporation of trivalent metal ions into the frame-work of MCM-41.

4. Cu-MCM-41

The Cu-MCM-41 materials can be prepared by I) impregnation II) ion-exchange and III) hydrothermal methods. Up to 2004, impregnation of copper to MCM-41 was limited to 5 weight% that limited the application of the material; a reported work has dealt with a copper content upto 30 weight% for wider applications [56]. It has been found that Cu-MCM-41 can be used as a catalyst for various reactions such as oxidation of phenol, benzene and alcohols. It is also active for NO decomposition and NO reduction over CO. It has been reported that Nb and Cu containing MCM-41 are applied for oxidizing organic compounds more efficiently [57]. Thelarge pore surface area of MCM-41is favorable for loading of Cu-metal and make the diffusion easier and increases the catalytic activity. Moreover, the localization of the metal in the active sites in the MCM-41 frame-work minimizes the leaching of metal during solid phase catalysis [58].

The most effective adsorbent for decomposition of HCN and CNCl is the functionalized Si/MCM-41-en-Cu(II). The explanation for this lies in the combination of a high surface area support, well-dispersed Cu(II) sites, discrete and accessible amine centers chemically bound to the support [59]. The oxidation of cyclohexane by Fe/Cu-MCM-41 has been effectively carried out [60]. Fujiyamaet al. reported the oxidation of 2,6-Di-tert-butylphenol by alkali metal supported, K/Cu-MCM-41 where the activity was much higher than pure Cu-MCM-41[61]. The selective hydroxylation employing Cu-MCM-41 was reported by Franco et al. [62]. The heterogeneous catalysis of photochemically enhanced oxidation of phenol by hydrogen peroxide was performed efficiently by varying the copper loading by impregnation method [63]. The wet air oxidation of aniline was experimented by Cu-MCM-41 with other transitional metals incorporates in addition to copper. The catalytic reduction of NO by Cu-MCM-41was studied by in situ EXFAS and XANES [64]. The adsorption study of CO over Cu-MCM-41 was characterized by FTIR that indicated the well-dispersion of active Cu sites on MCM-41[65].

Figure 9. SEM image of Fe-MCM-41.

5. Fe-MCM-41

Incorporation of trivalent atoms (B, Al, Ga, In and Fe) into the walls of MCM-41 develops negative centers or Bronsted acid sites that enhance the activities of the material towards adsorption and catalysis [66]. In addition to modification of acidity Fe-MCM-41 containing Fe(III) or Fe(II) can also create isolated redox centers, which can act as a very good oxidizing agent in presence of mild oxidants like hydrogen peroxide [30]. The incorporation of trivalent Fe(III), that is tetrahedrally arranged in the silica matrix of MCM-41, generates very strong acid siteswhich is regulated by the amount of iron loading [67]. The formation of iron-oxide and iron-hydroxide is very much expected during sol-gel synthesis as alkali is added during preparation. This can be prevented by the mode of addition of alkali, selecting the appropriate iron source and by hydrothermal treatment so that iron can be successfully and tetrahedrally incorporated into the silica matrix [68].

The incorporation of iron into MCM-41 can achieved through direct hydrothermal synthesis, template ion exchange, incipient wetness impregnation, solid-state impregnation, sol-gel method and other post-synthesis method. Thus, iron-rich ferrisilicates having predominantly tetrahedral Fe(III) and highly ordered mesoporous frameworks, but devoid of iron oxide impurities are most desirable. In 1997, Unger and Co-workers reported the synthesis of submicron spherical MCM-41. Spherical particles are often considered ideal for the adsorption study, but unfortunately the pores vary along with radius of the spherical MCM-41[69]. They further concluded that the addition of alcohol to the reacting mixture led to homogeneous crystallization system favoring the formation of spherical MCM-41. The synthesis and characterization of iron incorporated MCM-41(Fe-MCM-41) and its catalytic properties have been investigated by Wang and his coworkers [54]. During synthesis, the loading of iron into MCM-41 was up to 1.8% by both direct hydrothermal method and template ion exchange method. Recently, iron rich Fe-MCM-41 was synthesized with a higher loading around 8-10% [29].

The formation of iron hydroxides and oxides was restricted and the loading of iron into the framework was also highlighted for Fe-MCM-41 by Szegedi et al. [48]. The higher stability of Fe-MCM-41 was verified under temperature programmed reduction studies that showed the reduction of Fe^{3+} to Fe^0 was only 3%. Though the catalytic study and conversion into β-iron by Fe-MCM-41 has been reported [70], the negative centers were rarely exploited for catalytic activity in chemical reactions after removing heavy metal toxic ions. The oxidation of cyclohexaneat 373K and sulfur dioxideat 800-1100Kusing Fe-MCM-41 as catalyst, has established the thermal stability (up to1023K) of the catalyst [27, 71].

5.1. Preparation of Spherical Fe-MCM-41

Different amounts of iron incorporated spherical MCM-41 were prepared by slightly modifying the original procedure in order to avoid the formation of iron oxide and iron hydroxide precipitates in alkaline medium at pH = 8.0-8.5.Tetraethylorthosilicate (TEOS, Aldrich, India) was used as the silica source for all syntheses. The cationic surfactant Cetyl trimethylammonium bromide (CTAB, Aldrich, India) was used as the structure-directing agent. Fe $(NO_3)_3.9H_2O$ (Acros) salt was used for iron modification. CTAB, water, ethanol and TEOS were mixed in the proportion of 1 TEOS: 0.3 CTAB: 144 H_2O:58 EtOH. To that milky solution, 0.46, 0.18, 0.13 and 0.101 g of $Fe(NO_3)_3.9H_2O$ were added (where the Si/Fe ratios were 20, 50, 70 and 90 respectively.) and stirred for 30 min. resulting in a light yellow, clear solution. To that 3.7 ml of ammonia was added at one time causing immediate gel precipitation. The color of the precipitate was pale beige. Template removal was performed by calcined at 790 K.

6. CATALYTIC ACTIVITY OF METAL INCORPORATED MCM-41

The various works that report the catalytic application of metal incorporated MCM-41 are epoxidation of cyclohexane by Nb-MCM-41, liquid phase oxidation by V-MCM-41, photocatalytic degradation of organic substances and Cr(VI) by Ti-MCM-41 and epoxidation of styrene, ozonation of toluene by Co-MCM-41 and oxidation of adamantane by Fe-MCM-41[24,51,55]. Some fine chemicals like production of p-Cymene by isopropylation of toluene by Al-MCM-41, Photo-Fenton-like catalyst from Cu-MCM-41 are also reported [21, 57]. The environmental concern regarding removal of heavy metal toxic ions from surface water was attended by adsorption with metal doped MCM-41 such as removal of Cu(II) by Fe-MCM-41, adsorption citric acid by Al-MCM-41 have been well documented [59,64]. Table1 shows the surface properties of transition metal incorporated MCM-41.

7. BIMETAL MODIFIED MCM-41

7.1. Preparation of Bimetal Incorporated MCM-41

The synergistic effect of the two metals can usually improve the catalytic activity, selectivity and stability of a reaction compared to that of the single component metal catalysts. In general, supported bimetallic catalysts are very interesting materials because one metal can tune and/or modify the catalytic properties of the other metal as a result of both electronic and structural effects. Bimetallic catalysts supported on high surface area carriers, such as, silica and alumina, have attracted considerable attention recently because of their better performance in catalytic reactions which differs significantly from that of the corresponding monometallic counterparts. Additionally, the preparation of supported bimetallic catalysts (by deposition-precipitation) may lead to catalysts with new characteristics, where a specific interaction between the two metals could produce a hybrid catalyst whose behavior may differ from that of the catalysts prepared by conventional methods.

The synthesis of the bimetal modified MCM-41 follows the same methods as that of the single metal systems which are discussed above. Among the various methods the wetness impregnation and co-condensation (hydrothermal) methods are mostly followed. Study of different bimetal modified samples showed that the samples prepared by hydrothermal method exhibit remarkable redox properties than that of the impregnated samples. Their electron paramagnetic resonance (EPR) studies indicated that the Cu^{2+} ions in the ion-exchanged Cu-MCM-41 materials are buried within the mesoporous walls and hence they are less accessible to the foreign reactant molecules penetrating through the mesopores. The method of direct insertion of metal ion as precursors in the initial stage of the synthesis is scarce, although Hartmann et al. [72] have reported the hydrothermal synthesis of MCM-41 and MCM-48 materials containing Cu and Zn. With the help of temperature programmed reduction (TPR) experiments the authors have shown that the CuMCM-41 materials prepared by direct synthesis exhibit remarkable redox properties compared to the material prepared by the impregnation method. The materials have been synthesized via a classical method of hydrothermal treatment for at least 24 h. In some cases, NaOH has been used in the mother solution, which would result in the precipitation of $Cu(OH)_2$ and $Zn(OH)_2$ as impurities. The Cu/Zn-MCM-41 samples were prepared by direct insertion of metals as precursors at the initial stage of the synthesis using a H_2O/ethanol system at room temperature [73]. In case of noble metal modified MCM-41, the cost of synthesis became very high. But at the same time the non-noble metals are associated with rapid deactivation of the catalyst by carbon deposition and/or metal sintering. Hence the Pt/Co or Pt/Ni-MCM-41 [74] were synthesized by hydrothermal method with increased metallic dispersion and decreased catalytic deactivation.

The synergistic effect of the two metals can usually improve the catalytic activity, selectivity and stability of a reaction compared to that of the single component metal catalysts. The synthesis of the bimetal modified MCM-41 follows the same methods as that of the single metal systems (discussed above). The wetness impregnation and co-condensation methods are mostly followed.

7.2. Cu/Ni-MCM-41

The synthesis, physico-chemical characterization and study of catalytic activity towards hydrodehalogenation of chlorobenzene were done by Parida et al. [75]. Mesoporous silica containing different amounts of Cu and Ni were synthesized at room temperature using CTAB as template and TEOS as silica source in H_2O and ethyl alcohol medium. $Cu(NO_3)_2 \cdot 3H_2O$ and $Ni(NO_3)_2 \cdot 6H_2O$ were used as the Cu and Ni sources respectively. The as-synthesized samples were calcined in flowing air at 540 °C for 18 h to obtain modified MCM-41. Keeping the amount of Cu fixed, the Ni amount can be varied to get different wt% of Cu/Ni-MCM-41 samples.

Table 4. Surface properties of bimetal incorporated MCM-41

Sample	Method of synthesis	BET Surface area (m^2/g)	Pore volume (cm^3/g)	Pore diameter (A^0)	Ref.
Cu/Ni-MCM-41	Hydrothermal	733	0.85	22.1	[75]
Cu/Zn-MCM-41	Co-precipitation	758	0.41	22.8	[76]
Pt/Co-MCM41	-do-	869	1.33	25.1	[77]
Pt/Ni-MCM41	-do-	616	1.09	23.1	[78]
Cu/Ni-MCM-41	Impregnation	106.6	0.21	45.2	[79]
La/Mn-MCM-41	Hydrothermal	789	NA	26.5	[80]

NA-Not available

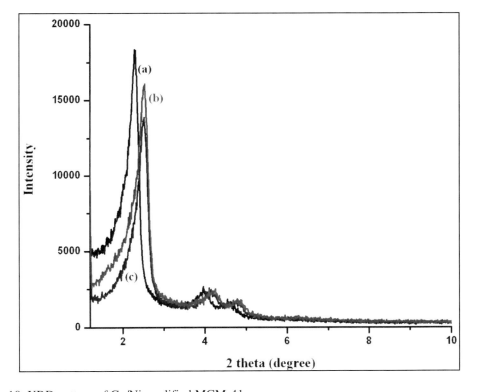

Figure 10. XRD pattern of Cu/Ni modified MCM-41.

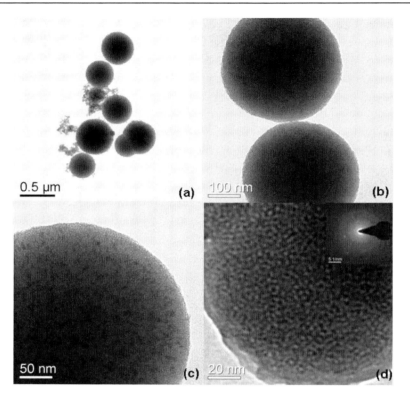

Figure 11. TEM image of Cu/Ni modified MCM-41.

The LAXRD and N₂ adsorption-desorption isotherms show the retention of mesoporousity after modification of MCM-41 surface. From FTIR study it is confirmed that the substitution of silicon by copper and nickel causes shifts of the lattice vibration bands to lower wave numbers. The wave number of the asymmetric Si–O–Si vibration band of (10)Ni-MCM-41 and (10)Cu/Ni-MCM-41 samples decreases. Theses shifts should be due to the increase in the mean Si–O distance caused by the substitution of the small silicon (radius 40 pm) by the larger size of Ni^{+2} (69 pm) and Cu^{+2} (73 pm). The observed shifts, which depend as well on the change in the ionic radii as on the degree of substitution, are comparatively small. In addition, the rocking motion of bridging oxygen perpendicular to the Si-O-Si plane can be correlated with the 445 cm^{-1} band which is common to all the spectra.

TEM images of the bimetallic mesoporous molecular sieves are characteristics for the mesoporous materials with hexagonal channel array, showing high quality in organization of channels of these catalysts. From the figure, it is confirmed that the particles are spherical in nature. The metal particles are well dispersed throughout the silica framework, which is clearly seen in the figure. The particle size can be confirmed from the TEM images. The particle size is calculated to be in between 0.5 μm-0.2 μm.

The reducibility of copper in the MCM-41materials has been investigated by the TPR measurements. The reduction temperature of bare NiO is 370 °C in Ni-MCM-41. The improvement of the reducibility of Ni (lowering of reduction temperature) when Cu is present in the sample is due to the synergistic electronic and structural interaction of copper and nickel [76]. This finding satisfies the property of the group 11 elements of the periodic table to decrease the reduction temperature of other metals.

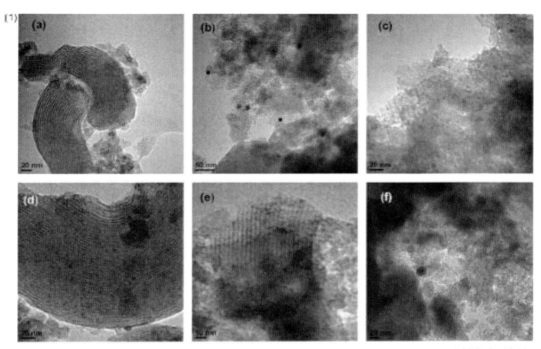

(Reprinted with permission from Chenite, A.; Le Page, Y.; Sayari, A. Chem. Mater., 1995, 7, 1015–1019. Copyright (1995) American Chemical Society)

Figure12. TEM image of a) MCM-41, b) Pt-MCM-41, c) Ni-MCM-41, d) Co-MCM-41 e) Pt/Ni-MCM-41 and f) Pt/Co-MCM-41.

8. Catalytic Activities of Bimetal Modified MCM-41

Utilization of bimetallic catalysts in catalytic reactions like oxidation, hydrogenation, hydrodehalogenation processes has drawn much attention in recent decades due to their high reducible performance. It was found that the catalyst deactivation rate in monometallic catalysts is significant and hence the low stability predominates. Therefore, in the field of hydrogenation, bimetallic catalysts are often used in order to improve selectivity and stability of a single component catalyst. The disposal of chlorinated organic wastes is a serious environmental problem. Polychlorinated biphenyls (PCBs), polychlorinated dibenzodioxins (PCDDs) and polychlorinated dibenzofurans (PCDFs) are carcinogenic, mutagenic, cumulative and stable chemicals. Also in the recent years, dechlorination in water has attracted much attention due to pollution problems of organic halides in water system.

Among the methods proposed for destruction of chlorinated organic compounds [77], the catalytic hydrodechlorination (HDC) is receiving more and more attention because it is simple, effective and safe. Noble metal catalysts, like Pd, Rh and Pt on various supports are active for the reaction under mild conditions [78-80]. Chlorobenzene and its derivatives were hydrodechlorinated on Pd/C catalyst in a flow system, Pd/AlPO$_4$–SiO$_2$ and Pd/C catalysts [81, 82] were tested in liquid phase. A few of these catalysts are introduced to large-scale applications because of their high cost [83, 84]. Transition metal catalysts, such as Ni, Ni–Mo on γ-alumina, silica or carbon composite required high temperature (>473 K) or high hydrogen pressure (2 MPa) to reach significant activity in the gas phase dechlorination process [85, 86]. An alternate to all these expensive methods for dehalogenation is to use the heterogeneous as catalysts. Under mild experimental conditions Cu/Ni-MCM-41 (Si/Cu+Ni=10) sample showed highest chlorobenzene (98%) conversion with 100% selectivity for benzene.

Some combination of Co-V, Co-Nb, Co-La, Ni-Ru, Ni-Cr, Ni-Ti, V-Ti, and V-Co are experimented [87]. Bimetal modified MCM-41 can be effectively used as a heterogeneous catalyst for oxidation [88], epoxidation [89] and alkylation reactions [90]. In hydrogenation reactions, bimetallic catalysts are often used to improve selectivity and stability of a single component catalyst. Cu and Cu/Zn mixed oxides are of great importance for several industrial processes such as low-pressure methanol synthesis, steam reforming of methanol for producing hydrogen for fuel cells, low temperature water-gas shift (WGS) reactions, CO oxidation. Cu/Ni supported catalysts can be used in steam reforming reactions of ethanol using various supports and SiO_2 are found very efficient.

CONCLUSION

This chapter deals with the various synthetic procedures for preparation of transition metal incorporated mesoporous materials with emphasis on MCM-41. The various transition metal modified MCM-41 materials are synthesized by ion-exchange, impregnation, co-condensation and solid state vaporization methods. The low angle XRD study concluded that the parent material retained its identity even after modification. There is a decrease in surface area, average pore size, specific pore volume of MCM-41 by the transition metal modification. The uniform distribution of metal oxides over the support was concluded from the surface characterizations like SEM and TEM. The application of the transition metal modified MCM-41 as catalysts were attributed to the reduction potential of the metals. The industrially important reactions like oxidation and epoxidation of aromatic substrates, addition and removal of H_2 and halogens, Fenton's and photo Fenton's reactions are carried out using the Co, Cr, V, Cu, Ni and Fe etc. modified MCM-41. The removal of heavy toxic metal ions like Cu (II) from surface water by adsorption method was done by Fe-MCM-41and their catalytic activity was reported. The similar type of behavior was shown by bimetal modified MCM-41 samples. The transition metal and bimetal modified mesoporous silica are in the fore-front of catalysts in the manufacture of chemicals, isolation of reaction intermediates and industrial products.

REFERENCES

[1]   Corma, A; Navarro, MT; Pariente, JP. *J. Chem. Soc., Chem. Commun.*, 1994, 147-148.
[2]   Kozhevnikov, IV; Sinnema, A; Jansen, RJJ; Pamin, K; van Bekkum, H. *Catal. Lett.*, 1995, 30, 241-252.
[3]   Chen, CY; Li, HY; Davis, ME. *MicroporousMater.*, 1993, 2, 27-36.
[4]   Davis, ME; Lobo, RF. *Chem. Mater.*, 1992, 4, 756-768.

[5] Kresge, CT; Leonowicz, ME; Roth, WJ; Vartuli, JC; Beck, JS. *Nature*, 1992,359,710-712.
[6] Beck, JS;Vartuli, JC; Kennedy, GJ;Kresge, CT; Roth, WJ; Schramm, SE. *Chem. Mater.*, 1994, 6, 1816-1821.
[7] Huo, Q; Margolese, DI; Ciesla, U; Feng, P; Gier, TE; Sieger, P; Leon, R; Petroff, PM; SchEth, F; Stucky, GD. *Nature*, 1994, 368, 317 – 321.
[8] Huo, Q; Margolese, DI; Ciesla, U; Demuth, DG; Feng, P; Gier, TE; Sieger, P; Firouzi, A; Chmelka, BF; Scheth, F; Stucky, GD. *Chem. Mater.*, 1994, 6, 1176 – 1191.
[9] Steel, A; Carr, SW; Anderson, MW. *J. Chem. Soc., Chem. Commun.*, 1994, 1571-1572.
[10] Tanev, PT; Chibwe, M; Pinnavaia, TJ. *Nature* 1994, 368, 321-323.
[11] Meléndez-Ortizn, HI; Garcı́a-Cerda, LA; Olivares-Maldonado,Y; Castruita, G; Mercado-Silva, JA; Perera-Mercado, YA. *Ceramics International*, 2012, 38, 6353-6358.
[12] Zimmermann, H; Kababya, S; Vega, S; Goldfarb, D. *Chem. Mater.*, 2005, 17, 3723-3727.
[13] Selvaraj, M; Pandurangan, A. *Eng. Chem. Res.*, 2004, 43, 2399-2412.
[14] Ozaydin, Z; Yasyerli, S; Dogu, G; *Ind. Eng. Chem. Res.*, 2008, 47, 1035-1042.
[15] Kirik, SD; Parfenov, VA; Zharkov, SM. *Micropor. and Mesopor. Mater.*, 2014, 195, 21-30
[16] Chenite, A; Page, Y.Le. *Chem. Mater.*, 1995, 7, 1015-1019.
[17] Schmidt, R; Stocker, M; Hansen, E; Akporiaye, D; Ellestad, OH. *Microporous Materials*, 1995, 3, 443-448.
[18] Zheng, S; Gao, L; Zhang, Q; Guo, J. *J. Mater. Chem.*, 2000, 10,723-727.
[19] Feuston, BP; Higgins, JB. *J. Phys. Chem.*, 1994, 98, 4459-62.
[20] Bhattacharyya, KG; Talukdar, AK; Das, P; Sivasanker, S. *J. Mol. Catal. A: Chem*, 2003, 197, 255–262.
[21] Lam, FLY; Yip, ACK; Hu, X. *Ind. Eng. Chem. Res.*, 2007, 46, 3328-3333.
[22] Zhai, QZ;Wang, P. *J. Iran. Chem. Soc.*, 2008,5,268-273.
[23] Srinivas, N; Radha Rani, V; Kulkarni, SJ; Raghavan, KV. *J. Mol Catal A: Chem* 2002, 179, 221–231
[24] Zheng, S; Gao, L; Zhang, Q; Guo, J. *J. Mater.Chem.*, 2000, 10, 723-727.
[25] Lam, KF; Chen, X; Mckay, G; Yeung, K. L. *Ind. Eng. Chem. Res.*, 2008, 47, 9376-9378.
[26] Wang, Ye; Zhang, Q; Shishido, T; Takehira, K. *J. Catal*, 2002, 209,186-196.
[27] Carvalho, W. A; Wallau, M; Schuchardt, U. *J. Mol. Cat.A*, 1999, 1, 144 ,91-99.
[28] Sobczac, I; Ziolek, M; Renn, M; Decyk, P; Nowak, I; Daturi, M; Lavalley, J-C. *Micro. Mesopor. Mater.*, 2004, 74, 23-36.
[29] Trejda, M; Daturi, M; Lavalley, JC; Nowak, I; Ziolek, M. *Stud. Sur. Sci. Catal.*, 2004, 154, 1490-1497.
[30] Parvulescu, V; Su, BL. *Catalysis Today*, 2001, 69, 315–322.
[31] Pârvulescu, V; Anastasescu, C; Su, BL. *J. Mol. Catal. A: Chem.*, 2004, 211, 143–148.
[32] Reddy, EP; Sun, B; Smirniotis, PG. *J. Phys. Chem. B.*, 2004, 108, 17198–17205.
[33] Reddy, EP;Davydov, L;Smirniotis, PG. *J. Phys. Chem. B.*, 2002, 106, 3394–3401.
[34] Poyraz, AS; Kuo, CH; Kim, E; Meng, Y; Seraji, MS; Suib, SL; *Chem. Mater.* DOI: 10.1021/cm501216c.

[35] Antonakou, E; Lappas, A; Nilsen, M. H; Bouzga, A; Stöcker, M. *Fuel.*, 2006, 85, 2202–2212.
[36] Shylesh, S; Singh, AP. *J. Catal.*, 2004, 228, 333–346.
[37] Kadgaonkar, MD; Laha, SC; Pandey, RK; Kumar, P; Mirajkar, SP; Kumar, R.*Catal. Today*, 2004, 97, 225–231.
[38] Samanta, S; Giri, S; Sastry, PU; Mal, NK; Manna, A; Bhaumik, A. *Ind. Eng. Chem. Res.*, 2003, 42, 3012–3018.
[39] Selvaraj, M; Sinha, PK; Lee, K; Ahn, I; Pandurangan, A; Lee,T.G. *Micropor. and Mesopor. Mater.*, 2005, 78, 139–149.
[40] Goscianska, J; Ziolek, M. *Stud. Sur. Sci. Catal.*, 2007, 165, 215-218.
[41] Luan, Z; Cheng, CF; Zhou, W; Klinowsi, J. *J. Phys. Chem.*, 1995, 99, 1018-1024.
[42] Samanta, S; Giri, S; Sastry, P.U.; Mal, NK; Manna, A; Bhaumik, A. *Ind. Eng. Chem. Res.*, 2003, 42, 3012-3018.
[43] Notari, B; Elley, DD; Hagg, WO; Gates, BC. *Adv. Catal*; Eds.: *American Press*: San Diego, CA, 1996, 41, 253.
[44] Vaudry, F; Renzo, F; Fazula, F; Schulz, P. *J. Chem. Soc. Faraday Trans.*, 1998, 94 617-618.
[45] Velu, S; Wang, L; Okazaki, M; Suzuki, K; Tomura, S. *Micropor. Mesopor. Mater.*, 2002, 54, 113–126.
[46] Ziolek, M; Lewandowska, A; Renn, M; Nowak, I. *Stud. Sur. Sci. Catal.*, 2004, 154, 2610-2617.
[47] Ying, JY; Mehnert, CP; Wong, M.S. *Angew. Chem. Int. Ed.*, 1999, 38, 56-77.
[48] Szegedi, A; Konya, Z; Mehn, D; Solymar, E; Pal-Borbely, G; Horvath, Z; Biro, LP; Kiricsi, I. *Appl. Catal. A: Gen.*, 2004, 272, 257-266.
[49] Lang, N; Delichere, P; Tuel, A. *Micropor. Mesopor. Mater.*, 2002, 56, 203–217
[50] Velu, S; Wang, L; Okazaki, M; Suzuki, K; Tomura, S. *Micropor. Mesopor. Mater.*, 2002, 54, 113–126.
[51] Chien, SH; Chun-Long Chen, MCK. *Journal of the Chinese Chemical Society*, 2005, 52, 733-740.
[52] Jiang, TS; Li, YH; Zhou, XP; Hao, QZ; Yin, HB. *J. Chem. Sci.*, 2010, 122, 371–379.
[53] Rana, RK; Viswanathan, B. *Catalysis Letters.*, 1998, 52, 25–29.
[54] Wang, Y; Zhang, Q; Shishido, T; Takehira, K. *J. Catal*, 2002, 209, 186-196.
[55] Selvaraj, M; Pandurangan, A. *Ind.eng.Chem.Res.*, 2004, 43, 2399-2412.
[56] Parida, KM; Rath, Dharitri. *Applied Catalysis A: General.*, 2007, 321, 101-108.
[57] Parida, KM; Dash, SK. *J.Hazard.Mater.*, 2010, 179, 642-647.
[58] Gokulakrishnan, N; Pandurangan, A; Sinha, PK. *Chemosphere*, 2006, 63, 458–468.
[59] vanDillen, AJ;Terörde, RJAM;Lensveld, DJ;Geus, JW; de Jong, KP. *J.Catal.*, 2003.
[60] de Jong, KP; van Dillen, AJ. Eds. *Heterogeneous Catalysis*,pub.; Department of Inorganic Chemistry andCatalysis, Utrecht University, Utrecht, The Netherlands, 1998, chapter 2.
[61] Terörde, RJAM. *Ph.D. Thesis*, Utrecht University, The Netherlands, 1996.
[62] Junges, U; Disser, S; Schmid, G;Schüth, F. *Stud. Surf. Sci. Catal*.1998, 117, 391-392.
[63] Kong, Y; Zhu, H; Yang, G; Gou, X; Hou, W; Yan, Q; Gu, M; Hu, C. *Adv.Func. Mater.*, 2004, 14,816-817.
[64] Schumacher, K; Grun, M; Unger, KK. *Micropor. Mesopor. Mater.*, 1999,27, 201-209.
[65] Evans, A. *J.Chem. Ind.*, 2000, 702.

[66] Naderi, M; Pickett, JL; Chinn, MJ; Robert Brown, D. *J. Mater. Chem.*, 2002,12, 1086–1089.
[67] Carvalho, WA; Wallau, M; Schuchardt, U. *J. Molecular Catal.A: Chemical.*, 1999, 144,91–99.
[68] Fujiyama, H; Kohara, I; Iwai, K; Nishiyama, S; Tsuruya, S; Masai, M. *J. Catal.* 1999, 188,417-425.
[69] Franco, LN; Perez, IH; Pliego, JA; Franco, AM. *Cataly. Today,*2002, 75, 189–195.
[70] Hartmann, M; Racouchot, S; Bischof, C. *Micropor. Mesopor. Mater.* 27 (1999) 309.
[71] Velu, S; Wang, L; Okazaki, M; Suzuki, K; Tomura, S. Microporous and Mesoporous Materials 54 (2002) 113-126.
[72] Liu, D; Cheo, W.N.E; Lim, Y.W.Y; Borgna, A; Lau, R; Yang, Y. *Cat.Today*, 2010, 229-236.
[73] Rath, Dharitri; Parida, KM. *Ind. Eng. Chem. Res.*, 2011, 50 (5), 2839–2849.
[74] Chenite, A; Le Page, Y; Sayari, A. *Chem. Mater.*, 1995, 7, 1015–1019.
[75] P^arvulescu, V; Anastasescu, C; Su, BL. *J.Mol.Catal. A: Chem.*, 2004, 211, 143–148.
[76] Caudo, S; Centi, G; Genovese, C; Perathoner, S. *Applied Catalysis B: Environmental* 2007, 70,437.
[77] Alonso, F; Beletskaya, I.P; Yus, M. *Chem. Rev.*, 2002, 102,409-429.
[78] Aramendia, MA; Borau, V; Garcia, IM; Jimenez, C; Marinas, A; Marinas, JM; Urbano, FJ. *Appl. Catal. B: Environ.*, 2003, 43, 71-79.
[79] Sajiki, H; Kume, A; Hattori, K; Nagase, H; Hirota, K. *Tetrahedron Lett.*, 2002, 43,7251-7254.
[80] Gopinath, R; Lingaiah, N; Sreedhar, B; Suryanarayana, I; Sai Prasad, PS; Obuchi, A. *Appl. Catal. B: Environ.*, 2003, 46,587-594.
[81] Konuma, K; Kameda, N. *J. Mol. Catal. A: Chem.*, 2002, 178,239-251.
[82] Murena, F; Gioia, F. *Appl. Catal. B: Environ.*, 2002, 38,39-50.
[83] Park, C; Menini, C; Valerde, JL; Keane, MA. *J. Catal.*, 2002, 211, 451-463.
[84] Pina, G; Louis, C; Keane, MA. *Phys. Chem. Chem. Phys.*, 2003, 5, 1924-1931.
[85] Lingaiah, N; Uddin, MA; Muto, A; Iwamoto, T; Sakata, Y; Kusano, Y. *J. Mol.Catal. A.*, 2000, 161, 157-162.
[86] Parvulescu, V; Constantin, C; Su, BL. *J.Mol.Catal A,* 2003, 202,171-175.
[87] Parvulescu, V; Anastacescu, C; Constantin, C; Su, BL. *Stud.Surf. Sci.Catal.*, 2002, 142, 1204-1205.
[88] Parvulescu, V; Anastacescu, C; Su, BL. *J.Mol.Catal A, Chem.*, 2004, 211, 165-167.
[89] Velu, S; Wang, L; Okazaki, M; Suzuki, K; Tomura, S. *Micropor. Mesopor. Mater.*, 2002, 54, 113-115.
[90] Sartori, G; Maggi, R. *Chem. Rev.*, 2011, 111(5), 181-214.

Chapter 4

MESOPOROUS ACID CATALYSTS FOR RENEWABLE RAW-MATERIAL CONVERSION INTO CHEMICALS AND FUEL

M. Caiado, J. Farinha and J. E. Castanheiro
Centro de Química de Évora, Departamento de Química,
Universidade de Évora, Évora, Portugal

ABSTRACT

Mesoporous catalysts, such as silica (PMOs, MCM-41, SBA-15) and activated carbons, have been used in heterogeneous catalysis, due to a combination of high surface areas and controlled pore sizes. These mesoporous materials have been used as catalyst in a wide range of chemical reactions. Due to environmental pressure and a decrease in fossil fuel sources, alternative fuel sources, such as biomass or renewable feedstock sources, have become increasingly popular. Traditionally, the biomass conversion is carried out over homogeneous catalysts. However, homogeneous catalysts have some disadvantages, such as difficulty in separations and the production of toxic waste. Solid catalysts can replace the homogeneous ones in order to make the processes simpler and more environmentally benign. Heterogeneous catalysts have been used in different reactions biomass conversion. In this work, the use of mesoporous acid catalysts (silica and activated carbons) for the biomass conversion into chemicals and fuel will be reviewed.

INTRODUCTION

Biomass is a promising alternative for fossil resources to produce chemicals, fuels and materials because it is inedible, renewable, abundant, and locally available in most areas. Lignocelluloses are generally composed of hemicellulose (25–35%), cellulose (40–50%), and lignin (15–20%) [1, 2].The main problem is how to efficiently remove the abundant oxygen content from biomass-derived products and convert it into a hydrophobic molecule with the appropriate combustion or chemical properties. Many efforts have been devoted to the search of heterogeneous catalytic systems, more selective, safe and environmentally friendly [3-6]. Figure 1 represents a simplified scheme of biomass transformation into biofuel and chemical over mesoporous catalysts. Ordered mesoporous catalysts could open the door for new catalytic processes, based partly on novel principles, owing to their hither to unprecedented intrinsic features. The development of catalysts (SBA-15, MCM-41 and activated carbons) in the field of conversion of biomass to fuels, chemicals requires knowledge of the complex nature of the substrates to be converted [7-9]. Table 1 summarizes the information of previous works. This chapter focuses on mesoporous acid catalysts as heterogeneous catalysts for biomass conversion into chemicals and fuel. The first topic discussed is the application of silica-based mesostructured material (like MCM-41, SBA-15) with sulfonic acid groups and silica-based mesostructured material with heteropolyacids, as acid catalysts, for the conversion of triglycerides, glycerol, terpenes and carbohydrates into chemicals and fuel. The second topic is the renewable sources conversion (triglycerides, glycerol) over mesoporous carbons, as solid acid catalysts.

1. SILICA-BASED MESOSTRUCTURED MATERIAL AS ACID CATALYSTS FOR THE RENEWABLE SOURCES CONVERSION

1.1. Silica-Based Mesostructured Material with Sulfonic Acids

This type of catalyst showed, as an advantage over the sulfonic resins and analogues, that its textural properties are fixed and non-dependent on the swelling behaviour because of the rigidity of the silica matrix [7].

Table 1. Conversion of biomass into fuel and chemicals over heterogeneous catalysts

	Ref.
Conversion of biomass into biofuels over catalysts	[1]
H-USY zeolites as catalysts in hydrolysis of hemicellulose	[2]
Conversion of hemicellulose over solid acid catalyst	[3]
Transformation of biomass into chemicals by different routes	[4]
Transformations of biomass into fuel and chemical products	[5]
Heterogeneous catalysts for biomass conversion in fine and specialty chemical	[6]
Mesostructured materials with catalytic applications	[7]
Conversion of biomass to fuels in the presence of zeolite and mesoporous materials	[8]
Ordered mesoporous materials in catalysis	[9]

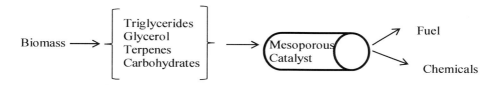

Figure 1. Converison of biomass into fuel and chemical over mesoporous catalysts.

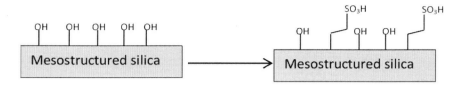

Figure 2. Preparation of sulfonic-acid-modified mesostructured silica

Conversion of Triglycerides

The transesterification of refined and crude vegetable oils was carried out over a propyl-sulfonic acid-modified mesostructured SBA-15 silica catalyst [10]. The reaction conditions were optimized by an experimental design methodology. It was observed that at temperature equal 180°C, methanol to oil molar ratio, 10, and catalyst loading 6 wt% referred to oil, the catalyst produced a FAME purity over 95 wt% for oil conversions close to 100%. It was also observed that high methanol concentrations led to a detrimental effect on the catalyst activity in the range of studied reaction parameters. Regardless of the presence of free fatty acids, the propyl-SO3H-modified mesostructured silica catalyst showed high activities towards the simultaneous esterification of FFAs and transesterification of triglycerides. It was demonstrated that these sulfonated mesostructured materials are a promising catalysts for the acid-catalysed preparation of biodiesel.

The esterification of fatty acids (oleic and lauric) with glycerol was studied in the presence of MCM-41 with sulfonic acid groups. The MCM-41-SO$_3$H materials were synthesized using an improved method of preparation. This method included the partial replacement of the surfactant used previously — hexadecyltrimethylammonium bromide, for a shorter chain surfactant such as dodecyltrimethylammonium bromide. It also included the use of tetramethylammonium hydroxide as co-structuring agent instead of NaOH. The materials obtained were characterized by different techniques and they presented a higher order in the channels arrangement confirmed by X-ray diffraction and transmission electron microscopy measurements. These materials have high amount of thiol groups which were oxidized with H_2O_2 in order to obtain the corresponding sulfonic groups (very strong acidity). Materials obtained with this new method presented a surface area higher than that of the standard preparation. A decrease in the pore size of these materials was observed, which can be attributed to the close packing of the propylthiol chains protruding from the walls into the channels. These materials showed that the catalysts prepared with mixtures of surfactants are more selective to the monoglycerides than the ones synthesized only with hexadecyltrimethylammonium bromide, probably due to the higher order in the channels packing. The catalytic activity is comparable for both type of samples probably due to the amount of sulfur in the different samples is quite similar [11].

The esterification of free fatty acids was studied over silica-based mesostructured material, SBA-15, functionalized with propyl sulfonic groups covalently bound onto the inner surface of the mesopores system [12]. Beef tallow with approximately 7 wt% FFA concentration was used as feedstock in order to condition it for the transesterification step using alkaline catalysts. The objective was to reduce the acidity below 0.5 wt%. It was observed that the catalytic activity of this material decreased when reusing the SBA-15-based catalyst. This poisoning effect was ascribed to the adsorption of polar impurities coming from the beef tallow which caused pore blocking. Other possibilities, such as the poisoning by formed water, were discarded since no poisoning effect was observed when esterifying palmitic acid in soybean oil. In order to avoid or partially reduce the deactivation phenomena, modification of the SBA-15 with sulfonic acid groups was carried out. Thus, simultaneous functionalization of the mesoporous silica was carried out with both the sulfonic acid groups and alkyl chains in order to provide hydrophobicity to the surface of the catalyst. The catalytic activity was enhanced.

Conversion of Glycerol

Glycerol can be converted by different reactions into products, which can be used as fuel additives [13-19]. The esterification of glycerol with acetic acid was studied over sulfonic acid functionalized mesoporous SBA-15 [20]. The catalysts were characterized with different techniques. It was observed that the optimized conditions were a temperature of 125°C and an acetic acid to glycerol molar ratio of 9:1. Under these reaction conditions, glycerol conversion over 90% and combined selectivities toward di- and triacetylglycerol of over 85% were achieved after 4 h of reaction over sulfonic acid-modified SBA-15. The activity (expressed per acid site) showed that the increasing catalyst activity trends agreed with the increasing acid strength of sulfonic group propylsulfonic<arenesulfonic<fluorosulfonic. It was also observed that the catalyst can be reused. A similar catalytic activity to the fresh catalyst was observed. The activities and selectivities of sulfonic acid functionalized mesostructured materials were comparable to those displayed by conventional acid catalysts.

Sulfonic-acid functionalized mesostructured silica was also used in etherification of glycerol with isobutylene to yield tert-butylated derivates [21]. The reaction conditions were optimized. Under optimized conditions (75°C and isobutylene to glycerol molar ratio of 4:1), a glycerol conversions up to 100% and combined selectivities towards di- and tri-tert-butylglycerol about 92% were achieved, after 4 h of reaction over arenesulfonic-acid-modified SBA-15. Under this reaction conditions, no isobutylene oligomerization was observed. The activity and selectivity obtained over sulfonic acid-functionalized mesostructured silica were comparable to those displayed by widely used macroporous commercial acid resins.

Arenesulfonic acid-functionalized mesostructured silica was used in etherification of glycerol with anhydrous ethanol [22]. The reaction conditions were optimized. It was observed that the maximize glycerol conversion and yield towards ethyl-glycerols were obtained at T = 200°C, ethanol/glycerol molar ratio = 15/1, and catalyst loading = 19 wt%. Under these reaction conditions, 74% glycerol conversion and 42% yield to ethyl ethers have been achieved after 4 h of reaction. However, under these reaction conditions, different by-products of glycerol were obtained. Under low ethanol concentration, secondary reactions of glycerol are enhanced. However, the gradual increase of ethanol concentration, promotes the glycerol etherification reactions towards ethyl-glycerols.

The acetalisation of glycerol with acetone was studied over SBA-15 with sulfonic acid groups [23]. The main product of acetalisation of glycerol was 2, 2-dimethyl-1, 3-dioxolane-4- methanol (solketal). It was observed that these modified catalysts (arene- and propyl-sulfonic mesostructured materials) have low surface area, which can limit the access of highly polar glycerol and acetone molecules to the active sites. The optimize reaction conditions over arenesulfonic acid SBA-15 has been established. The use of lower grades of glycerol, such as technical (purity of 91.6 wt%) and crude (85.8 wt%) glycerol, has also provided high conversions of glycerol over sulfonic acid-modified heterogeneous catalysts (84% and 81%, respectively). The SBA-15 with sulfonic acid groups has been reused. It was observed high initial activity without any regeneration treatment, up to three times for refined and technical glycerol. However, the high sodium content in crude glycerol deactivated the sulfonic acid sites by cation exchange. However, the deactivation is reversed by simple acidification of the catalyst after reaction.

Sulfonic-acid grafted silica have demonstrated high activity and selectivity at temperatures below 373 K for the etherification of glycerol with a series of alkyl aromatic and allylic alcohols but proved inactive with aliphatic alcohols at these temperatures [24].

The etherification of glycerol with ethanol has been investigated over sulfonic-acid grafted silica acid catalysts with the aim of producing selectively mono ethers. The only reaction observed at 433 K was the etherification of ethanol into diethylether while at 473K a high conversion of glycerol was obtained, confirming that glycerol etherification is a more difficult reaction than ethanol etherification. The passivation of the surface of the grafted silica by SiMe3 groups, which conferred a hydrophobic character to the surface, led to a significant decrease of the catalyst activity, more pronounced for the glycerol etherification than for the formation of diethylether. This finding is in line with the conclusion reached above in the case of the acidic resins [25].

Gas-phase dehydration of glycerol was carried out over sulfonic-functionalized mesoporous silica (SBA-15) at moderate temperature (275 and 300°C). For the tested conditions all the samples show a very high catalytic activity. At the lowest temperature and for 140 h on stream a nearly constant selectivity toacrolein of ca. 80% could be obtained. Both the pore size and density of acid sites play an important role in the deactivation rate and selectivity: larger pore size may extend significantly the catalytic activity and high density of acid sites may reduce acrolein production. Increasing the temperature to 300°C is beneficial to the acrolein production, but it also leads to an increase of the deactivation rate [26].

Conversion of Terpenes

MCM-41 and Aerosil-200 functionalized with different quantity of SO_3H groups were used as catalysts for the heterogeneous synthesis of camphene from alpha-pinene. The surface was successfully modified, controlling the acidity degree by means of the quantity of MPTS used during the treatment. The catalysts were characterized by thermodesorption of NH_3 and physisorption of nitrogen. Three types of acid sites were found on the functionalized MCM-41, whereas on Aerosil-200 mainly strong acid sites were found. The conversion of α-pinene increased with the quantity of medium strength acid sites. The highest conversion was 100% with yield to camphene of 39.3% for catalyst with the higher content of SO_3H. The highest yield to camphene was 42.2% with a conversion of 95% and was reached for moderate acid treatment of catalyst. The Aerosil-200 based catalyst did not show conversion of α-pinene. Other terpenes as limonene, terpinolene and δ-terpinene were obtained as by products. The maximum yield obtained was of 47%, which is attractive from the industrial point of view. The reuse of catalyst shown an activity loss of 30–40% after each reaction, and removal of adsorbed products can reduce the loss of activity, nevertheless more studies must be made to determine the best way to minimize the catalyst deactivation. [27]

The acetoxylation of a-pinene was carried out over SBA-15 with sulfonic acid groups. The products of acetoxylation of a-pinene are acetates (a-terpinyl acetate, bornyl acetate and b-fenchyl acetate) and hydrocarbons (camphene, tricyclene, limonene, g-terpinene, terpinolene and a-terpinene). Catalysts with different amount of sulfonic acid groups were prepared. It was observed that the activity increases with increase of the surface area and porous volume. Catalytic stability of the sample that showed the highest activity (C1) was evaluated by performing consecutive batch runs with the same catalyst sample. After the fifth batch, the catalyst exhibited a good initial activity [28].

Mn(Salen) complexes immobilized on sulfonic acid-functionalized SBA-15 molecular sieves (SBA-15-pr-SO$_3$-Mn(Salen)) catalyze the Mukaiyama-type oxidation of R-(+)-limonene selectively to the 1, 2-epoxide with molecular oxygen at 298 K (Salen = N, N-ethylenebis(salicylidenaminato)). The diasteriomeric excess for the endo-enantiomer was 39.8%. The Mn complexes immobilized on the sulfonated surface showed more enhanced catalytic activity than the "neat" complexes and zeolite-Y-encapsulated Mn(Salen). Immobilization modified the oxidation state of Mn and the conformational geometry of the complex, thereby increasing their catalytic activity. Theactive sites were also better dispersed and more easily accessible for the substrate molecules in the case of sulfonic acid-functionalized, mesoporous SBA-15 than in zeolite-Y. Thecatalysts were, however, not stable during the oxidation reaction; Mn ions were leached out of the solid phase during their action [29].

SBA-15 molecular sieves were functionalized with propylamine, propylthiol and propylsulfonic acid groups. Mn(Salen)Cl complexes were grafted on these organo-functionalized SBA-15. The support and the type of organo-functional group influenced the electronic structure (oxidation state and redox behavior) and chemoselectivity of the Mn-complexes in the oxidation of limonene. The Mn ions were reduced from +3 to +2, the extent of this reduction on different supports decreasing in the order: SBA-15-pr-SH > SBA-15-pr-SO$_3$H > SBA-15-pr-NH$_2$. Mn(Salen)Cl supported on propylthiol-functionalized SBA-15 yielded the 1, 2-limonene epoxide with 100%chemo- and regioselectivity. Higher electron density at the site of Mn ions and the consequent lower redox potential of the Mn-complexes on immobilization are the probable causes for their efficient and selective catalytic activity. Solvents, additives (N-MeIm) and co-reagents (iso-butyraldehyde), which facilitated formation of Mn^{2+} ions, enhance the catalytic activity. A part of the Mn complexes was leached out of the solid phase during the reaction and the extent of this with different catalysts decreased in the order: SBA-15-pr-NH2-Mn(Salen)Cl> SBA-15-pr-SO3H-Mn(Salen)Cl> SBA-15-pr-SH-Mn(Salen)Cl.Efficient, chemo- and regioselective, mesoporous catalysts for the oxidation of R(+)-limonene were preparedby modifying the surface of SBA-15 molecular sieves withorgano-functional groups and Mn Schiff base complexes [30].

The methoxylation of α-pinene was studied over sulfonic acid-functionalized mesoporous silica (MCM-41, PMO) at 60°C. The support functionalization was achieved by the introduction of 3-(mercaptopropyl)trimethoxysilane onto the surface of these materials either by grafting or by co-condensation. The thiol groups were oxidized to SO$_3$H by treatment with H$_2$O$_2$. All the catalysts were active in the studied reaction being the PMO-SO$_3$H-g the best one. Good values of selectivity to α-terpinyl methyl ether were obtained with these catalysts. Catalytic stability of the PMO-SO$_3$H-g was evaluated by performing consecutive batch runs with the same catalyst sample. After the third batch it was observed a stabilisation of the activity [31].

Carbohydrates

Propyl- and arene-SO$_3$H-modified mesostructured SBA-15 were used as catalysts in the esterification of levulinic acid with ethanol. Under optimized reaction conditions (T = 117°C, ethanol/levulinic acid molar ratio = 4.86/1 and catalyst/levulinic acid = 7 wt%) almost 100% of levulinic acid conversion was achieved after 2 h of reaction, being negligible the presence of levulinic acid by-products or ethers coming from intermolecular dehydration of alcohols. Despite their lower acid strength, propyl-SO$_3$H acid sites provided higher activity than arene-SO$_3$H groups as a consequence of the lower hydrophilicity of the sulfonic-acid site microenvironment of the former, which reduces the poisoning effect by water molecules coming from the esterification reaction. The catalyst has been reused, without any regeneration treatment, up to three times keeping almost the high initial activity. Interestingly, a close catalytic performance to that achieved using ethanol has been obtained with bulkier alcohols [32].

Mesoporous silica was modified with sulfonic acid groups either by one-pot or by grafting method. The hydrolysis of sucrose and starch was carried out over sulfonated mesoporous silicas. The catalysts can be recyclable and there producible activity is achieved. The catalysts showed higher conversion and turnover frequency than conventional Amberlyst-15, Nafion-silica and HZSM-5 catalysts [33].

1.2. Silica-Based Mesostructured Material with Heteropolyacids (HPA)

Figure 3 represents the immobilization of heteropoliacids on surface of mesostructured silica.

Conversion of Triglycerides

Tungstophosphoric acid (PW), molybdophosphoric acid (PMo) and tungstosilicic acid (SiW) immobilized on SBA-15 were used, as catalysts, in the esterification of palmitic acid with methanol, at 60°C. All catalysts exhibited high catalytic activity in palmitic acid esterification with methanol. It was observed that the catalytic activity decreases in the follow series: PW1-SBA-15 > SiW-SBA-15 >PMo- SBA-15. A series of PW immobilized on SBA-15 with different PW loadings from 2.7 wt% to 8.3 wt% were prepared. It was observed high catalytic activity with low amount of tungstophosphoric acid immobilized on SBA-15. In order to optimize the reaction conditions, the effect of different parameters, such as catalyst loading, carbon length of the alcohol and temperature, molar ratio of fatty acid to methanol in the presence of PW3-SBA-15 were studied. PW3-SBA-15 catalyst was also used in the esterification of stearic and oleic acid with methanol. High catalytic activity was observed [34].

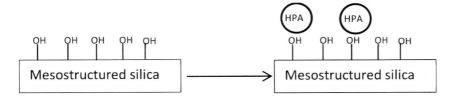

Figure 3. Immobilization of HPAs on mesostructured silica.

The simultaneous esterification/transesterification of FFA-containing soybean oil was carried out over tungstophosphoric acid immobilized on a mixture of Ta_2O_5 and SiO_2. The hydrophobic/hydrophilic balance can be tuned to enhance the catalytic activity and reusability of the catalyst. These materials displayed good catalytic activity in the simultaneous esterification/transesterification of FFA-containing soybean oil [35].

Esterification of fatty acid, lauric acid with 1-butanol was successfully carried out over a solid acid comprising 12-tungstophosphoricacid and MCM-41. 95% yield was obtained for butyl laurate. Influence of various reaction parameters such as catalyst concentration, acid/alcohol molar ratio and reaction time were studied to optimize the conditions for maximum yields. Also the catalyst was regenerated and reused up to four cycles [36].

Esterification of oleic acid with methanol was carried out over 12-tungstophosphoric acid (TPA) anchored to two different mesoporous silica supports, MCM-41 and SBA-15. The activation energy was found to be 44.6 and 52.4 $kJ.mol^{-1}$ for 12-tungstophosphoric acid anchored to SBA-15 and MCM-41, respectively. Based on the catalytic as well as kinetic studies, 12-tungstophosphoric acid anchored to SBA-15 was found to be a better catalyst. The large surface area, high value of total acidity and availability of more Bronsted acid sites, all together imply excellent catalytic activity of TPA3/SBA-15 as compared to that of TPA3/MCM-41. Catalytic results demonstrated that the size of the support channels plays a crucial role in catalyst performance [37].

12-Tungstophosphoric acid anchored to SBA-15 was synthesized and used for biodiesel production by esterification of free fatty acid, oleic acid with methanol. Influence of various reaction parameters (such as catalyst concentration, acid/alcohol molar ratio and reaction temperature) on catalytic performance was studied. The catalyst shows high activity in terms of conversion of oleic acid as well as high turnover frequency of 9.3 min^{-1}. The catalyst can be reused after simple regeneration. It was also observed that the catalyst can be used for biodiesel production from waste cooking oil under mild conditions [38].

Conversion of Glycerol

Different silica-, alumina-, and aluminosilicate-supported heteropolyacid catalysts were prepared using phosphomolybdic acid $H_3PMo_{12}O_{40} \cdot xH_2O$, phosphotungstic acid $H_3PW_{12}O_{40} \cdot xH_2O$, silicotungstic acid $H_4SiW_{12}O_{40} \cdot xH_2O$, and ammonium phosphomolybdate $(NH_4)_3PMo_{12}O_{40} \cdot xH_2O$ as precursor compounds. The catalysts, in particular tungsten-based materials, were interesting acid catalysts for the dehydration of glycerol in the gas phase. In particular, the influence of selected support materials, catalyst loading, and temperature on acrolein formation was studied at standardised reaction conditions (10% by weight of glycerol in water, 225–300°C, modified contact time 0.15 $kg.hmol^{-1}$). Tungsten based heteropolyacids showed outstanding performance and stability. Acrolein was always the predominant product with maximum selectivity of 75% at complete conversion over silicotungstic acid supported over alumina and aluminosilicate [39].

Dehydration of glycerol in gas phase was carried out over mesoporous silica supported silicotungstic acid ($H_4SiW_{12}O_{40} \cdot xH_2O$) led to an 86.2% selectivity of acrolein and a 98.3% conversion of glycerol at an ambient pressure and 548K [40].

Dehydration reaction of glycerol was carried out over solid catalysts, such as zeolites, heteropolyacids, mixed metal oxides and (oxo)-pyrophosphates, often suffer from quick deactivation under reaction conditions because of carbon deposition [41].

1.3. Other Mesostructure

Liquid phase a-pinene isomerization was carried out over various aluminosilicate catalysts. MSU-S(BEA) and MSU-S(Y) mesoporous molecular sieves with different Si/Al ratios were prepared and characterized by XRD, XRF, N_2 adsorption, 27Al MAS NMR, NH3-TPD and 2, 6-di-tert-butyl-pyridineadsorption. The activity correlates well with the amount of the accessible acid sites on the catalyst surface. MSU-S(BEA) with Si/Al ratio of 67 has the highest activity in comparison to others. Ninty-seven percent conversion of a-pinene and 91% yield for main products like camphene, limonene, tricylene and terpinolene can be obtained at 70°C. The catalyst is stable and reusable, and the product yield is only reduced by 10% after four runs, which is probably caused by the slow dealumination in the framework wall during the reaction [42].

The kinetics of the nopol synthesis by Prins condensation of b-pinene and paraformaldehyde over Sn-MCM-41 synthesized by impregnation was evaluated using the initial reaction rate method. The reaction rate equation obtained from a kinetic model based on the Langmuir–Hinshelwood formalism with the surface reaction of adsorbed reactants on catalytic sites of the same nature as the limiting step, gave a good prediction of the experimental data. The effect of temperature on the kinetics of nopol synthesis over Sn-MCM-41 obtained by impregnation was studied between 75 and 100° C [43].

Esterification of glycerol with lauric acid in supercritical CO_2 medium was carried out in the presence of aluminum and zirconium-containing mesoporous MMS-H catalysts. The catalyst gave a 93% conversion of glycerol with a 48% selectivity of lauric acid monoglyceride and a 41% selectivity of diglyceride [44].

Table 2 summarizes the different reaction studied in the presence of mesostructured silica.

2. MESOPOROUS CARBONS AS SOLID ACID CATALYSTS FOR THE RENEWABLE SOURCES CONVERSION

Mesoporous carbon catalysts have received considerable interest in the scientific community because of their tunable porosity, hydrophobic surface, and ease of functionalization with active groups such as–SO_3H. The formation of silica templates is a typical method to prepare a mesoporous carbon, but little is understood on its role for effective sulfonation [45]. Figure 4 shows the functionalization of mesoporous carbons with sulfonic groups.

Conversion of Triglycerides

The esterification of oleic acid showed that the reactivity of the carbon-based catalysts is dependent on the total acidity, but independent of the surface area. These findings show that carbon-based catalysts are suitable for esterification which is useful for biodiesel production. The role of silica template in the preparation of sulfonated mesoporous carbon catalysts has been investigated based on the surface area, pore size, pore volume, total acidity, sulfur content, and esterification of oleic acid. The silica template method via the confined activation process technique produced disordered mesoporous sugar char at low pyrolysis temperature (400°C). Sulfonation before removing the silica templates retained 67% of the surface area, increased by 0.5 nm in average pore diameter, and decreased 33% in pore volume relative to the sugar char (CMK-w). On the other hand, sulfonation after removing the silica template totally collapsed the internal pores. Silica templates provided support to the internal pores of the char, but prevented sulfuric acidfrom effectively reaching the internal surface during the sulfonation process. The esterification of oleic acid was dependent on the total acidity, but independent of the surface area. In general, the development of mesoporosity and high surface area carbon-based catalysts is not useful without having the high acidity [46].

Table 2. Renewable sources conversion over mesostructured silica

Catalyst	Reaction studied	Ref.
SBA-15-SO$_3$H	Transesterification of refined and crude vegetable oils with methanol	[10]
MCM-41-SO$_3$H	Esterification of fatty acids (oleic and lauric) with glycerol	[11]
SBA-15-SO$_3$H	Esterification of free fatty acids	[12]
SBA-15-SO$_3$H	Esterification of glycerol with acetic acid	[20]
SBA-15-SO$_3$H	Etherification of glycerol with isobutylene	[21]
SBA-15-SO$_3$H	Etherification of glycerol with anhydrous ethanol	[22]
SBA-15-SO$_3$H	Acetalisation of glycerol with acetone	[23]
Silica-SO$_3$H	Etherification of glycerol with alcohols	[24]
Silica-SO$_3$H	Etherification of glycerol with ethanol	[25]
SBA-15-SO$_3$H	Dehydration of glycerol	[26]
MCM-41-SO$_3$H	Pinene isomerization	[27]
SBA-15-SO$_3$H	Acetoxylation of alpha-pinene	[28]
SBA-15-pr-SO$_3$-Mn(Salen)	oxidation of R-(+)-limonene	[29, 30]
MCM-41-SO$_3$H and PMO-SBA-15	Methoxylation of alpha-pinene	[31]
SBA-15-SO$_3$H	Esterification of levulinic acid with ethanol	[32]
HMM-1 with SO$_3$H	Hydrolysis of sucrose and starch	[33]
SBA-15 with HPA	Esterification of palmitic acid with methanol	[34]
Ta$_2$O$_5$ and SiO$_2$	Esterification/transesterification of soybean oil	[35]
MCM-41-HPW	Esterification of lauric acid with 1-butanol	[36]
MCM-41-HPW	Esterification of oleic acid with methanol	[37]
SBA-15-HPW	Biodiesel production by esterification of free fatty acid	[38]
Silica-HPW	Dehydration of glycerol	[39]
Mesoporous silica - HSiW	Dehydration of glycerol	[40]
MSU-S(Y)	a-Pinene isomerization	[42]
Sn-MCM-41	Synthesis of nopol	[43]
MMS-H	Esterification of glycerol with lauric acid	[44]

Figure 4. Functionalization of mesoporous carbons with sulfonic groups.

Mesoporous carbon materials with thin pore walls (~1.7 nm) were synthesized using low-cost -Al$_2$O$_3$ as a hard template and in situ polymerized resorcinol–furfural resin as the carbon precursor. Compared with sugar, resin, a widely used carbon precursor, has higher carbon yield and simplifies the synthetic process. Ph-SO$_3$H modified mesoporous carbon was synthesized by covalent grafting of Ph-SO$_3$H groups on mesoporous carbon via the diazonium salt. The modified carbons were shown to possess high surface area (~1000 m^2/g), a bimodal pore size distribution and high strong acid density (1.86 mmol H$^+$/g). These sulfonated carbons were used as solid acid catalysts in the esterification of oleic acid and methanol, a key reaction in biodiesel production. Compared with the traditional solid acid Amberlyst-15, the optimized carbon catalyst exhibited much higher activity with a rate constant (1.34 h^{-1}) three times to that of Amberlyt-15 and a turnover frequency (TOF) of 128 h^{-1}eight times that of Amberlyst-15. The efficient catalytic ability was attributed to the high surface area and a proper mesopore texture. This carbon catalyst could then be easily separated from the product by filtration. The catalyst was reused six times, and no distinct activity drop was observed after the initial deactivation [47].

A carbon-based solid acid catalyst was prepared by the sulfonation of carbonized vegetable oil asphalt. This catalyst was employed to simultaneously catalyse esterification and transesterification to synthesis biodiesel when a waste vegetable oil with large amounts of free fatty acids (FFAs) was used as feedstock. The physical and chemical properties of this catalyst were characterized by a variety of techniques. The maximum conversion of triglyceride and FFA reached 80.5 wt.% and 94.8 wt.% after 4.5 h at 220 C, when using a 16.8 M ratio of methanol to oil and 0.2 wt.% of catalyst to oil. The high catalytic activity and stabilityof this catalyst was related to its high acid site density (–OH, Brönsted acid sites), hydrophobicity that prevented the hydration of –OH species, hydrophilic functional groups (–SO3H) that gave improved accessibility of methanol to the triglyceride and FFAs, and large pores that provided more acid sites for the reactants [48].

The synthesis of highly functionalized porous silica–carbon composites made up of sulfonic groups attached to a carbon layer coating the pores of three types of mesostructured silica (i.e., SBA-15, KIT-6 and mesocellular silica) is presented. The synthesis procedure involves the following steps: (a) removal of the surfactant, (b) impregnation of the silica pores with a carbon precursor, (c) carbonization and (d) sulfonation. The resulting composites have a high BET surface area (ever 590 m^2 g^{-1}), a large pore volume (1–1.8 cm^3 g^{-1}) and a porosity made up of uniform mesopores with a width of~10–29 nm. In addition, these sulfonated materials are characterized by a high density of acidic groups (SO$_3$H, COOH and OH) attached to the deposited carbon layer. The combination of the textural and chemical properties of these materials gives rise highly effective solid acids with a large number of strong acid sites (i.e., SO$_3$H) located inside wide and accessible mesopores. This ensures high mass transfer rates. These solid acids were investigated as catalysts for the esterification of maleic anhydride, succinic acid and oleic acid with ethanol. The results obtained show that these materials have a high intrinsic catalytic activity (TOF), which is superior to that of commercial solid acids such as Amberlyst-15. The reusability of these sulfonated composites has been conformed in the case of the esterification of maleic anhydride with ethanol. The synthesis procedure reported in this study provides a new family of mesoporous silica–carbon materials that, due to their special properties, can be used not only as solidacid catalysts but also as catalytic supports and adsorbents for the immobilization of biomolecules and removal of heavy metals, by selecting the appropriate functional groups for the carbon layer [49].

Esterification of oleic acid was carried out over a variety of sulfonated carbons. All sulfonated materials show some loss in activity associated with the leaching of active sites. Exhaustive leaching shows that a finite amount of activity is lost from the carbons in the form of colloids. Fully leached catalysts show no loss in activity upon recycling. The best catalysts; 1, 3, and 6; show initial TOFs of 0.07 s^{-1}, 0.05 s^{-1}, and 0.14 s^{-1}, respectively. Significantly, the leachate solutions obtained from catalysts 1, 3, and 6, also showed excellent esterification activity. The results of TEM and catalyst poisoning experiments on the leachate solutions associate the catalytic activity of these solutions with carbon colloids [50].

Esterification of oleic acid with methanol was tested over a series of carbon-based solid acid catalysts. The material was prepared by the sulfonation of mesoporous carbon substrates with thin pore walls. The highest turnover frequency (TOF) observed was 78 h^{-1}, five times that of Amberlyst-15. The high catalytic cactivity may be attributed to the good dispersion of the catalysts in methanol. Catalysts with improved dispersion were obtained by a modified preparation, after which the highest TOF observed was 109 h^{-1}, seven times, that of Amberlyst-15 [51].

A carbon-based acid catalyst was prepared by the sulfonation of partially carbonized de-oiled canola meal (DOCM), a by-product of canola seed processing. Partial carbonization of DOCM was carried out at 300 and400°C and the partially carbonized material prepared at 400°C was steam activated to improve its surface properties. Four types of carbon catalysts were prepared using the partially carbonized material and DOCM by concentrated sulfuric acid treatment. The catalyst partially carbonized at 300°C established various functional groups such as aromatic carbon atoms, phenolic-OH, COOH, and carbonyl groups on the surface. The catalyst produced by partial carbonization at 300°C followed by sulfuric acid treatment showed better performance towards esterification of oleic acid and high FFA canola oil. The maximum conversion reached 93.8% after 24 h at 65°C using 1:60 mol:mol of FFA to methanol and 7.5wt.% of catalyst to FFA. The catalyst deactivated gradually over four recycles, and the activity could be regenerated by sulfuric acid treatment [52].

Esterification of the fatty acids was carried out in the presence of solid acid carbon supported catalysts. The materials were generated from biomass by pyrolysis (400–500 °C)to generate a soft to hard carbon backbone (i.e., biochar) for addition of acidic functional groups. Acid catalysts were synthesized by sulfonating the biochar and wood derived activated carbon using concentratedH_2SO_4 at 100, 150 and 200°C (12 h) and gaseous SO_3 (23°C). The sulfonated carbons presented SO_3H groups on the 100°C sulfonated biochar and activated carbon (AC), with higher active site densities (SO_3H density) for the SO_3sulfonated material. The sulfonated carbons were tested for their ability to esterify free fatty acids with methanol in blends with vegetable oil and animal fat (5–15 wt.% FFA). Esterification of the fatty acids was typically complete (~90–100% conversion) within 30–60 min at 55–60°C (large methanol excess), but decreased with lower methanol to oil ratios using the biochar catalysts (e.g., 70%, 6 h, 20:1). Solid acid catalysts derived from wood based activated carbon had significantly higher activity compared to the biochar derived catalysts (e.g., 97%, 6 h, 6:1). Wood based activated carbon sulfonated using SO_3 generated a solid acid catalyst with the highest esterification activity and reuse capability due to a combination of particle strength, hydrophobicity (difference between total acid density and SO_3H group density), and high surface area (1137 m^2/g) and sulfonic acid group density (0.81 mmol/g). For the SO_3 sulfonated AC catalyst, such a regeneration step was not required, as the fractional conversion of palmitic and stearic acid (5% FFA, 10:1, 3 h) remained >90% after 6 cycles [53].

Acids solids sulfonated carbons Starbons-300 were prepared by using a mixture of $ClSO_3H/H_2SO_4$ and H_2SO_4 (98%) as sulfonating agents. Depending on the sulfonation method used important effects were found on the acidity, textural properties and activity. SO_3H groups linked to the Starbons-300 surface were observed by FTIR photoacoustic spectroscopy. Total number of acid sites was determined by acid-base back titration. The content of SO_3H groups was estimated by measuring the cationic exchange capacity. The elemental analysis was carried out by XPS and SEM. The sulfonated Starbons-300 showed very high activity for the esterification of free fatty acids with ethanol. Conversions up to 60% of oleic acid and selectivity to the ester of 100% were reached after 3 h of reaction at 80°C. The results showed that the sulfonated carbon Starbons-300 catalysts are promising acids catalysts to be used in the esterification of free fatty acids with ethanol. The sulfonated carbons Starbons-300 catalysts showed a high selectivity (100%) to the esterification of free fatty acids [54].

Transesterification of crude Jatropha oil was carried out in presence of tungstophosphoric acid (TPA) supported on activated carbon (AC) using ultrasound-assisted process. The catalysts were characterized for physical and chemical properties to examine the effects of different TPA loadings (15%, 20%, 25% w/w). The catalysts were then used in the transesterification of Jatropha oil with high free fatty acid content. The catalyst with 20% TPA loading achieved the best methyl ester yield, achieving 87.33% in just 40 min. The quality of the feed stock was varied by increasing the water content and FFA content to test the tolerance of the catalysts towards these parameters separately. The catalyst reusability and leaching study performed on TPA20-AC catalyst revealed that the reaction was mainly heterogeneous in nature with an appreciable contribution of homogeneous reaction [55].

Conversion of Glycerol

Glycerol etherification with isobutylene was carried out over carbon-based solid acid catalyst. The materials were prepared by sulfonation of partially carbonized peanut shell, and characterized by SEM, EDS, BET analysis, FTIR spectroscopy, NH3-TPD, and TGA. The analytic results indicated that sulfonated peanut shell catalyst has an amorphous porous structure with a high acid capacity and good thermal stability and exhibits better catalytic activity for the glycerol etherification reaction than cation-exchange resin. With a molar ratio of isobutylene to glycerol of 4:1, a catalyst-to-glycerol mass ratio of 6 wt%, a reaction temperature of 343 K, and a reaction time of 2 h, glycerol was completely transformed into a mixture of glycerol ethers including mono-tert-butylglycerols (MTBGs), di-tert-butylglycerols (DTBGs), and tri-tert-butylglycerol (TTBG), and the selectivity toward the sum of the desired DTBGs and TTBG of 92.1% was obtained. Excellent reusability of the catalyst was also confirmed by repeated experiments [56].

The glycerol acetylation reaction was carried out over sulfonation of carbon-based materials prepared by controlled porous structure by sucrose carbonization produced a highly active, and stable solid acid catalyst. The catalysts prepared exhibited an appropriated porous system of the interconnected micro- and mesoporosity that allowed the surface sulfonation using fuming sulfuric acid to form sufficient but variety catalytic active centers; type C–SO$_3$H. The combination of low temperature carbonization at 673 K using silica as template material and the sulfonation with fuming sulfuric acid produced the best active catalyst. The use of this catalyst in the esterification reaction of glycerol with acetic acid allowed conversions above 99% with selectivity towards triacetin around 50%, while the esterification at the same conditions using a conventional catalyst like Amberlyst-15 also gave over 99% conversion but with a selectivity to triacetin of about 10%. [57]

Acetylation of glycerol was carried out over activated carbon (AC) treated with sulphuric acid. The treatment was carried out with sulphuric acid at 85 C for 4 h to introduce acidic functionalities to its surface. The unique catalytic activity of the catalyst, AC-SA5, was attributed to the presence of sulfur containing functional groups on the AC surface, which enhanced the surface interaction between the glycerol molecule and acyl group of the acetic acid. A 91% glycerol conversion with a selectivity of 38%, 28% and 34% for mono-, di- and triacetyl glyceride, respectively, was achieved at 120 C and 3 h of reaction time. The activity of the material, AC-SA5, was attributed to the presence of sulfur containing functional groups on the AC surface, which enhanced the surface interaction between the glycerol molecule and acyl group of the acetic acid. It was observed the AC-SA5 catalyst showed good stability in its catalytic activity when reused in up to four consecutive batch runs [58].

Table 3. Renewable sources conversion over mesoporous carbon catalysts

Catalyst	Reaction studied	Ref.
CMK with sulfonic acid	Esterification of oleic acid	[46]
mesoporous carbon with sulfonic groups	Esterification of oleic acid and methanol	[47]
mesoporous carbon with sulfonic groups	Esterification and transesterification to synthesis biodiesel	[48]
mesoporous carbon with sulfonic groups	Esterification of maleic anhydride, succinic acid and oleic acid with ethanol	[49]
Sulfoneted carbon	Esterification of oleic acid	[50]
mesoporous carbon with sulfonic groups	Esterification of oleic acid with methanol	[51]
sulfonated carbon	Esterification of oleic acid and high FFA canola oil	[52]
sulfonated carbons	Esterification of the fatty acids	[53]
sulfonated carbons Starbons-300	Esterification of free fatty acids with ethanol	[54]
HPW supported on activated carbon	Transesterification of crude Jatropha oil	[55]
sulfonated carbon	Glycerol etherification with isobutylene	[56]
sulfonated of carbon	Glycerol acetylation	[57]
sulfonated of carbon	Acetylation of glycerol	[58]
sulfonated multi-walled carbon nanotubes	Transesterification cottonseed oil with methanol	[59]

The transesterification cottonseed oil with methanol was carried out over a carbon-based solid acid catalyst, which was prepared by the sulfonation of carbonized vegetable oil asphalt. This catalyst was characterized by scanning electron microscopy/energy dispersive spectroscopy, BET surface area and pore size measurement, thermogravimetry analysis and Fourier transform infrared spectroscopy. The sulfonated multi-walled carbon nanotubes (s-MWCNTs) was also prepared and used to catalyse the same transesterification as the asphalt catalyst. The asphalt-based catalyst shows higher activity than the s-MWCNTs for the production of biodiesel, which may be correlated to its high acid site density, its loose irregular network, and large pores, can provide more acid sites for the reactants. The conversion of cottonseed oil 89.93% was obtained (using the asphalt based catalyst) when the methanol/cottonseed oil molar ratio was 18.2, reaction temperature at 260°C, reaction time 3.0 h and catalyst/cottonseed oil mass ratio of 0.2%. The catalyst can be re-used [59].

Table 3 summarizes the different reaction studied in the presence of mesoporous carbons.

CONCLUSION

Biomass can be obtained all over the world and it generally occurs in the form of organic materials such as grass, wood, agricultural crops and their residues and wastes. These materials come from the biological photosynthesis from CO_2, water and sunlight, thus making them sustainable and green feedstocks with zero carbon emission. Traditionally, the conversion of biomass to chemicals and fuels is carried out in the presence of homogenous catalysts. In order to transform these reactions into a "green process", these reactions have been carried out over different heterogeneous catalysts. Mesoporous materials, such as MCM-41, SBA-15 and activated carbons, have been used in heterogeneous catalysis as catalyst supports, due to a combination of high surface areas and controlled pore sizes. The functionalization of organic groups onto the surface of these materials could be by grafting on the surface or by co-condensation. These modified mesoporous materials have been used as catalyst in a wide range of chemical reactions

REFERENCES

[1] Alonso, DM; Bond, JQ; Dumesic, JA. *Green Chem.*, 12, 1493 (2010).
[2] Zhou, L; Shi, M; Cai, Q; Wu, L; Hu, X; Yang, X; Chen, C; Xu, J. *Micropor. Mesopor. Mater.*, 169, 54 (2013).
[3] Dhepe, PL; Sahu, R. *Green Chem.*, 12, 2153 (2010).
[4] Corma, A; Iborra, S; Velty, A. *Chem. Rev.*, 107, 2411 (2007).
[5] Serrano-Ruiz, JC; Luque, R; Sepulveda-Escribano, A. *Chem. Soc. Rev.*, 40, 5266 (2011).
[6] Maki-Arvela, P; Holmbom, B; Salmi, T; Murzin, D. *Catal.Rev. Sci. Eng.* 49, 197 (2007).
[7] Melero, JA; van Grieken, R; Morales, G. *Chem. Rev.*, 106, 3790(2006).
[8] Perego, C; Bosetti, A. *Micropor. Mesopor. Mater.*, 144, 28 (2011).
[9] Taguchi, A; Schüth, F. *Micropor. Mesopor. Mater.*, 77, 1 (2005).

[10] Melero, JA; Bautista, LF; Morales, G; Iglesias, J; Briones, D. *Energy Fuels,* 23, 539 (2009).
[11] Díaz, I; Mohino, F; Pérez-Pariente, J; Sastre, E. *Appl. Catal. A: Gen.,* 205, 19 (2001).
[12] Mbaraka, IK; McGuire, KJ; Shanks, BH. *Ind. Eng. Chem. Res.*, 45, (2006) 3022.
[13] Frusteri, F; Arena, F; Bonura, G; Cannilla, C; Spadaro, L; Blasi, OD. *Appl. Catal.A Gen.,* 367, 77 (2009).
[14] Goncalves, VLC; Pinto, BP; Silva, JC; Mota, CJA. *Catal. Today,* 133, 673 (2008).
[15] Reddy, PS; Sudarsanam, P; Raju, G; Reddy, BM*Catal. Commun.,* 11, 1224 (2010).
[16] Jerome, F; Pouilloux, Y; Barrault, J. *Chemsuschem,* 1, 586 (2008).
[17] Rahmat, N; Abdullah, AZ; Mohamed, AR. *Review. Renew. Sust. Energ. Rev.,* 14, 987 (2010).
[18] Klepacova, K; Mravec, D; Kaszonyi, A; Bajus, M. *Appl. Catal. A Gen.,* 328, 1 (2007).
[19] Lee, HJ; Seung, D; Jung, KS; Kim, H; Filimonov, IN. *Appl. Catal. A Gen.,* 390, 235 (2010).
[20] Melero, JA; van Grieken, R; Morales, G; Paniagua, M. *Energy Fuels,* 21, 1782 (2007).
[21] Melero, JA; Vicente, G; Morales, G. *Appl. Catal A: Gen.,* 346, 44 (2008).
[22] Melero, JA; Vicente, G; Paniagua, M; Morales, G; Muñoz, P. *Bioresour. Technol.,* 103, 142 (2012).
[23] Vicente, G; Melero, JA; Morales, G; Paniagua, M; Martín, E. *Green Chem.,* 12, 899 (2010).
[24] Gu, Y; Azzouzi, A; Pouilloux, Y; Jérôme, F; Barrault, J. *Green Chem.*, 2, 164 (2008).
[25] Pariente, S; Tanchoux, N; Fajula, F. *Green Chem.*, 11, 1256 (2008).
[26] Lourenço, JP; Macedo, MI; Fernandes, A. *Catal. Commun.,* 19, 105 (2012).
[27] Román-Aguirre, M; Gochi, YP; Sánchez, AR; Torreand, L; Aguilar-Elguezabal, A. *Appl. Catal. A: Gen.*, 334, 59 (2008).
[28] Machado, J; Castanheiro, JE; Matos, I; Ramos, AM; Vital, J; Fonseca, IM. *Micropor. Mesopor.Mater.,* 163, 237 (2012).
[29] Saikia, L; Srinivas, D; Ratnasamy, P. *Appl. Catal. A: Gen.*, 309, 144 (2006).
[30] Saikia, L; Srinivas, D; Ratnasamy, P. *Micropor. Mesopor. Mater.*, 104, 225 (2007).
[31] Castanheiro, JE; Guerreiro, L; Ramos, AM; Fonseca, IM; Vital, J. *Stud. Surf. Sci. Catal.,* 174, 1319 (2008).
[32] Melero, JA; Morales, G; Iglesias, J; Paniagua, M; Hernández, B; Penedo, S. *Appl. Catal. A: Gen.,* 466, 116 (2013).
[33] Dhepe, PL; Ohashi, M; Inagaki, S; Ichikawa, M; Fukuoka, A. *Catal.Lett.,* 102, 163 (2005).
[34] Tropecêlo, AI; Casimiro, MH; Fonseca, IM; Ramos, AM; Vital, J; Castanheiro, JE. *Appl. Catal. A Gen.,* 390, 183 (2010).
[35] Xu, L; Wang, Y; Yang, X; Hu, J; Li, W; Guo, Y. *Green Chem.,* 11, 314 (2009).
[36] Brahmkhatri, V; Patel, A. *Fuel,* 102, 72 (2012).
[37] Patel, A; Brahmkhatri, V. *Fuel Processing Technol.,* 113, 141(2013).
[38] Brahmkhatri, V; Patel, A. *Appl. Catal. A: Gen.,* 403, 161 (2011).
[39] Atia, H; Armbruster, U; Martin, A. *J. Catal.,* 258, 71 (2008).
[40] Atia, H; Armbruster, U; Martin, A. *Appl. Catal. A Gen.*, 393, 331 (2011).
[41] Katryniok, B; Paul, S; Belliere-Baca, V; Rey, P; Dumeignil, F. *Green Chem.*, 12, 2079 (2010).
[42] Wang, J; Hua, W; Yue, Y; Gao, Z. *Bioresource Technol.*, 101, 7224 (2010).

[43] Villa, AL; Correa, LF; Alarcón, EA. *Chem. Eng. J.,* 215–216, 500 (2013).
[44] Sakthivel, A; Nakamura, R; Komura, K; Sugi, Y. *J. Supercrit. Fluids*, 42, 219 (2007).
[45] Xing, R; Liu, Y; Wang, Y; Chen, L; Wu, H; Jiang, Y; He, Mand Wu, P. *Micropor.Mesopor.Mater.*, 105, 41(2007).
[46] Janaun, J; Ellis, N. *Appl. Catal. A Gen.,* 394, 25 (2011).
[47] Geng, L; Yu, G; Wang, Y; Zhu, Y. *Appl. Catal. A Gen.,* 427– 428, 137 (2012).
[48] Shu, Q; Gao, J; Nawaz, Z; Liao, Y; Wang, D; Wang, J. *Appl. Energy,* 87, 2589 (2010).
[49] Valle-Vigón, P; Sevilla, M; Fuertes, AB. *Appl. Surf. Sci.,* 261, 574(2012).
[50] Deshmane, CA; Wright, MW; Lachgar, A; Rohlfing, M; Liu, Z; Le, J; Hanson, BE. *Bioresource Technol.,* 147, 597 (2013).
[51] Geng, L; Wang, Y; Yu, G; Zhu, Y. *Catal.Commun.,* 13, 26 (2011).
[52] Rao, BVSK; Mouli, KC; Rambabu, N; Dalai, AK; NPrasad, RB. *Catal.Commun.,* 14, 20 (2011).
[53] Kastnera, JR; Millera, J; Gellera, DP; Locklinb, J; Keithc, LH; Johnson, T. *Catal.Today,* 190, 122 (2012).
[54] Aldana-Pérez, A; Lartundo-Rojas, L; Gómez, R; Niño-Gómez, ME. *Fuel,* 100, 128 (2012).
[55] Badday, AS; Abdullah, AZ; Lee, K-T. *Renew.Energy,* 62, 10 (2014).
[56] Zhao, WQ; Yang, BL; Yi, CH; Lei, ZO; Xu, J. *Ind. Eng. Chem. Res.,* 49, 12399 (2010).
[57] Sánchez, JA; Hernández, DL; Moreno, JA; Mondragón, F; Fernández, JJ. *Appl. Catal. A: Gen.,* 405, 55 (2011).
[58] Khayoon, MS; Hameed, BH. *Bioresour. Technol,* 102, 9229 (2011).
[59] Shu, Q; Zhang, Q; Xu, G; Nawaz, Z; Wang, D; Wang, J. *Fuel Processing Technol.,* 90, 1002 (2009).

Chapter 5

MESOPOROUS TiO₂ AS EFFICIENT PHOTOCATALYSTS AND SOLAR CELLS

Adel A. Ismail[1,2]

[1]Advanced Materials Department, Central Metallurgical R & D Institute,
CMRDI, Helwan, Cairo, Egypt
[2]Advanced Materials and NanoResearch Centre,
Najran University, Najran, Saudi Arabia

ABSTRACT

Structuring of titania on a nanometer scale has attracted considerable attention in the past few years, since nanostructured titania shows outstanding properties and has widespread application potentials: e.g., in photovoltaics, photocatalysis, and gas sensing. Increasing attention has recently been focused on the simultaneous achievement of high bulk crystallinity and the formation of ordered mesoporous TiO_2 frameworks with high thermal stability. Mesoporous TiO_2 have continued to be highly active in photovoltaics and photocatalytic applications because it is beneficial for promoting the diffusion of reactants and products, enhancing photovoltaics and photocatalytic efficiency by facilitating access to the reactive sites on the surface of mesoporous TiO_2 films. This steady progress has demonstrated that mesoporous TiO_2 nanoparticles are playing and will continue to achieve an important role in the protections of the environment and in developing progress of dye-sensitized solar cells (DSSC). This chapter focuses on the preparation and characterisation of mesoporous titania as efficient photocatalysts and solar cells.

1. INTRODUCTION

Mesoporous materials have attracted considerable interest for their applications as catalysts, gas separators, sensors, and energy converters.[1-4] TiO_2 mesoporous is an interesting material for photocatalytic applications due to it being continuous, which may be beneficial compared to separated individual nanoparticles, in particular for catalyst recovery.

The reasons for the low number of studies made on ordered mesoporous titania as a photocatalyst are likely related to the difficulties in making it as an ordered material. Ever since Antonelli et al. [5] first synthesized mesoporous titania in 1995, many efforts have been made to control the crystallization and to make more crystalline material with maintenance of mesoscale order. Usually, mesoporous TiO_2 is prepared by template-based methods using soft templates (surfactant and block polymers) and hard templates (porous silica, polystyrene spheres, porous carbon).[6-9] However, the wall of these materials is normally amorphous, and under heat treatment, crystallization results in collapse of the uniform mesoporous structure. Another approach to titania-containing mesoporous materials for photocatalysis has been to modify existing mesoporous silica with titania. This approach has had good results and, thus, made it even more interesting to prepare ordered mesoporous (highly) crystalline TiO_2. [10,11] TiO_2 has been widely used as a photocatalyst for the removal of hazardous organic substances and as an electrode material for dye-sensitized solar cells[12, 13] due to its strong oxidizing and reducing ability under UV light irradiation. Two of the most important factors affecting the photocatalytic activity of TiO_2 are its specific surface area in a continuous structure rather than in discrete particles and crystallinity. This continuity can be expected to make the electron transfer within the material easier, resulting in higher activity. If mesoporous titania could be prepared with an anatase crystalline wall, it would be a useful material applicable to high performance photocatalyst. In the last decade, research efforts have been directed to enhance the activity of the mesoporous TiO_2 photocatalysts using various methods such as increasing catalyst surface-to-volume ratio, sensitization of the catalyst using dye molecules, [14-15] doping the catalyst with nonmetals such as nitrogen, carbon, floride and iodine [16-19] and transition metals.[20-22] On the other hand, dye-sensitized solar cells (DSCs) using mesoporous TiO_2 electrodes are regarded as a promising alternative to silicone-type solar cells because of their low-cost fabrication.[23-26] Recently, plastic DSSCs have attracted much more attention from the view point of the fabrication of less weighty solar cells, and low-temperature sintering of mesoporous TiO_2 on plastic transparent substitutes made from Sn-doped In_2O_3 (also called ITO) has been reported.[25-26]. In the present review, the recent developments in the syntheses of mesoporous TiO_2 by the surfactant assembly will be focused. Also, this review focuses on the preparation and characterisation of mesoporous TiO_2 as efficient photocatalysts and solar cells.

2. SYNTHETIC METHODS OF PURE MESOPOROUS TITANIA

Mesoporous TiO_2 as active photocataylsts could be prepared by different methods such as sol-gel, hydrothermal, sonochemical, microwave and electrodeposition. In the past decade, mesoporous TiO_2 have been well synthesized with or without the use of organic surfactant templates. Template was used as structure-directing agents for organizing network forming TiO_2 and mixed oxides species in nonaqueous solutions. The most commonly used organic templates were amphiphilic poly(alkylene oxide) block copolymers, such as $HO-(CH_2CH_2O)_{20}(CH_2CH(CH_3)O)_{70}(CH_2CH_2O)_{20}H$ (designated $EO_{20}PO_{70}EO_{20}$, called Pluronic P-123)[24-27,33,37,38,53,70,71,84,87,90,106, 112, 123] and $HO(CH_2CH_2O)_{106}$-$(CH_2CH(CH_3)O)_{70}(CH_2CH_2O)_{106}H$ (designated $EO_{106}PO_{70}$-EO_{106}, called Pluronic F-127).[73,79,98,100,106,110,124,128] Besides triblock copolymers as structure-directing agents, diblock polymers were also used such as [C$_n$H$_{2n-1}$(OCH$_2$CH$_2$)$_y$OH, Brij 56 (B56, n/y)16/10)[49] Other surfactants employed to direct the formation of mesoporous TiO$_2$ include tetradecyl phosphate by Antonelli and Ying [5] and commercially available dodecyl phosphate by Stone and Davis[83], Tween 80[57] and cetyltrimethylammonium bromide (CTAB) TiO$_2$.[29,30-32,43,45-47, 85,98,103]

TiO$_2$ is an interesting material for photocatalytic applications and it is regarded as the most efficient and environmentally benign photocatalyst, and it has been most widely used for photodegradation of various pollutants.[61] The principle of the semiconductor photocatalytic reaction is straightforward. When photons with energies > 3.2 eV, i.e., exceeding the band gap energy of TiO$_2$, are absorbed by the anatase particles in the mesoporous TiO$_2$ photocatalysts electrons are rapidly promoted from the valence band to the conduction band leaving holes behind in the valence band.[61,107] The thus formed electrons and holes participate in redox processes at the semiconductor/water interface. The valance band holes migrate to the surface of the particles where they react with adsorbed hydroxide ions (or water molecules), generating adsorbed˙OH radicals. This photodecomposition process usually involves one or more radicals or intermediate species such as.OH, O$_2^-$•, H$_2$O$_2$, or O$_2$, which play important roles in the photocatalytic reaction mechanisms.[61] The photocatalytic activity of a semiconductor is largely controlled by (i) the light absorption properties, e.g., light absorption spectrum and coefficient, (ii) reduction and oxidation rates on the surface by the electron and hole, (iii) and the electron-hole recombination rate.

2.1. Sol- Gel Method

Sol-gel process, a colloidal suspension, or a sol, is formed from the hydrolysis and polymerization reactions of the precursors, which are usually inorganic metal salts or metal organic compounds such as metal lakesides. Complete polymerization and loss of solvent leads to the transition from the liquid sol into a solid gel phase. First, a homogeneous solution is obtained by dissolving the surfactant(s) in a solvent. TiO$_2$ precursors are then added into the solution where they undergo hydrolysis catalyzed by an acid catalyst and transform to a sol of Ti-O-Ti chains.[27-34, 37] Chen et al. [27] prepared mesoporous TiO$_2$ using TiCl$_4$ and Ti(OBu)$_4$ as the precursors and P123 as the template by the nonhydrolytic evaporation-induced self-assembly (EISA) method with ordered 3-D TiO$_2$ with a uniform pore size and high surface area was obtained. Photodegradation of phenol shows that the sample has a better photoactivity than commercial TiO$_2$ P25. This is attributed to the well ordered 3D open-pore structure, which, combined with its relatively large surface area and pore volume, can facilitate the mass transport of the organic pollutants. Therefore, both the regular open pore morphology and the biphase structure are playing crucial roles in determining the sample's photoactivity.

Beyers et al., reported the preparation TiO$_2$ mesoprous using CTAB as surfactant.[29]In typical, TTIP was added to an ethanolic HCl solution, resulting in a titanium precursor solution, which was added to an ethanolic CTAB solution. After vigorous stirring, the resulting solution was transferred into an open Petri dish for 7 days at 60°C to evaporate the solvent. The molar ratios were Ti: CTAB: HCl: H$_2$O: EtOH) 1:0.16:1.4: 17:20. The photocatalytic activity of mesoporous TiO$_2$ for decomposition of rhodamine 6G could be increased by changing the synthesis medium from basic to acidic conditions. The slower condensation of the TiO$_2$ precursor lead to better formed mesoporous structure, with better accessibility for photocatalysis. However, TiO$_2$ anatase have prepared by using the sol-emulsion gel method in presence of both CTAB[30-32] and cyclohexane. The as-prepared anatase powders exhibited high photocatalytic activity and could be effectively used as the catalyst for photodegradation of methyl orange, bromopyrogallol red, and methylene blue. Shiraishi et al. [33] have developed highly selective methods for photochemical organic synthesis, driven by a mesoporous TiO$_2$, which enables a transformation of benzene into phenol, with very high selectivity (>80%). Briefly, as follows 1g of P123 was dissolved in dry ethanol (20 g). TiCl$_4$; 0.6 g and TTIP; 2.5 g were added to the solution, and the mixture was stirred for 2 h at room temperature.[34] The system proposed here exhibits significant advantages for organic synthesis: (i) additive-free, (ii) cheap source of oxidant (H$_2$O), and (iii) mild reaction condition. Liu et al. [35] have prepared nanostructure anatase TiO$_2$ monoliths using 1-butyl-3-methylimidazolium tetrafluoroborate (BMIM$^+$BF4$^-$) ionic liquids as template solvents by a simple sol-gel method with a peptization process at ambient temperature. The as-prepared products showed wormhole-like mesoporous structures and mesoporous structures of the product with a surface area of ca. 260 m^2 g^{-1} were retained upon calcining to 450°C, showing excellent thermal stability. Furthermore, the products revealed photodegradation ability towards rhodamine B than that of the commercially TiO$_2$ P25. Lu et al. [36] have synthesized well-defined, crystalline TiO$_2$ nanoparticles at room temperature by using spherical polyelectrolyte brush particles as a template. The template particles consist of a polystyrene core from which long chains of poly(styrene sodium sulfonate) are grafted. Ti(OC$_2$H$_5$)$_4$ is hydrolyzed in the presence of brush particles leading to the formation of well-dispersed TiO$_2$ nanoparticles (Figure 1). The as-prepared TiO$_2$ nanocomposites present high photocatalytic activity for the degradation of Rhodamine B under UV irradiation.

Zhan et al. [37] have fabricated long TiO$_2$ hollow fibers with mesoporous walls with the sol-gel combined two-capillary spinneret electrospinning technique using a P123. In a typical synthesis, 1.48 mmol P123 was dissolved in 20.0 mL absolute ethanol, and 40.0 mmol Ti(OC$_4$H$_9$)$_4$ was added into the mixture of 10.0 mL ethanol and 2.0 mL concentrated HCl. The two solutions were mixed at room temperature and aged at 45°C for 16 h under N$_2$ to give a spinnable sol.[38] The photodegradation rate of methylene blue and gaseous formaldehyde for the TiO$_2$ hollow fibers was faster than that for P25 and mesoporous TiO$_2$ powders. Yu et al. [39] have prepared TiO$_2$ hollow microspheres, based on template-directed deposition and in situ template-sacrificial dissolution, is developed in pure water by using SiO$_2$ microspheres as templates and TiF$_4$ as the precursor at 60°C (Figure 2). In a typical synthesis, theTiF$_4$ powder is dissolved in distillated water. Subsequently, 0.01 g of the SiO$_2$ microspheres is added into a 40 mL TiF$_4$ solution, which is then maintained at 60°C for 12 h. It is found that the prepared TiO$_2$ hollow microspheres show a stronger absorption in the UV-visible region (310–700 nm) than P25. Thus, this also leads to the enhanced photocatalytic activity of the TiO$_2$ hollow spheres. The prepared TiO$_2$ hollow spheres exhibit hierarchically nanoporous structures and a high photocatalytic activity.

2.2. Hydrothermal Method

Hydrothermal synthesis is normally conducted in steel pressure vessels called autoclaves with teflon liners under controlled temperature and/or pressure with the reaction in aqueous solutions. The temperature can be elevated above the boiling point of water, reaching the pressure of vapor saturation. The temperature and the amount of solution added to the autoclave largely determine the internal pressure produced. Many groups have used the hydrothermal method to prepare mesoporous TiO₂ nanoparticles. [40-51]

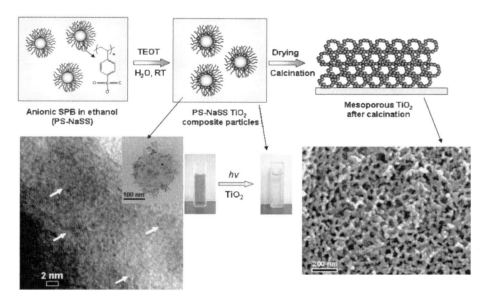

Figure 1. Scheme of macropores structured mesopores TiO₂ nanomaterials, where the macropores are formed from PS core particles and the mesopores are formed by removing the polyelectrolyte brushes. Reprinted with permission from ref. 36, Copyright 2009 ACS.

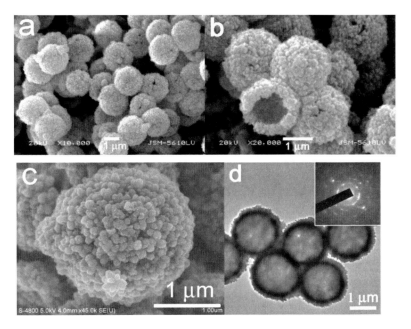

Figure 2. (a and b) SEM, (c) FESEM, and (d) TEM images of TiO₂ hollow spheres obtained in a 0.02 M TiF4 aqueous solution at 60°C for 12 h. Inset in (d) shows the SAED pattern of individual TiO₂ hollow sphere. Reprinted with permission from ref. 39, Copyright 2008 ACS.

Formation Mechanism of Stable Porous TiO₂: TiO$_2$ crystallites with low crystallinity, are first prepared by a hydrolysis process in the acid system, and it can be expected that they easily form agglomerations because of the existence of some amorphous phase resulting from Ti(OH)n. [40] Subsequently, the CTAB introduction during the hydrothermal process on the basic condition can effectively disperse the agglomeration and further induce the assembly of the as-prepared crystallites. On the basic condition, the CTA$^+$ groups are always positively charged, while the nanoparticles are negatively charged (Figure3). Thus, the strong interaction between the nanoparticles and CTAB, resulting from the electrostatic attraction, results in the dispersion of nanoparticles together with the transformation from amorphous to TiO$_2$ crystallites and further induces the assembly of the as-prepared crystallites to form mesoporous TiO$_2$.[41-43, 47,48]

Mesoporous TiO$_2$ with an amorphous wall can be prepared by sol-gel reaction of titanium oxysulfate sulfuric acid hydrate (TiOSO$_4$.xH$_2$SO$_4$.xH$_2$O)[45] and Ti(SO$_4$)$_2$ [43] in the presence of the cationic surfactant CTAB at room temperature. The results suggest that the obtained mesoporous material thus prepared has good adsorbability as well as photocatalytic activity for 2-propanol to yield acetone. The large surface area, small crystalline size, and well-crystallized anatase mesostructure can explain the high photocatalytic activity of mesoporous TiO$_2$ nanoparticles calcined at 400°C for degaradtion of Rhodamine B. [48] Synthesize mixed-phase TiO$_2$ nanocrystals with tunable brookite-to-rutile ratios using titanium tetrachloride as the titanium source and triethylamine [46] and using CTAB as the template, followed by a post treatment in the presence of ethylenediamine.[47] The high crystallinity, large specific surface area, and heterojunction microstructure between anatase and brookite may be responsible for the high photocatalytic activity in terms of the degradation of phenol and rhodamine B under UV irradiation. Interestingly, the prepared TiO$_2$ photocatalysts showed higher photocatalytic efficiency than the Degussa P25 in the degradation of phenol and Rhodamine B in water solution under UV irradiation. Trimodally sponge-like macro-/mesoporous titania was prepared by hydrothermal treatment of precipitates of (Ti(OC$_4$H$_9$)$_4$) in pure water.[44] The prepared TiO$_2$ samples exhibit a disordered worm-like macroporous frameworks with continuous nanocrystalline titania particles). The hierarchically porous titania prepared at 180°C for 24 h displayed an especially high photocatalytic activity for acetone decomposition probably due to its special pore-wall structure, and its photocatalytic activity was about three times higher than that of Degussa P-25.

Wang et al. [49] have synthesized mesoporous TiO$_2$ using TTIP and nonionic poly(alkylene oxide)-based surfactant (decaoxyethylene cetyl ether,C$_{16}$(EO)$_{10}$, Brij56) as the structural-directing agent. In a typical synthesis, TTIP (4.8 mL) was added dropwise to 30 mL of an aqueous solution containing C$_{16}$(EO)$_{10}$ (10 wt%) under very gentle stirring. The mixture was transferred to a Teflon-lined autoclave to age at 80°C for 24 h. The catalyst which calcined at 350°C possessed an intact macro/mesoporous structure and showed photocatalytic reactivity for ethylene oxidation 60% higher than that of P25. Further heating at temperatures above 600°C destroyed both macro- and mesoporous structures, accompanied by a loss in photocatalytic activity.[49] The high photocatalytic performance of the intact macro/mesoporous TiO$_2$ may be explained by the existence of macrochannels that increase photoabsorption efficiency and allow efficient diffusion of gaseous molecules. Liu et al. [50] have synthesized porous TiO$_2$ hollow aggregates on a large scale by means of a simple hydrothermal method without using any templates (Figure 4). The Rhodamine B degradation rate of the porous TiO$_2$ hollow aggregates is more than twice that of P25.

The higher photocatalytic activity of the porous TiO$_2$ hollow aggregates can be explained by considering several factors: The larger specific surface area (porous TiO$_2$ hollow aggregates ca. 168 m^{2}g^{-1} versus Degussa P25 powder ca. 45 m^{2}g^{-1}); hence, there are more reactant adsorption/desorption sites for catalytic reaction. 2) The prevention of the unwanted aggregation of the nanoparticles clusters, which is also helpful in maintaining the high active surface area. 3) The highly porous structure, which allows rapid diffusion of various reactants and products during the reaction. 4) The smaller crystal size, which means more powerful redox ability owing to the quantum-size effect; moreover, the smaller crystal sizes are also beneficial for the separation of the photogenerated hole and electron pairs.[50,51]

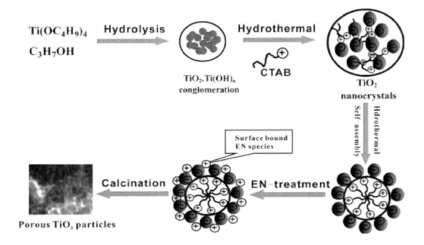

Figure 3. Scheme of idealized formation model of stable porous TiO$_2$ nanoparticles. Reprinted with permission from ref. 47, Copyright 2007 ACS.

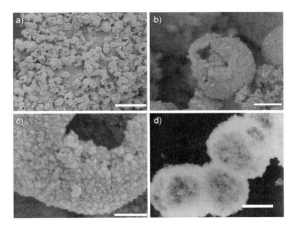

Figure 4 a–c). FESEM and d) TEM images of the porous TiO$_2$ hollow aggregates prepared by one-step hydrothermal treatment at 160°C for 6 h, recorded at different magnifications. The scale bars for a)–d) are 5 mm, 500 nm, 100 nm, and 500 nm, respectively. Reprinted with permission from ref. 50, Copyright 2007 Wiley-VCH.

2.3. Microwave Method

A dielectric material can be processed with energy in the form of high-frequency electromagnetic waves. The principal frequencies of microwave heating are between 900 and 2450 MHz. Microwave radiation is applied to prepare various mesoprous TiO2 nanoparticles. Crystalline anatase mesoporous nanopowders 100–300 nm in size with worm hole-like pore sizes of 3–5 nm were prepared by a modified sol–gel of TTIP, accelerated by a microwave hydrothermal process.[52] The organic surfactant, tetradecylamine, which is used as a self-assembly micelle in the sol–gel and microwave hydrothermal process, enables to harvest crystallized mesoporous anatase nanoparticles with a high-surface area. Mesoporous worm hole-like and crystalline powders with surface areas of 243–622 m^2/g are obtained. It is shown that crystallization by calcination at 400°C/3 h inevitably reduced the surface area, while the microwave hydrothermal process demonstrated a rapid formation of crystalline mesoporous TiO_2 nanopowders with a high-surface area and excellent photocatalytic effects for methylene blue photodegradation.

2.4. Sonochemical Method

Ultrasound has been very useful in the synthesis of mesoporous TiO_2 active photocatalysts.[53, 54] Yu et al. [53] applied the sonochemical method in preparing highly photoactive TiO_2 nanoparticle photocatalysts with anatase and brookite phases using the hydrolysis of TTIP in pure water or in a 1:1 $EtOH-H_2O$ solution under ultrasonic radiation. In this study mesoporous TiO_2 with a bicrystalline (anatase and brookite) framework was synthesized directly under high-intensity ultrasound irradiation. Typically, 0.032 mol TTIP and 3.2 g of P123 were dissolved in 20 mL of absolute ethanol. After the mixture was stirred for 1 h, the resulting solution was added to 100 mL of deionized water, drop by drop, under sonication. The suspension was sonicated for 3 h by a high-intensity probe and the powder was collected and dried in an oven at 373 K. The as-prepared sample was calcined at 673 K for 1 h. Both as-prepared samples exhibited better activities than the commercial photocatalyst P25 in the degradation of *n*-pentane in air. The degradation rate of mesoporous TiO_2 synthesized in the presence of triblock copolymer was about two times greater than that of P25. The high activities of the mesoporous TiO_2 with a bicrystalline framework can be attributed to the combined effect of three factors: high brookite content, high surface area, and the existence of mesopores. Yu et al. [54] prepared mesoporous TiO_2 nanocrystalline powders ultrasonic-induced hydrolysis reaction of $(Ti(OC_4H_9)_4)$ in pure water without using any templates or surfactants. It was found that the as-prepared products by the ultrasonic method were composed of anatase and brooktie phases. The photocatalytic activity of the samples prepared by ultrasonic method is higher than that of commercial Degussa P25 and the samples prepared by conventional hydrolysis method.

2.5. Electrodeposition

Electrodeposition is commonly employed to produce a coating, usually metallic, on a surface by the action of reduction at the cathode. TiO_2/benzoquinone hybrid films have been electrodeposited anodically from basic Ti(IV)-alkoxide solutions containing hydroquinone in presence of tetramethylammoniumhydroxide.[55] The photodegradation of methylene blue (MB) as a representative for organic pollutants have been studied. The results revealed that the amorphous film calcined at 350°C is quite inactive but the activity increases with increasing calcination temperature with the exception of the film calcined at 500°C. Matsumoto et al., reported a new method to prepare a mesoporous TiO_2 photocatalyst film onto alumite using an electrochemical technique[56], where the initial electrodeposition was carried out by electrolysis in $(NH_4)_2[TiO(C_2O_4)_2]$ solution, followed by pulse electrolysis in $TiCl_3$. This film had a high catalytic activity for the decomposition of acetaldehyde, CO_2 with the corresponding concentration was detected in the cell after 45 min even under fluorescent lamp illumination.

2.6. Mesoporous TiO$_2$ Thin Film Active Photocatalysts

Various titania sols containing poly(oxyethylenesorbitan monooleate) (Tween 80) surfactant to tailor-design the porous structure of TiO_2 using dip-coating at different molar ratios of Tween 80/isopropyl alcohol/acetic acid/TTIP=R:45:6:1 have been synthesized[57]. The prepared TiO_2 photocatalytic membrane has great potential in developing highly efficient water treatment and reuse systems, for example, decomposition of organic pollutants, inactivation of pathogenic microorganisms, physical separation of contaminants, and self-antifouling action because of its multifunctional capability. Wang at al. [58] have obtained 3D ordered mesoporous sulfated-TiO_2 reacting a cubic mesoporous amorphous TiO_2 film with sulfuric acid at high temperature to produce sulfur-containing mesoporous TiO_2 with nanocrystalline frameworks. The resulting 3D ordered mesoporous sulfated-TiO_2 superacids are found to be attractive photocatalysts for degradation of bromomethane

High-quality mesostructured titania thin films have been prepared on silicon substrates by spin coating [59] Post-treatment of the films in supercritical carbon dioxide, in the presence of small amounts of either TTIP, tetramethoxysilane, or other precursors, greatly improved the thermal stability of the mesoporous coatings without affecting their optical transparency or integrity. A 0.02 M solution of stearic acid in methanol was first coated on the titania-coated silicon wafers by spin coating. The decrease in the stearic acid concentration with the sc-CO_2/TTIP-treated film is consider-ably faster compared to the untreated and sc-CO_2/TMOStreated films. The higher efficiency can be explained by its high porosity and the presence of its highly photoactive anatase nanocrystalline structure. Without sc-CO_2 treatment calcination at temperatures above 550°C inevitably leads to collapse of the ordered mesoporous structure and formation of rutile crystallites in the walls of the TiO_2 thin films from the transformation of anatase nanocrystals. [59] Kim et al. [60] prepared monodisperse spherical mesoporousTiO_2 with a morphology size of approximately 800 nm via the sol-gel approach using a triblock copolymer surfactant and titanium TTIP with 2,4-pentanedione in aqueous solution. It coated onto glass substrates without cracking by using the doctor blade method with various amounts of polyethylene oxide (PEO) and polyethylene glycol (PEG) (Figure 5). The results revealed that found that the efficacy of photocatalytic disinfection with the film adhesion method is strongly dependent on the surface area and crystallite size.

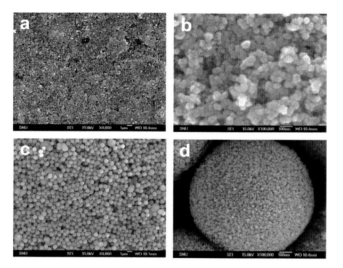

Figure 5. SEM images of TiO$_2$ coated with 40 wt.% of PEG and PEO and calcined at 400°C. (a) a P25 TiO$_2$ coated sample, (b) a magnified image of the P25 TiO$_2$ coated sample, (c) a monodisperse spherical mesoporous TiO$_2$ coated on glass, (d) a magnified image of the coated mesoporous TiO$_2$. Reprinted with permission from ref. 60, Copyright 2009 ACS.

3. TRANSIOTION METALS DOPED TiO$_2$ AS ACTIVE PHOTOCATALYTS

TiO$_2$ anatase can only be excited by UV irradiation (λ<380 nm) because of its large band gap energy of 3.2 eV. Moreover, the rapid recombination of photoinduced electrons and holes greatly lowers the quantum efficiency. [61] Therefore, it is of great interest to improve the generation and separation of photoinduced electron-hole pairs in TiO$_2$ for further applications. The manipulation of semiconductor heterostructures is one of the effective methods for photoinduced electron-hole generation and separation in recent years. [62, 63] Multiple-semiconductor devices can absorb a larger fraction of the solar spectrum, which is beneficial for the excitation of the semiconductor and thus the photoinduced generation of electrons and holes. Moreover, the coupling of two different semiconductors could transfer electrons from an excited small band gap semiconductor into another attached one in the case of proper conduction band potentials. This favors the separation of photoinduced electrons and holes and thus improves the photocatalytic efficiency of semiconductor heterostructure dramatically.

Fe (III) Doped TiO$_2$

Xuan et al. [64] have prepared well-defined magnetic separable, hollow spherical Fe$_3$O$_4$/TiO$_2$ hybrid photocatalysts through a poly(styrene-acrylic acid) (PSA) template method. In typical, 0.5 g sample of monodispersed PSA latex spheres was dispersed in 150 mL of aqueous ethanol (1:2, v/v) solution via ultrasound. Then, 50 mL of the above solution was added to 100 mL of water containing 6×10^{-4} M FeCl$_3$. The suspension was allowed to stand for 1 day to ensure the surface of the PSA spheres was sufficiently adsorbed by Fe^{3+}

ions. A 0.1 g sample of as-prepared PSA/Fe$_3$O$_4$ composite particles was dispersed in a 55 mL mixture of *n*-butyl alcohol and ethanol (10:1, v/v) in an ultrasonic bath. A predetermined amount of TBOT (0.2 mL) was dissolved in 10 mL of ethanol and added to the above mixture. Then, the solution was stirred overnight to allow a saturated adsorption of TBOT on the shell of the composite particles. Fe$_3$O$_4$/TiO$_2$ hybrid hollow spherical with hollow nature exhibit good photocatalytic activity for degaradtion of RhB under UV light irradiation and can be recycled six times by magnetic separation without major loss of activity.

Kim et al. [65] have synthesized mesoporous iron oxide-layered titanate nanohybrids through a reassembling reaction between exfoliated titanate nanosheets and iron hydroxide nanoclusters, in which an electrostatic attraction between both nanosized species could be achieved at low pH of 1.5. The photocatalytic activity revealed that the present nanohybrids could induce the photodegradation of methylene blue and DCA under visible light illumination (λ > 420 nm). Fe-doped nanocrystalline TiO$_2$ with a mesoporous structure was prepared via a facile nonhydrolytic sol-gel route.[66,67] 1.0 mL of TiCl$_4$ was added dropwise into 20 mL of a *t*-butyl alcohol solution containing the required amount of Fe(NO$_3$)$_3$ at 313 K and allowed to stir for 3 h at 333 K until the formation of a translucent alcogel occurred and then calcined for 5 h at 773K. During photodegradation of methylene blue under visible light irradiation, as-prepared Fe/TiO2 exhibited a higher activity than either the undoped TiO$_2$ or the Fe/TiO$_2$ obtained via the traditional hydrolytic sol-gel route. The promoting effect of the Fe-doping on the photocatalytic activity for methylene blue decomposition could be attributed to the formation of intermediate energy levels that allow Fe/TiO$_2$ to be activated easily in the visible area. The nonhydrolytic sol-gel method is superior to the traditional hydrolytic sol-gel method owing to the controllable reaction rate and lack of surface tension, which ensures the formation of mesopores and well-crystallized anatase in the Fe/TiO$_2$ sample, leading to a higher activity since the reactant molecules are easily adsorbed and the recombination between the photoelectrons and the holes is effectively inhibited.[66] A new multifunctional nanocomposite (FexOy@Ti-hexagonal mesoporous silica (HMS)) involving superparamagnetic iron oxide nanoparticles, ordered mesoporous channelshas been developed via the coating of as-synthesized iron oxide nanoparticles with an amorphous silica layer followed by the sol-gel polymerization using TEOS, tetrapropyl orthotitanate (TPOT), and a structure-directing reagent.[68] The FexOy@Ti-HMS acted as an efficient heterogeneous catalyst for the liquid-phase selective oxidation reactions of organic compounds using hydrogen peroxide (H$_2$O$_2$) as an oxidant. The bifunctional composites were synthesized by wet impregnation, drying, ethanol washing, and calcinations process.[69] The meso-TiO$_2$/α-Fe$_2$O$_3$ composites possess synergy of the photocatalytic ability of meso-TiO$_2$ for oxidation of As (III) to As (V) and the adsorption performance of α-Fe$_2$O$_3$ for As (V). The results show that the *meso*-TiO$_2$/α -Fe$_2$O$_3$ composites can oxidize higher toxic As (III) to lower toxic As (V) with high efficiency at various pH values in the photocatalysis reaction (Figure 6). At the same time, As (V) is effectively removed by adsorption onto the surface of composites. Tayade et al. [22] have prepared mesoporous nanocrystalline TiO$_2$ by hydrolysis of TTIP, and the band gap of the TiO$_2$ was modified with transition metal ions Ag, Co, Cu, Fe, and Ni having different work functions by the wet impregnation method. The investigations were carried out to demonstrate the effect of ionic radius and work function of metal ions on photocatalytic activity of mesoporous TiO$_2$ for degradation of acetophenone (AP) and nitrobenzene (NB) in aqueous medium under ultraviolet light irradiation. The initial rate of the photocatalytic degradation of AP and NB varies due to the change in band gap of the

Bi(III) Doped TiO$_2$

nanocrystallines with ordered mesoporous structure are synthesized by EISA method using P123 surfactant as a template.[70] In a typical synthesis, 1.7 g of TiCl4, 3.0 g of (Ti(OBu)$_4$), and a certain amount of Bi(NO3)3.5H$_2$O were added into 12.0 mL of ethanol solution containing 1.0 g of P123. The Bi$_2$O$_3$-photosensitization of TiO$_2$ could extend the spectral response from UV to visible area, making the Bi$_2$O$_3$/TiO$_2$ photocatalyst easily activated by visible lights for degradation of pchlorophenol. The ordered mesoporous channels facilitate the diffusion of reactant molecules. Meanwhile, the high surface area could enhance the Bi$_2$O$_3$ dispersion, the light harvesting, and the reactant adsorption. Furthermore, the highly crystallized anatase may promote the transfer of photoelectrons from bulk to surface and thus inhibit their recombination with photoholes, leading to enhanced quantum efficiency. Besides the excellent activity, the recycling test also demonstrated that this catalyst was quite stable during liquid-phase photocatalysis since no significant decrease in activity was observed even after being used repetitively for 10 times, showing a good potential in practical application. Kong et al. [71] have been prepared visible-light-driven mesoporous bismuth titanate photocatalyst, which possesses wormlike channels, mixed phase mesostructured frameworks, large pore diameter (~6.1 nm), and low band gap energy (2.5 eV) via a modified EISA. The calcined sample exhibited visible-light photocatalytic reactivity valued by the degradation of 2,4-DCP in aqueous media. However Zhang et al. [72] have synthesized BiOI/TiO$_2$ heterostructures with different Bi to Ti molar ratios through a simple soft-chemical method at a temperature as low as 80°C. The photocatalytic activities of these BiOI/TiO$_2$ were evaluated on the degradation of methyl orange under visible-light irradiation ($\lambda > 420$ nm). The results revealed that the BiOI/TiO$_2$ heterostructures exhibited much higher photocatalytic activities than pure BiOI and TiO$_2$, respectively, and 50% BiOI/TiO$_2$ showed the best activity among all these heterostructured photocatalysts. The visible-light photocatalytic activity enhancement of BiOI/TiO$_2$ heterostructures could be attributed to its strong absorption in the visible region and low recombination rate of the electron-hole pairs because of the heterojunction formed between BiOI and TiO$_2$.

Cr(III) Doped TiO$_2$

Yu et al. [73] have fabricated an ordered and well-crystallized cubic Im3m mesoporous Cr–TiO$_2$ photocatalyst To synthesize the cubic Im3m mesoporous Cr–TiO$_2$, 1.5 g of F-127 dissolved in 19 mL of ethanol containing 0.15 g Cr(NO$_3$)$_3$.9H$_2$O. To this solution was added 0.015 mol of TiCl$_4$ with vigorous stirring for 0.5 h. The as-prepared transparent sample was then calcined at 400°C for 5 h in air to remove the surfactant species. The pure mesoporous TiO$_2$ is ineffective but the mesoporous Cr–TiO$_2$ shows a very high decomposition rate.[73,74] This must be due to the Cr^{3+} doping, which allows activation of the mesoporous

TiO$_2$ sample in the visible light region. The excellent photocatalytic performance is also related to the open mesoporous architecture with a large surface area, good anatase crystallinity and a 3D-connected pore system. It is known that chemical reactions are most effective when the transport paths through which molecules move into or out of the nanostructured materials are included as an integral part of the architectural design.[75, 76] The 3D-interconnected mesochannels in the cubic mesoporous Cr–TiO$_2$ composite serve as efficient transport paths for reactants and products in photocatalytic reactions. [77] However the Ti-Cr-MCM-48 photocatalyst prepared in a single step exhibits far superior photocatalytic activity compared to the TiO$_2$-Cr-MCM-48 prepared by a post-impregnation method. The high activity of the Ti-Cr-MCM-48 photocatalyst is attributed to a synergistic interaction between Cr ions dispersed in the silica framework and the nanocrystalline nature of titania crystallites anchored onto the pore walls. The catalysts were examined for the photocatalytic degradation of acetaldehyde. The photoactivity of the catalysts was evaluated by measuring the loss of acetaldehyde and the production of CO$_2$ in the gas phase upon visible-light irradiation. Ti-Cr-MCM-48 prepared in a single step showed the highest activity for CO$_2$ production. The high activity of Ti-Cr-MCM-48 arises from the synergistic interaction of the Cr ions dispersed in the MCM-48 framework and the titania nanocrystallites anchored onto the pore walls of MCM-48. The highly dispersed chromium ions can be excited by visible-light radiation to form a CT excited state, involving an electron transfer from O^{2-} to Cr^{6+}.[78]

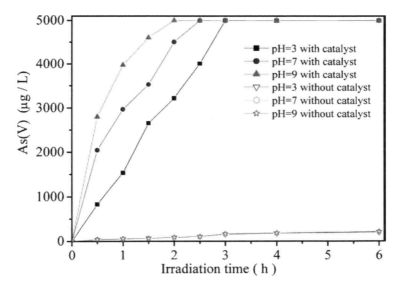

Figure 6. Time profiles of photocatalytic oxidation As (III) to As (V) with and without *meso*-TiO$_2$/R-Fe$_2$O$_3$ composites (sample A) at pH 3, 7, and 9, respectively). Reprinted with permission from ref. 69, Copyright 2008 ACS.

WO$_3$ and ZrO$_2$ Doped TiO$_2$

Pan et al. [79] have prepared highly ordered cubic mesoporous WO$_3$/TiO$_2$ thin films by spin coating via an EISA process, employing metal alkoxides as inorganic precursors and Pluronic F127 as a structure-directing agent. With the incorporation of WO$_3$ into TiO$_2$, the ordering of the mesopore structure has been appreciably improved. Titanium TTIP was dissolved and stabilized in ethanol solution acidified with aqueous HCl for 10 min at room temperature. Tungsten(V) pentaethoxide, stabilized by 2 equiv of 2,4-pentadione, was then added with vigorous stirring. After the solution was stirred for an additional 10 min, the resultant W-Ti sol was added to the Pluronic F127 ethanol solution dropwise. The molar ratio of the final composition in the Ti-W sol was as follows: (Ti + W)/F127/HCl/H$_2$O/ethanol) 1:0.005:1.75:10:25. The transparent sol was aged under mild agitation for 3 h and then deposited on Pyrex glass substrate by spin coating. The removal of F127 used as a liquid-crystal template was carried out by heating at 400°C for 4 h. The characterization results suggest that the majority of incorporated W species are located at the surface of the mesopore wall instead of doping into the TiO$_2$ lattice., the photocatalytic activity of WO$_3$/TiO$_2$ thin films in decomposing 2-propanol in the gas phase was optimized at 4 mol% of WO$_3$ concentration. Its photocatalytic activity was 2.2 times that of a mesoporous TiO$_2$ film and 6.1 times that of a nonporous TiO$_2$ film derived from a typical sol-gel method. The enhanced photocatalytic activity of WO$_3$/TiO$_2$ is ascribed to the increase in surface acidity.

Liu et al. [80] have produced of codoped Zr^{4+} and F$^-$ ions within anatase hollow microspheres by a fluoride mediated self-transformation strategy. Urea was used to catalyze the hydrolysis of aqueous mixtures of titanium sulfate (Ti(SO$_4$)$_2$) and zirconium oxide chloride (ZrOCl$_2$) in the presence of ammonium fluoride (NH$_4$F) under hydrothermal conditions. The concomitant participation of F$^-$ promotes lattice substitution of Ti^{4+} ions by Zr^{4+} and facilitates the transformation of surface-segregated amorphous ZrOx clusters into Zr-F species. The better photocatalytic activity of fluorinated samples may be at least partially attributed to the presence of well-crystallized anatase with retention of small grain size, high specific surface area and porosity, as well as hollow microarchitecture. Codoping is associated with electron transfer-mediated charge compensation between the Zr/F impurities, which reduces the number of both bulk and surface defects and provides a stabilizing effect on the local structure. Moreover, these synergetic interactions influence the textural characteristics and surface states of the TiO$_2$ host, such that the photocatalytic activity with regard to the decomposition of gaseous toluene is enhanced. Also The formation of porous TiO$_2$/ZrO$_2$ networks using mixed titania/zirconia precursor solutions was achieved by a polymer gel templating technique.[21] The mixed TiO$_2$/ZrO$_2$ network structures exhibit higher surface areas than a corresponding pure titania network, and in a certain range of metal oxide compositions. The photocatalytic efficiencies of the TiO$_2$ and TiO$_2$/ZrO$_2$ networks have been assessed by monitoring the photodecomposition of two organic molecules: salicylic acid and 2-chlorophenol. The TiO$_2$ network was found to exhibit an efficiency of ~60% and ~65% of the standard P25 TiO$_2$ for the salicylic acid and 2-chlorophenol reactions, respectively. For both photocatalytic reactions the presence of zirconia in the titania network (at a molar ratio of 1:9) resulted in enhanced photocatalytic activity relative to the pure TiO$_2$ network (80% and 100% the efficiency of Degussa P25), which is believed to be due to a number of factors including an increased surface area and a decrease of the anatase to rutile crystal phase transformation. A further increase in the zirconia-to-titania ratio leads to decreased activity as amorphous materials are obtained and zirconia itself does not act as a photocatalyst under the experimental conditions used.

Ce(III) and Zn(II) Doped TiO$_2$

Visible light induced metal-to-metal charge transfer (MMCT) for hetero-bimetallic Ti(IV)-O-Ce(III) assemblies on the pore of mesoporous silica, MCM-41 have been achieved.[81] The bimetallic Ti/Ce assembly exhibited the intense MMCT absorption up to 540. The synthesis of Ti(IV)-O-Ce(III) assemblies on the pore of MCM-41 was made by applying the nucleophilic character of titanol groups of (OH)Ti(OSi)3 sites. Namely, isolated (OH)Ti(OSi)3 sites were first synthesized on the pore surface of MCM-41. Subsequently, dehydrated Ti-MCM-41 crystalline was added to a dry acetonitrile solution of Ce(III)NO$_3$ and stirred at 55°C for 18 h. The pale yellow crystallites thus obtained were filtered, washed, and dried under dynamic vacuum at 40°C. It was concluded that the catalytic oxidation of 2-propanol is driven by the visible-light induced MMCT of Ti(IV)-O-Ce(III) assemblies. Sinha et al. [20] have prepared a thermally stable mesoporous ceria-titania using hexadecylamine as structure-directing reagent and triethanolamine as an additive in mixed propanol-water medium. These novel mesoporous ceria-titania materials showed high performance for the removal of toluene. The toluene removal performance was further enhanced for Pt impregnated mesoporous ceria-titania.

Kim et al. [82] have investigated the chemical bonding character and physicochemical properties of mesoporous zinc oxide-layered titanate nanocomposites synthesized by an exfoliation-restacking route. Upon hybridization with zinc oxide nanoparticles, the photocatalytic activity of layered titanate is enhanced with respect to the oxidative photodegradation of phenol and dichloroacetate. But of greater importance is that the chemical stability of guest zinc oxide against acidic corrosion is greatly improved by hybridization with layered titanate.

Nb(V) and PO$_4$(III) and Doped TiO$_2$

Stone et al[83] have prepared mesoporous titania and niobia molecular sieves by a ligand-assisted templating method. TTIP precursor (5.0 g) was mixed with acetylacetone (0.9 mL) in a separate flask and immediately added to the surfactant solution dodecyl phosphate. The resulting slurry was placed into a Teflon-lined autoclave and heated to 353 K for at least 5 days without agitation. The transition metal oxides were tested as photocatalysts in the liquid-phase oxidative dehydrogenation of 2-propanol to acetone. The observed quantum yield of the reaction was 0.45 over P25. However, mesoporous titania converted 2-propanol with a very low quantum yield of 0.0026. A very low quantum yield was also found for the mesoporous niobia sample compared to a crystalline standard. Apparently, the surface reactivities of the poorly crystallized samples were suppressed by defects that act as electron-hole traps. The relative inactivity of our mesoporous titania samples can be attributed to a high surface concentration of defects which can act as surface electron-hole recombination sites and/or the poisoning of catalytic surface sites by the phosphorus remaining from the surfactant. Yu et al. [84] synthesized a phosphated mesoporous TiO$_2$ by incorporating phosphorus from phosphoric acid directly into the framework of TiO$_2$ via a surfactant-templated approach. Briefly, 3 grams of P123 and 0.03 mol of TTIP were dissolved in 30 mL of absolute ethanol. After the solution was vigorously stirred for 1 h, 0.003 mol of H$_3$PO$_4$ and 40 mL of H$_2$O were added into the solution. The yellow powder was obtained after slow thorough vaporization of water and ethanol. The surface area of phosphated mesoporous TiO_2 exceeded 300 m^2/g after calcination at 400°C. It was found that the incorporation of phosphorus could stabilize the TiO_2 framework and increase the surface area significantly. This stabilization is attributed to two reasons: the more complete condensation of surface Ti-OH in the as-prepared sample and the inhibition of grain growth of the embedded anatase TiO_2 by the interspersed amorphous titanium phosphate matrix during thermal treatment. Both pure and phosphated mesoporous TiO_2 show significant activities on the oxidation of n-pentane. The higher photocatalytic activity of phosphated mesoporous TiO_2 can be explained by the extended band gap energy, large surface area, and the existence of Ti ions in a tetrahedral coordination.

Mechanism of Phosphated Mesoporous TiO_2: It is well-known that a large amount of uncondensed Ti-OH exists on the surface of the as-prepared amorphous mesoporous TiO_2. During calcination, the rapid reactions between the uncondensed Ti-OH would cause the walls of mesoporous TiO_2 to collapse. This is why pure mesoporous TiO_2 has a relatively poor thermal stability. Reactions between phosphoric acid and uncondensed Ti-OH in the as-prepared phosphated mesoporous TiO_2 may result in more completely condensed walls, which effectively prevent the collapse of the mesoporous structure during calcination.[84]

Ni(III) and La(III) Doped TiO_2

Jing et al. [23] have prepared Ni-TiO_2 by using of TBOT and acetylacetone in presence of laurylamine. Then, various amounts of $Ni(NO_3)_2$ were added into the sol to give doping level of Ni^{2+} from 0 to 3 wt%. The mixture was further aged at 313 K for 6 days and then transferred into a stainless steel autoclave and heated at 353 K for 6 days and 373 K for another day. The sample was finally calcined in flow air at 723 K for 4 h. XRD and EDX results indicated that Ni^{2+} was incorporated into the framework of the mesoporous TiO_2 in a highly dispersed way. The results of photocatalytic hydrogen evolution in aqueous methanol solution under UV–Vis light irradiation showed that activity of hydrogen production strongly depended on the amount of Ni doped. The highest activity was achieved with Ni doping of 1%. The results were rationalized by assuming that Ni^{2+} serves as shallow trapping sites, greatly enhancing the activity of the mesoporous photocatalyst.

Dai et al. [85] have prepared TiO_2 nanoparticles as follows: 1.1402 g $Ti(SO_4)_2$ was dissolved into 3.5 ml H_2O water. The obtained solution was added into CTAB solution under stirring. The molar ration of $Ti(SO_4)_2$:$La(NO_3)_3$:CTAB:H_2O is 0.95:0.05:0.12:100, then adjusted the pH 0.6. After stirring for 30 min, the resulting mixture was transferred into an autoclave at 100 °C for 72 h. Photocatalytic degradation of commercial phoxim emulsion in aqueous suspension was investigated by using La-doped mesoporous TiO_2 as the photocatalyst under UV irradiation. La-doped TiO_2 with mesostructures showed much better photoactivity than that of undoped ones and the P25 due to its large surface area, highly crystallized mesoporous wall, and more active sites for concentrating the substrate.[86]

Multi-cations Doped TiO$_2$

Yang et al. prepared silver and indium oxide codoped titania nanocomposites by a one-step sol-gel-solvothermal method in the presence of P123.[87] The resulting Ag/In$_2$O$_3$-TiO$_2$ three-component systems mainly exhibited an anatase phase structure, high crystallinity, and extremely small particle sizes with Ag particles well-distributed on the surface. Compared with pure anatase TiO$_2$, the Ag/In$_2$O$_3$-TiO$_2$ systems showed narrowing of the band gap due to the change in the band position caused by the contribution of the In 5s5p orbits to the conduction band. At 2.0% Ag and 1.9% In$_2$O$_3$ doping, the Ag/In$_2$O$_3$-TiO$_2$ system exhibited the highest UV-light photocatalytic activity for degaradtion rhodamine B (RB) and methyl ter-butyl ether (MTBE), and nearly total degradation of dye RB (25 mg L^{-1}) or MTBT (200 mg L^{-1}) was obtained after 120 min UV-light irradiation. In addition, the UV-light photocatalytic activity of three-component systems exceeded that of pure TiO$_2$ and two-component (Ag/TiO$_2$ or In$_2$O$_3$-TiO$_2$) systems as well as the commercial photocatalyst, Degussa P25. These results indicate that (i) codoped TiO$_2$ system is more photoactive than single-doped TiO$_2$ system, (ii) single doped TiO$_2$ system is more photoactive than pure TiO$_2$, and (iii) Ag/In$_2$O$_3$-TiO$_2$ with P123 is more photoactive than that prepared without P123. This is explained by (i) Enhanced quantum efficiency of Ag and In$_2$O$_3$ codoped TiO$_2$ systems (ii) Decreased band gap due to coupling of semiconductor In$_2$O$_3$ with TiO$_2$, which results from formation of dopant energy level within the band gap of TiO$_2$. (iii) Size and morphology of Ag/In$_2$O$_3$-TiO$_2$ are also responsible for its enhanced photocatalytic activity. Chu et al., [89] have synthesized 3D highly porous TiO$_2$-4%SiO$_2$-1%TeO$_2$/Al$_2$O$_3$/TiO$_2$ composite nanostructures (30-120 nm) directly fixed on glass substrates by anodization of a superimposed Al/Ti layer sputterdeposited on glass and a sol-gel process. The porous composite nanostructures exhibited enhanced photocatalytic performances in decomposing acetaldehyde gas under UV illumination. Specially, the composite nanostructure showed the highest photocatalytic activity that is 6-10 times higher than commercial P-25 TiO$_2$.

4. HIGHLY ORDER MESOSTRUCTURED TiO$_2$-SiO$_2$ PHOTOCATALYSTS

The mesoporous TiO$_2$-SiO$_2$ composites exhibit a binary function for the degradation of organic pollutants derived from their high surface areas and highly crystalline anatase nanoparticles. Large surface areas and pore volumes can enrich organic molecules in the channels to contact with anatase nanocrystals which act as the catalytic sites to degrade the molecules under UV light irradiation. This unique feature makes the composites exhibit excellent photodegradation activity. Dong et al. [90] have prepared highly ordered 2-D hexagonal mesoporous crystalline TiO$_2$-SiO$_2$ nanocomposites with variable Ti/Si ratios (0 to ∞). These mesostructured TiO$_2$-SiO$_2$ composites were obtained using TTIP and TEOS as precursors and P123 based on the solvent evaporation-induced co-self-assembly process under a large amount of HCl.

The amorphous silica acts as a glue linking the TiO$_2$ nanocrystals and improves the thermal stability. As the silica contents increase, the thermal stability of the resulting mesoporous TiO$_2$-SiO$_2$ nanocomposites increases and the size of anatase nanocrystals decreases. The mesoporous TiO$_2$-SiO$_2$ nanocomposites exhibit excellent photocatalytic activities more than P25 for the degradation of rhodamine B in aqueous suspension due to the bifunctional effect of highly crystallized anatase nanoparticles and high porosity (Figure 7).

Wang et al. [91] have synthesized TiO$_2$/SBA-15 composites through a postsynthetic approach with the assistance of ethylenediamine. While Li et al. [92] have prepared the same composite for photodecomposition of Orange II. Ethylenediamine plays a double role: (1) etching the surface of silica SBA-15 and directing its regeneration and (2) introducing nitrogen atoms to the lattice of TiO2. The excellent photocatalytic activity of the composites is evaluated via the photodecomposition of phenol in the liquid phase under visible- and ultraviolet-light illumination. The conversion of phenol varies with the content of TiO$_2$ in the composites, and the optimal value is up to 46.2% under illumination in the visible region. Li et al. [93] have prepared a core/shell structure of a nano titania/Ti-O-Si species modified titania embedded in mesoporous silica by the sol-gel method. In typical experiment, 0.165 mol/L TiOCl$_2$ aqueous solution was obtained by dropping known amounts of TiCl$_4$ into 100 mL H$_2$O in an ice-water bath. With strong magnetic stirring, 1.37 g of H$_2$SO$_4$ was added to 110 mL of the 0.165 mol/L TiOCl$_2$ aqueous solution, and then a known amount of a 2.0 mol/L Na$_2$SiO$_3$ solution and 33 g of CTAB were added dropwise to the mixture. The mixture was aged at 90°C for 48 h and heated at 480°C for 2 h. The as-synthesized TiO$_2$-xSiO$_2$ composites exhibit both much higher absorption capability of organic pollutants and better photocatalytic activity for the photooxidation of benzene than pure titania and P25. The better photocatalytic activity of as-synthesized TiO$_2$-xSiO$_2$ composites than pure titania is attributed to their high surface area, higher UV absorption intensity, and easy diffusion of absorbed pollutants on the absorption sites to photogenerated oxidizing radicals on the photoactive sites. Xuzhuang et al. [94] have fabricated a new composite Ti/clay by reaction between TiOSO$_4$ and a synthetic layered clay laponite. The large number of the anatase crystals and better accessibility to the sites by UV light and reactant molecules are the major factors enhancing the photocatalytic activity. The high photocatalytic activity for phenol decomposition resulted from the unique structure of the composite, in which anatase nanocrystals were attached on leached laponite fragments. The performance of the catalysts is related to their structural features, and it is found that the catalytic activity increased with increasing size of the anatase crystals in the catalysts, specific surface area, and mesopore size. Li et al. [95] have synthesized monodispersed concentric hollow nanospheres with mesoporous silica shell and anatase titania core by the combination of sol-gel reaction and distillation-precipitation polymerization. The first synthesis step involved the preparation of cross-linked poly(methacrylic acid) (PMAA) core nanospheres via distillation-precipitation polymerization in the presence of ethylene glycol dimethylacrylate. The next step involved the synthesis of PMAA/TiO$_2$ composite nanospheres via the sol-gel process, using the cross-linked PMAA nanospheres as the template cores. In the subsequent step, the PMAA/TiO$_2$ composite nanospheres were coated with a uniform PMAA layer via distillation-precipitation polymerization to produce the PMAA/TiO$_2$@PMAA core-shell particles. Finally, coating of the core-shell particles with an outer silica shell, derived from the sol-gel reaction of the TEOS precursor, produced the PMAA/TiO$_2$@PMAA@SiO$_2$ trilayer hybrid nanospheres (Figure 8). Photocatalytic decomposition of methyl orange in the concentric hollow reactors is followed an apparent first-order rate constant. The observed rate constant for the concentric hollow nanospheres as photocatalysts seems to be lower than those reported for naked and doped TiO$_2$ nano- and microparticles.[96] The lower reaction rate observed in the present work is probably due to the low content of mesoporous anatase titania in the hollow nanospheres.

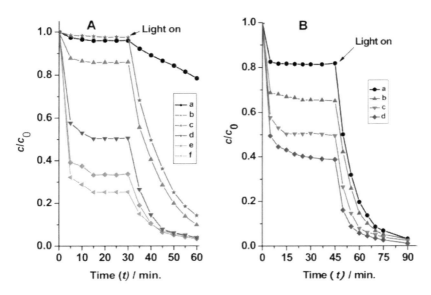

Figure 7. Photocatalytic degradation of RhB monitored as the normalized concentration change versus irradiation time in the presence of (A) mesoporous TiO$_2$-SiO$_2$ composites prepared with different Ti/Si ratios. (a) Mesoporous TiO$_2$ calcined at 400°C for 2 h; (b) commercial photocatalyst P25; (c) mesoporous 90TiO$_2$-10SiO$_2$ composite calcined at 700°C for 2 h; (d) mesoporous 80TiO$_2$-20SiO$_2$,(e) 70TiO$_2$-30SiO$_2$, and (f) 60TiO$_2$-40SiO$_2$ composites calcined at 850°C for 2 h. (B) Photocatalytic degradation for mesoporous 80TiO$_2$-20SiO$_2$ composites calcined at 700°C for 4 h (a), 800°C for 2 h (b), 850°C for 2 h (c), and 900°C for 2 h (d). Reprinted with permission from ref. 90, Copyright 2007 ACS.

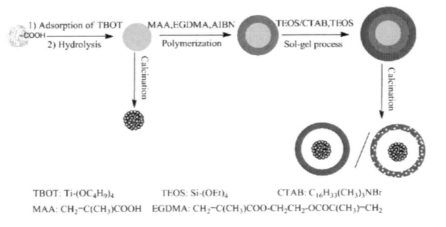

Figure 8. Schematic of combined polymerization and pol-gel reactions for the preparation of nearly monodispersed concentric hollow nanospheres composed of mesoporous silica shells and anatase titania inner cores. Reprinted with permission from ref. 95, Copyright 2009 ACS.

Aronson et al. [97] have grafted TiO$_2$ onto the pore surface of MCM-41 and FSM-16 by reacting TiCl4 in hexanes with the as-synthesized mesostructured silicate. It was found that titania forms well-dispersed isolated (TiO$_2$)n clusters ($n \sim$ 30- 70) within the channel structure. These are attached to the silicate walls via Si-O-Ti bonds. A minor second phase consisting of anatase crystallites 10-25 nm in diameter on the external surface of the mesoporous silicate crystals was sometimes obtained. It is concluded that an organic moiety, such as the surfactant present in the pores, or a physical constraint, such as the pore walls, is necessary to prevent the creation of large TiO$_2$ agglomerates and enable the formation of nanosized TiO$_2$ clusters. The titania-grafted MCM-41 samples exhibited good catalytic activity for photobleaching of rhodamine-6G and for oxidation of α-terpineol; however, product selectivity was low. Alvaro et al. [98] and Maldotti et al. [99] reported the preparation a series of structured mesoporous silicas, starting from colloidal TiO$_2$ nanoparticles in combination with TEOS using neutral Pluronic or cationic CTAB as templates (Figure 9). Even though the activity of these new mesostructured materials for the degradation of phenol in aqueous solution is lower that those found for P-25 TiO$_2$, the turnover frequency of the photocatalytic activity is much higher for the mesoporous titania. Also, Both pure mesoporous TiO$_2$ and mixture of 50% TiO$_2$ and 50% SiO$_2$ can induce cyclohexane photooxidation to yield cyclohexanone.

Li et al. [100] have prepared a core/shell SiO$_2$/TiO$_2$ photocatalyst using a liquid-phase deposition method. The photocatalytic activity of the core/shell SiO$_2$/TiO$_2$ catalyst for decomposition of Orange II in liquid phase was observed to be comparable with that of P25. Kang et al. [101] have prepared mesoporous SiO$_2$-modified nanocrystalline TiO$_2$ photocatalysts by sol-hydrothermal processes, followed by post-treatment with F127-modified silica sol. In a typical process, 5 mL of Ti(OBu)$_4$ is dissolved in 5 mL of C$_2$H$_5$OH produce Ti(OBu)$_4$-C$_2$H$_5$OH solution. Meanwhile, 5 mL of water and 1 mL of HNO$_3$ are added to another 20 mL of C$_2$H$_5$OH in turn to form an ethanol-HNO$_3$-water solution. After the two resulting solutions are stirred for 30 min, respectively, the Ti(OBu)$_4$-C$_2$H$_5$OH solution is slowly added dropwise to the ethanol-HNO$_3$-water solution under vigorously stirring to carry out a hydrolysis. The sol-hydrothermal production obtained is transferred into a bottle. Subsequently, a desired volume of F127-modified SiO$_2$ sol, prepared in advance by hydrolysis of TEOS [102] is added to the weighing bottle. Mesoporous SiO$_2$-modified nanocrystalline TiO$_2$ samples can exhibit much higher photocatalytic activity for degrading rhodamine B than P25 TiO$_2$, which is explained mainly by the high photoinduced charge carrier separation rate resulting from the high anatase crystallinity and the large surface area related to the small nanocrystallite size and mesoporous SiO$_2$ as well as still possessing a certain amount of surface hydroxyl group. Morishita et al. [103] have employed Ti-containing mesoporous organosilicas (T-OS), synthesized by a surfactant templating method with an organosilane precursor, as the photocatalyst and have studied the effects on the olefin conversion and the epoxide selectivity. The T-OS catalysts demonstrate the same high epoxide selectivity as does T-S, but scarcely improve the olefin conversion. Photoluminescence measurement reveals that the T-OS catalysts with high surface hydrophobicity enhance the access of hydrophobic olefins to the photoexcited Ti-oxide species as expected, but destabilize the excited species themselves. ESR analysis demonstrates that the T-OS catalysts also destabilize the active oxygen radical (O$_2^-$), a crucial oxidant for olefin epoxidation, formed on the excited Ti-oxide species. These destabilizations counteract the enhanced olefin access to the excited species, resulting in almost no improvement in olefin conversion. Hu et al. [104] have prepared Ti-MCM-41 mesoporous molecular sieves using TEOS and TPOT as the starting materials and CTAB as a structure directing agent. A high Ti/Si gel ratio was used for this preparation due to the high solubility of Ti under strongly acidic conditions. The reaction mixture was stirred for 5 days at room temperature and then the solid product obtained was filtered, washed, dried, and calcined at 823 K for 6 h with airflow. It was found that an increase in the Ti content caused the structure of the Ti-oxides in Ti-MCM-41 to change from an isolated tetrahedral coordination to adjacent Ti-oxide species with Ti^{4+} of tetrahedral coordination. Ti-HMS (0.60 wt% as Ti) was prepared by a hydrothermal synthesis method from the following composition: 1.0 TEOS/0.0076 TPOT/0.2 dodecylamine/9.0 EtOH/160 H_2O.[105,106] The photocatalytic reactivity of these catalysts for the decomposition of NO into N_2 and O_2 was found to strongly depend on the local structure of the Ti-oxide species including their coordination and distribution, i.e., the charge transfer excited state of the highly dispersed isolated tetrahedrally coordinated Ti-oxides act as the active sites for the photocatalytic decomposition of NO into N_2 and O_2. Ti-MCM-41 showed higher photocatalytic reactivity than Ti-HMS for the decomposition of NO. Rohlfing et al. [107] have fabricated titania-silica composite films with a high content of crystalline titania phase and periodic mesoporous structure by a low temperature "brick and mortar" approach (Figure 10). Pre-formed titania nanocrystals were fused with surfactant-templated sol-gel silica, which acts as a structure-directing matrix and as a chemical glue. Using P123 as the structure-directing agent, the structure formation of the mesoporous silica is greatly disturbed as a result of the presence of TiO_2 nanoparticles. On the contrary, the use of F127, whose molecules are larger and whose poly(ethylene oxide) blocks are more hydrophilic, enables the preparation of composite TiO_2 - SiO_2 mesoporous architectures that can accommodate up to 50 wt% of nanocrystals and yet retain the periodicity of the porous structure. While films of pure silica are inactive for photooxidation of NO, the activity of those containing TiO_2 nanocrystals increases almost linearly with the TiO_2 content, approaching the conversion efficiency of 3.9-4% for the films composed solely of titania particles taken as a reference. This linearity confirms the homogeneous distribution of the particles and their good accessibility for molecules from the gas phase.

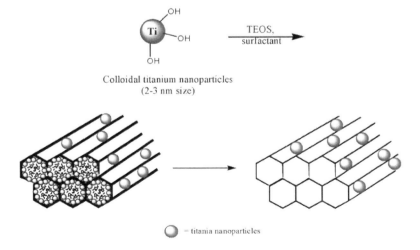

Figure 9. Schematic of general procedure used for the synthesis of mesoporous materials. Reprinted with permission from ref. 98, Copyright 2006 ACS.

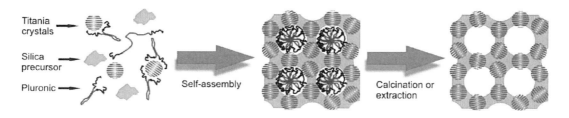

Figure 10. Formation of nanocomposite titania-silica mesoporous films using pre-formed titania nanocrystals stabilized by the pluronic polymer and amorphous sol-gel silica precursor. Reprinted with permission from ref. 107, Copyright 2009 ACS.

Allain et al. [108] have dispersed TiO_2 colloidal within a transparent silica binder with a mesoporous structure. In typical synthesis, 5.5 mL of a solution of TEOS in ethanol (2.25 molL^{-1}) is hydrolyzed by adding 2.25 mL of an aqueous acidic solution of HCl under vigorous stirring. [109] The obtained solution was then aged for 1 h at 60°C and cooled down to room temperature. The copolymer was then dissolved into the obtained sol. The TiO_2 particles were then added into the obtained sol by slow addition of a given volume of a commercial aqueous solution of acid-stabilized TiO_2 dispersion with a TiO_2 content of 220 g L^{-1}. Stearic acid was first deposited on the film by spin-coating from a solution in tetrahydrofuran. Studies of photodegradation kinetics show that such mesoporous films are at least 15 times more active than films synthesized with a usual microporous silica binder. Moreover, the measured quantum-yield efficiency is 1.1% and the improved photoactivity of the films is obtained as resulting from the closer proximity between the organic molecules and the surface of the TiO_2 crystallites as well as the improved diffusion rate of water and oxygen through the interconnected pore network. Ogawa et al. [110] have prepared transparent self-standing films of titanium-containing (Ti/Si ratio of 1/50) silica-surfactant mesostructured materials by the solvent evaporation method from tetramethoxysilane, vinyltrimethoxysilane, TTIP, and octadecyltrimethylammonium chloride. The films were converted to titanium-containing nanoporous silica films by subsequent calcination in air at 550°C, while their highly ordered mesostructures and macroscopic morphology were retained after the surfactant removal. Titanium-containing nanoporous silica films with hexagonal and cubic symmetry were obtained by changing the composition of the starting mixtures. The titanium ions exist in the silica network as a tetrahedrally coordinated species. UV irradiation of the titanium-containing nanoporous silica film in the presence of CO_2 and H_2O led to the evolution of CH_4 and CH_3OH, indicating high selectivity for the formation of CH_3OH, showing the characteristic reactivity of the charge transfer excited complexes of the tetrahedrally coordinated titanium oxide species.

5. NOBLE METALS/TiO$_2$ MESOPOROUS

Doping of noble metals with mesoprous TiO_2 photocatalysts was proposed to enhance the photocatalytic activity due to their different Fermi levels, characterized by the work function of the metals and the band structure of the semiconductors. Upon contact, a Schottky barrier can be formed between the TiO_2 and the noble metals, leading to a rectified charge carrier transfer. Ismail et al. [111] have suggested the mechanism of highly Pd, Pt and Au/TiO$_2$

mesoporuous for photooxidation of methanol (Figure 11). 3-D mesoporous TiO$_2$ network acts as an antenna system transferring the initially generated electrons from the location of light absorption to a suitable interface with the noble metal catalyst and subsequently to the location of the noble metal nanoparticle where the actual electron transfer reaction will take place. Within this antenna model, it can be envisaged that the overlap of the energy bands of the nanoparticles forming this network will result in unified energy bands for the entire system enabling a quasi-free movement of the photogenerated charge carriers throughout. Consequently, an electron generated by light absorption within one of the nanoparticles forming the network will subsequently be available to promote redox processes anywhere within the structure.

Bain et al. [24] have prepared mesoporous Au/TiO$_2$ nanocomposites. Briefly, Pluronic surfactant P123, TiCl$_4$, Ti(OBu)$_4$, and AuCl$_3$ were mixed in ethanol. Casting the mixture followed by an aging process resulted in homogeneous mesostructured nanocomposites. Calcination removed P123 and created crystalline mesoporous TiO$_2$ networks embedding gold nanoparticles. The conversion of phenol oxidation and chromium reduction continuously increases from 22% to 95% when Au content is increased from 0 to 0.5%. A near three-time improvement in phenol decomposition is achieved when 0.5% of Au was doped, unambiguously suggesting a significantly improved photocatalytic activity. However Ismail et al. prepared it by using TTIP with hydrogen tetrachloroaurate in the presence of a F127 for methanol oxidation.[111]

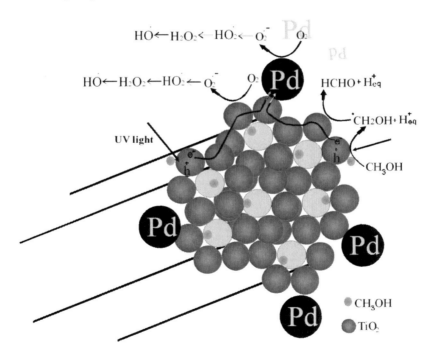

Figure 11. The proposed antenna and reaction mechanisms for methanol photooxidation to illustrate the enhanced photonic efficiency of mesostructured Pd/TiO$_2$ photocatalyst, absorption of UV light by the semiconducting nanoparticle promotes an electron from the valence band to the conduction band. The lines in the scheme show cut perpendicular to the c axis of the hexagonal pore system which extends infinitely in this direction. Reprinted with permission from ref. 111, Copyright 2009 ACS.

Srinvasn et al. [112] have synthesized three-dimensionally ordered macroporous (3DOM) TiO$_2$ by colloidal crystal templating against polystyrene spheres using a metal alkoxide precursor. The 3DOM TiO$_2$ walls which are predominantly anatase (>98%) were decorated homogeneously with gold nanoparticles (5-7 nm) by pH-controlled precipitation of Au from HAuCl$_4$ using sodium hydroxide. Macroporous 3DOM TiO$_2$ (Figure12) with pore diameter 0.5μm had the highest first-order rate constant of 0.042 min^{-1} for decomposition of MB, compared to 0.025 min^{-1} for P25 TiO$_2$. Deposition of gold on the 3DOM TiO$_2$ surfaces decreased the reaction rate by covering the surface active sites

Bannat et al. [113] have synthesized mesoporous Au/TiO$_2$ films by the EISA method. For mesoporous TiO$_2$ films deposited on an ITO layer, the photonic efficiency for NOx oxidation is higher than for films prepared on glass, because the pore structures are altered. Incorporation of Au results in a significant improvement in the photonic efficiency due to the generation of Schottky barriers, which inhibit the recombination of electron-hole pairs and thereby increase the concentration of photogenerated holes at the film surface reacting with NO. Li et al. [114] have developed a simple method to generate nanoporous organic–inorganic hybrid films and arrays of Au–TiO$_2$ nanobowls using di-block copolymers as templates in combination with a sol–gel process. The photocatalytic activity of a representative hybrid PS-b-PEO–HAuCl$_4$–TiO$_2$ film in terms of the decomposition of MB is similar nanostructured TiO$_2$.

Silver nanoparticles were introduced into the mesopores using wet impregnation followed by heat treatment.[25] The cubic structured mesoporous titania had higher stability than the hexagonal structure and could be formed with a high content of nanocrystalline anatase with conservation of the meso-order. In typical synthesis, 1 g sample of P123 was dissolved in 12 g of ethanol (magnetic stirrer, 1 h). Separately, 4.2 g of titanium ethoxide was plunged into 3.2 g of HCl, and the mixture was stirred vigorously for 10 min, resulting in a clear solution. [115] The slides were dipped into the solution and withdrawn into open air and were dried at room temperature in air for 24 h. The dried film was heated at 400°C. Silver nanoparticles were incorporated into the mesoporous titania by in situ heat-induced reduction after impregnation of the films in a Ag(I)-containing solution. It was found that the meso-ordered titania had a sufficiently high crystallite content to be photoactive for oxidation of stearic acid and that the reaction mechanism resembled that of a non-meso-ordered titania sample. It was shown that mesoordered titania was photocatalytically active and that the activity was influenced by the presence of silver nanoparticles. Stathatos et al. [88] reported a transparent mesoporous titania films which have been deposited on glass slides by a sol-gel procedure in the presence of Triton X-100 reverse micelles in cyclohexane. Benzothiazolium, 2-[[4-[ethyl(2-hydroxyethyl)- amino]phenyl]azo]-5-methoxy-3-methyl(T-4)-methoxysulfate (Basic Blue 41), has been adsorbed on these films from aqueous solutions, and the photodegradation of the dye by visible-light illumination has been monitored by absorption spectrophotometry. Films doped with silver ions, incorporated through the reverse micellar route, are more efficient photocatalysts than pure titanium films and become even more efficient when they are treated with UV radiation. Films doped with ruthenium ions are less efficient for photocatalysis but when they are treated with UV radiation, they also become more efficient photocatalysts than pure titania films. [88]

Wang et al. [26] have prepared highly dispersed Pt nanoparticles embedded in a cubic mesoporous anatase thin film. Briefly, A solution of TTIP (5.3 mL) in fuming HCl (2.7 mL) was added dropwise to a template solution prepared by dissolving P123 (1 g) in absolute ethanol (15 mL). The as-synthesized thin films were aged at 48°C for 24 h, and then aged again at room temperature for 1 h. The thin films were subsequently treated with NH_3 vapor for 5 s. After that, the films were heated at 200°C for 12 h and then at 350°C 4 h to remove the template and increase cross-linking of the inorganic framework. A calcined TiO_2 thin film was immersed in a bottle filled with a solution of metal precursor ($PtCl_4$). After sonication under vacuum for 3 min, the film was stored in a vacuum oven for 12 h. After drying at 30°C under vacuum, the film was irradiated with UV light in the presence of methanol vapor for 2 h. The diameter of the Pt cluster can be controlled to below 5 nm, and the high dispersion of these clusters gives rise to catalytic activity for the oxidation of carbon monoxide. Furthermore, the pore-stabilized Pt particles contact and interact with the anatase-TiO_2 nanocrystals embedded in the mesonetwork, forming semiconductor/metal nanoheterojunctions. These nanoheterojunctions promote the separation of charge carriers on UV-excited TiO_2, thus significantly improving the photocatalytic activity of porous Pt/TiO_2 composites toward killing bacteria cells of M. lylae. Lakshminarasimhan et al. [116] have syntheisized mesoporous TiO_2 consisting of compactly packed nanoparticles without surfactant. In typical synthesis, 4.4 mL of TTIP was added to 100 mL of ethanol containing 0.4 mL of 0.01 M aqueous KCl under vigorous stirring. After 6h stirring, the precipitate was collected by filtering and washed thoroughly with distilled water several times. The washed precipitate was dried overnight in an oven at 85°C. The dried powder was calcined at 450°C for 1 h in a muffle furnace. The activity of meso-TiO_2 exhibited a unique dependence on Pt cocatalyst loading. Under both UV and visible light irradiation, the highest activity for H_2 production was obtained around 0.1 wt% Pt and further increase reduced the activity, whereas other nonporous TiO_2 samples exhibited a typical saturation behavior with increasing Pt load. The enhanced photocatalytic activity of meso-TiO_2 is ascribed to the compact and dense packing of TiO_2 nanoparticles forming a uniform agglomerate, which enables efficient charge separation through interparticle charge transfer.

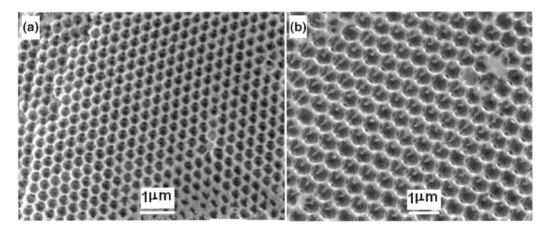

Figure 12. Secondary electron images (SEI) showing the pore morphology of (a) 0.5 μm (MT5) and (b) 1.0 μm three dimensionally ordered macroporous titania. Reprinted with permission from ref. 112, Copyright 2007 ACS.

6. NONMETALS DOPED/TiO$_2$ MESOPOROUS ACTIVE PHOTOCATALYSTS

Meanwhile, the relatively high rate of electron-hole recombination often results in a low quantum yield and poor efficiency of photocatalytic reactions. These fundamental problems prevent anatase TiO$_2$ from practical applications. Therefore, one of the endeavors to improve the performance of TiO$_2$ is to increase its optical activity by shifting the onset of its response from the UV to the visible region. A promising approach is the doping of TiO$_2$ with nonmetals. The rationale behind this approach is to sensitize TiO$_2$ toward visible light either by generating newly created midgap states or by narrowing the band gap. The observed band gap shift from the UV into the visible region has been attributed to (i) substitution of lattice oxygen by the anion or (ii) formation of interstitial species in vacancies or micro-voids that give rise to surface or near-surface states [117-119].

6.1. Nitorgen Doped TiO$_2$ Mesoporous

Fang et al. [120] have prepared visible-light-active mesoporous N-doped TiO$_2$ photocatalysts by the precipitation of titanyl oxalate complex ([TiO(C$_2$O$_4$)$_2$]2-). Briefly, 23.0 g of H$_2$C$_2$O$_4$ ·2H$_2$O was dissolved in 200 mL of distilled water kept at 15°C, into which 10 mL of titanium tetrachloride (TiCl$_4$) was drip under stirring, then ammonium hydroxide was added to adjust the pH of the solution to 8-9 to form precipitate. The precipitate was filtered out, washed with distilled water and ethanol, dried at 70°C, and finally calcined in air at the desired temperatures for 2 h. N-TiO$_2$ photocatalysts exhibit comparable UV-light activity and visible-light activity in the photodegradation of methyl orange. The doped N species locates at the interstitial sites in TiO$_2$, which leads to the band gap narrowing of TiO$_2$. A novel and interesting result is that N-doped TiO$_2$ calcined at 400°C has Bronsted acid sites arising from covalently bonded dicarboxyl groups, which greatly enhances the adsorption capacity for methyl orange. Chi et al. [121] prepared mesoporous N-doped TiO$_2$ microspheres a template-free solvothermal method. The N-doped TiO$_2$ mesoporous spheres show higher visible-light photocatalytic activity than the undoped TiO$_2$. The dual role of urea helps the formation of a mesoporous structure and the doping of nitrogen into TiO$_2$ to be completed simultaneously during the solvothermal process. The amount of urea shows the crucial effect on the mesoporous structure and nitrogen doping in TiO$_2$. The enhanced photocatalytic activities of N-TiO$_2$ in UV light may be due to the increase of the surface deficiency after the introduction of nitrogen into the TiO$_2$ structure. It also could be found that, with the increase of the nitrogen amount in the spheres, the visible-light photocatalytic activity would also be enhanced, which could be the evidence to confirm the role of nitrogen in the lattice for improvement of the visible-light response of N-TiO$_2$.

Nitrogen-containing surfactant dodecylammonium chloride (DDAC) was introduced as a pore templating material for tailor-designing the structural properties of TiO$_2$ and as a nitrogen dopant for its visible light response. [17] When the liquid molecular precursor of titanium is added to DDAC solution, TTIP is hydrolyzed and condensed around the self-assembled surfactants, forming a surfactant organic core/TiO$_2$ inorganic shell composite, as demonstrated in Figure 13. Red shift in light absorbance up to 468 nm, 0.9 eV lower binding energy of electrons in Ti 2p state, and reduced interplanar distance of crystal lattices proved nitrogen doping in the TiO$_2$ lattice. Due to its narrow band gap at 2.65 eV, N-TiO$_2$ efficiently degraded cyanobacterial Toxin Microcystin-LR (MC-LR) under visible spectrum above 420 nm. Acidic condition (pH 3.5) was more favorable for the adsorption and photocatalytic degradation of MC-LR on N-TiO$_2$ due to electrostatic attraction forces between negatively charged MC-LR and charged N-TiO$_2$. Even under UV light, MC-LR was decomposed 3-4 times faster using N-TiO$_2$ than control TiO$_2$. The degradation pathways and reaction intermediates of MC-LR were not directly related to the energy source for TiO$_2$ activation (UV and visible) and nature of TiO$_2$.

Cong et al. [122] have synthesized N-doped TiO$_2$ nanocatalysts with a homogeneous anatase structure hrough a microemulsion-hydrothermal method by using triethylamine, urea, thiourea, and hydrazine hydrate. In the microemulsion system, Triton X-100 was used as the surfactant, 1-hexanol as the cosurfactant, cyclohexane as the continuous oil phase, and TBOT dissolved in nitric acid (5 mol/L) as the aqueous phase. Analysis by Raman and X-ray photoemission spectroscopy indicated that nitrogen was doped effectively and most nitrogen dopants might be present in the chemical environment of Ti-O-N and O-Ti-N. The results of photodegradation of rhodamine B and 2,4-dichlorophenol in the visible light irradiation ($\lambda >$ 420 nm) suggested that the TiO$_2$ photocatalysts after nitrogen doping were greatly improved compared with the undoped TiO$_2$ photocatalysts and Degussa P-25. Nitrogen doping could inhibit the recombination of the photoinduced electron and thereafter increase the efficiency of the photocurrent carrier.

Wang et al. [123] prepared mesoporous TiO$_{2-x}$N$_x$/ZrO$_2$ visible-light photocatalysts by sol gel method. Results revealed that nitrogen was doped into the lattice of TiO$_2$ by the thermal treatment of NH$_3$-adsorbed TiO$_2$ hydrous gels, converting the TiO$_2$ into a visible-light responsive catalyst. The introduction of ZrO$_2$ into TiO$_{2-x}$N$_x$ considerably inhibits the undesirable crystal growth during calcination. Consequently, the ZrO$_2$-modified TiO$_{2-x}$N$_x$ displays higher porosity, higher specific surface area, and an improved thermal stability over the corresponding unmodified TiO$_{2-x}$N$_x$ samples. The photocatalytic activity of the samples was evaluated by the decomposition of ethylene in air under visible light ($\lambda >$ 450 nm) illumination. The activity of the TiO$_{2-x}$N$_x$ is initially quite high and decreases rapidly with increasing calcination temperature. At the sintering temperature of 400°C, the conversion of C$_2$H$_4$ on TiO$_{2-x}$N$_x$ is 28%, but it drops to 7% at 500°C. Soni et al. [124] have prepared N-doped TiO$_2$ mesoporous thin films templated using titanium alkoxide solution containing thiourea as the nitrogen source and P123 by EISA method. The as-prepared film, containing anatase nanocrystallites exhibited photocatalytic activity in the blue region of the visible spectrum. The shift in the bandgap was monitored by UV-vis spectroscopy and the photocatalytic properties were characterized by monitoring the photodegradation of methylene blue (MB) upon irradiation with visible light. Multiply coated thin films having different thicknesses were prepared to improve the efficiency of N-doped TiO$_2$ thin films. The adsorption capacity and photocatalytic activity of thin films in the visible region was bigger for multiply coated films than for thinner films as a result of their increased surface area. The photocatalytic activity improved significantly with the number of coating cycles. As expected, the thicker film coated five times shows the best response.

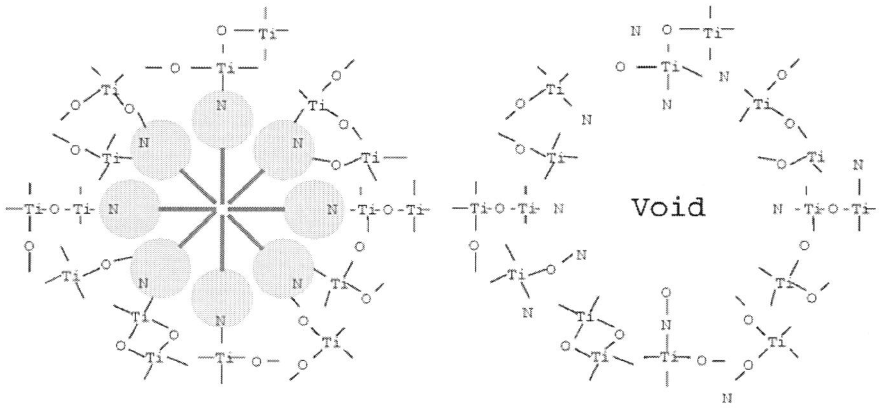

Figure 13. Schematic incorporation of Ti-O-Ti network onto self-organized DDAC surfactant micelles to form an organic core/inorganic shell composite, followed by the removal of the organic templates to form N-TiO$_2$ with mesoporous Structure. Reprinted with permission from ref. 17, Copyright 2007 ACS.

Martínez-Ferrero reported the preparation nanocrystalline mesoporous N-doped TiO$_2$. [125] Briefly, the initial solution, containing TiCl$_4$/EtOH/H$_2$O/F127 with molar ratios of 1:40:10:0.005, was stirred for 15 min at room temperature for homogenization. [126,127] The transparent and slightly viscous solution was then deposited by dip coating. The liquid layers were evaporated at a fixed relative humidity (RH) of 15%, transferred, and aged in a sealed environmental chamber with a RH fixed at 70% for 18 h. Then, the following thermal treatment was applied: 24 h at 130°C and 3 h at 350°C for densification of the amorphous TiO$_2$, 15 min flash heating at 500°C, and 5 min flash heating at 550°C. The introduction of nitrogen into the anatase structure starts at 500°C, with N bonding to titanium via oxygen substitution. Increasing the treatment temperature leads to the formation of TiN (TiN$_{1-x}$O$_x$) and N-doped rutile showing mixed-valence Ti states. Microstructural characterization shows that the ordered mesoporosity is maintained until 700°C, where TiN (TiN$_{1-x}$O$_x$) begins to form. Optical characterization shows that the discrete introduction of N is able to shift the titania absorption edge. The photocatalytic tests for degradtion of methylene blue give the best results under visible light excitation for the film nitrided at 500°C. At this temperature the concentration of nitrogen in the structure is optimal since oxygen vacancies are still not important enough to promote the recombination of the photogenerated electrons and holes.

6.2. Carbon Doped TiO$_2$ Mesoporous

Huang et al., prepared mesoporous nanocrystalline C-doped TiO$_2$ photocatalysts through a direct solution phase carbonization using TiCl$_4$ and diethanolamine as precursors. [16] In typical synthesis, 10 mmol of TiCl$_4$ was added dropwise into 30 mL of ethanol under stirring. After stirring for 1 h, a transparent yellowish sol was formed, and then 4 mL of diethanolamine was added into the sol. The solution turned colorless after the addition finished. Under stirring for 24 h at ambient temperature, the solution was maintained at 60°C for several hours, resulting in a vivid yellow gel. The resulting gels were then calcined at different temperatures for 5 h in a muffle furnace. After calcination, the powder photocatalysts were obtained XPS results revealed that oxygen sites in the TiO$_2$ lattice were substituted by carbon atoms and formed a C-Ti-O-C structure. The photocatalytic activities of the as-prepared samples were tested in a flow system on the degradation of NO at typical indoor air levels under simulated solar-light irradiation. The samples showed a more effective removal efficiency of NO than TiO$_2$ P25 on the degradation of the common indoor pollutant NO. However, for the C-doped TiO$_2$ calcined at 500°C, the removal rate reached the highest value after being irradiated for 30 min. Comparing to the C-doped TiO$_2$ calcined at 600°C, the carbon-doped TiO$_2$ calcined at 500°C showed superior photocatalytic activity on the degradation of NO at parts per billion levels, which can be explained by the band gap energy, the surface properties, as well as the mesoporous architecture. According to the above mechanism, inhibiting the undesirable electron-hole pair recombination is important to enhance the photocatalytic activity because it can improve the ability to produce hydroxyl radical group OH , which is possibly beneficial for oxidation of NO. A recent study revealed that the holes formed for carbon-doped TiO$_2$ photocatalysts under visible light irradiation were less reactive than those formed under UV light irradiation for pure TiO$_2$. [128] For the carbon-doped TiO$_2$ samples, the holes were trapped at midgap levels and showed less mobility, which was beneficial for the capture of surface hydroxyl to produce OH. However, the density and nature of the localized states in the band gap was significantly influenced by the carbon dopant concentration, [128] which may be used to explain the difference in photocatalytic activity of carbon-doped TiO$_2$ calcined at 500 and 600°C on the degradation of NO at typical parts per billion levels. Liu et al. [129] reported highly ordered mesoporous carbon-titania nanocomposites with nanocrystal-glass frameworks via the organic-inorganic-amphiphilic coassembly followed by the in situ crystallization technology. In typical synthesis, the carbon-titania nanocomposites were synthesized via the evaporation-induced triconstituent co-assembly followed by the in situ crystallization technology. First, a stock TiCl$_4$ solution (20 wt%) was prepared by dropping TiCl$_4$ into the mixture of ethanol and deionized water (7:1 EtOH:H$_2$O in mass ratio) at 0°C under vigorous stirring for 30 min. In parallel, 1.5 g of F127 was dissolved in 10.0 g of ethanol with 1.0 g of deionized water. Then, 7.0 g of TiCl$_4$ solution was slowly added with stirring for 1 h at room temperature. The resol precursor (Mw < 500) used as a carbon precursor was prepared accordingly.[130] The carbon-titania nanocomposites with controllable texture properties and composition can be obtained in a wide range from 20 to 80 wt% TiO$_2$ by adjusting the initial mass ratios. The C-TiO$_2$ nanocomposites with "bricked-mortar" frameworks exhibit highly ordered 2D hexagonal mesostructure and high thermal stability up to 700 °C. The nanocomposites have high surface area (465 m^2 g^{-1}) and large pore size (~4.1 nm). The carbon-titania nanocomposites show good photocatalytic activity for the photodegradation of Rhodamine B in an aqueous suspension, which may be attributed to the highly crystallized frameworks and high adsorptive capacity from the large surface areas.

Zhang et al. [131] have prepared hollow TiO$_2$ microparticles about 20-60 μm in size and hollow TiO$_2$/carbon composite microparticles about 30-90 μm in size by employing commercial Sephadex G-100 beads as the template as well as the carbon precursor. The cross-linked dextran gel template was first immersed in aqueous TiCl$_4$ solution to allow the surface mineralization of TiO$_2$, resulting in the formation of hollow microparticles of titania/G-100 hybrids. The photocatalytic activity of the obtained hollow microparticles of TiO$_2$ and TiO$_2$/carbon composites was investigated by monitoring the photodegradation of Rhodamine B. In both cases, the product calcined at an intermediate temperature exhibited the highest photocatalytic activity possibly because of a compromise between the anatase crystallinity and the surface area. Compared with the hollow TiO$_2$ microparticles, the hollow TiO$_2$/carbon composite microparticles exhibit remarkably enhanced photocatalytic activity. Lei et al. [132] prepared a 3D ordered macroporous TiO$_2$/graphitized carbon. The graphitized carbons were formed by catalytic graphitization of polystyrene arrays, which were used as both template and carbon source for the generation of macroporous composite. The graphitization degree and the content of graphitic carbon in the composite were dependent on the pyrolysis temperature and confinement effect of macroporous oxides skeleton. It was found the TiO$_2$/graphitized carbon showed higher activity in terms of degradation of Rhodamine B and eosin Y than TiO$_2$/amorphous carbon and TiO$_2$ P25.

6.3. Floride Doped TiO$_2$ Mesoporous

Pan et al. [133] synthesized monodisperse F-TiO$_2$ hollow microspheres by hydrothermal treatment of TiF$_4$ in H$_2$SO$_4$ aqueous solution at 160°C for 4 h. H$_2$SO$_4$ acts as an acid source to promote the HF etching. It also increases the ionic strength of the aqueous solution which governs the aggregation of hydrolyzed TiO$_2$ primary particles and the formation of porous microspheres. The absorption spectrum of the hollow microspheres exhibits a stronger adsorption in the UV-visible range of 325-800 nm than that of P25. The hollow inner structure associated with accessible mesopores at the spherical surface allow the light-scattering inside their pore channels as well as their interior hollows, enhancing the light harvesting and thus increase the quantity of photogenerated electrons and hole to participate in the photocatalytic decomposition of the contaminants. The removal rates of MB over the course of the photocatalytic degradation reaction are shown in Figure 14, which indicates that with identical UV-light exposure of 6 h, the mesoporous F-TiO$_2$ hollow microspheres show higher photocatalytic activity in the degradation of MB than that of P25.

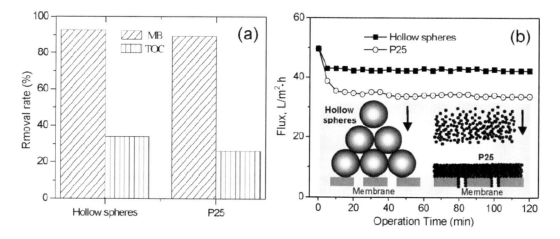

Figure 14. (a) MB, TOC removal, and (b) membrane flux over mesoporous F-TiO$_2$ hollow microspheres and P25. Insert of panel b: Schematic diagram of membrane fouling caused by photocatalysts. Reprinted with permission from ref. 133, Copyright 2008 ACS.

Zhou et al. [134] have doped F⁻ ions into crystalline TiO$_2$ by hydrothermal treatment of TiF$_4$ in an HCl solution. In a typical experiment, 60 mL of aqueous TiF4 solution (0.002 M) with a pH of 1.67 was placed in a Teflon-lined autoclave and heated at 180°C for 20 h in an oven. The white TiO$_2$ solid was obtained by centrifuging and washing with a large amount of deionized water. [135] Field-emission electron microscope and transmission electron microscope images showed that the products exhibited a flowerlike morphology with a hollow interior. The flowerlike F-doped TiO$_2$ hollow microspheres synthesized at 180°C showed the highest photocatalytic activity for the degradation of methylene blue under visible light irradiation. Yu et al. [18] have used a novel and simple method for preparing highly photoactive nanocrystalline F-doped TiO$_2$ photocatalyst with anatase and brookite phase was developed by hydrolysis of TTIP in a mixed NH$_4$F-H$_2$O solution. Fluoride ions not only suppressed the formation of brookite phase but also prevented phase transition of anatase to rutile. The F-doped TiO$_2$ samples exhibited stronger absorption in the UV-visible range with a red shift in the band gap transition. The photocatalytic activity for oxidation of acetone in air of by F-doped TiO$_2$ photocatalysts exceeded that of Degussa P25 when the molar ratio of NH$_4$F to H$_2$O was kept in the range of 0.5-3.

Yu et al. [136] have prepared mesoporous surface-fluorinated TiO$_2$ anatase phase by a one-step hydrothermal strategy in a NH$_4$HF$_2$-H$_2$O-C$_2$H$_5$OH mixed solution with TBOT as precursor. The photocatalytic activity of F-TiO$_2$ powders for decomposition of acetone is obviously higher than that of pure TiO$_2$ and P25 by a factor of more than 3 times due to the fact that the strong electron-withdrawing ability of the surface Ti-F groups reduces the recombination of photogenerated electrons and holes, and enhances the formation of free OH radicals. Yu et al. [137] prepared TiO$_2$ thin film using TTIP in the presence of a P123. The surface modification of the films was conducted by dipping the as prepared TiO$_2$ films in an aqueous 0.25M trifluoroacetic acid (TFA) solution at room temperature. The results show that TFA is chemisorbed on the surface. The photocatalytic activity of modified TiO$_2$ thin films for acetone oxidation in air is higher than that of unmodified TiO$_2$ thin films, and the modified film treated at 250°C shows the highest activity. This is ascribed to the fact that the TFA complex bound on the surface of TiO$_2$ acts as an electron scavenger and, thus, reduces the recombination of photogenerated electrons and holes. The enhancement is only temporary, however, as the TFA eventually decomposes under the strong oxidizing environment of photocatalysis.

7. DYE-SENSITIZED SOLAR CELLS

The dye-sensitized solar cells (DSC) provide a technically and economically credible alternative concept to present day p–n junction photovoltaic devices. In contrast to the conventional systems where the semiconductor assume both the task of light absorption and charge carrier transport the two functions are separated here. Light is absorbed by a sensitizer, which is anchored to the surface of a wide band semiconductor. Charge separation takes place at the interface via photo-induced electron injection from the dye into the conduction band of the solid. Since Grätzel introduced the nanoporous films into dye-derived wideband semiconductor research and made the breakthrough in the photoelectric conversion efficiency of dye-sensitized solar cells (DSSCs), the researchers interests have been paid on DSSCs for their high efficiency, their potential low-cost and simple assemble technology, especially in the past 6 years since Grätzel and coworkers were able to demonstrate the first 10% efficient cells[23, 24,138-144]. Dye-derivatized mesoporous titania film is one of the key components for high efficiency in such cells. They use bis(bipyridyl)Ru(II) complexes cis-di(isothiocyanato)bis(2,2-bipyridyl-4,4-dicarboxylate) ruthenium (II), known as the N3 dye, in conjunction with the nanocrystalline colloidal TiO_2 films and I_3^-/I^- solution in an organic volatile solvent mixture, to convert 10% of AM 1.5 solar radiation into electrical energy. The mesoporous TiO_2 layer in the dye-sensitized solar cells must exhibit efficient connectivities between nanocrystals to ensure the electron transfer to the collector electrode, extremely high surface area to increase the electron-hole pair density generated at the hybrid interface, and large open pores to promote hole-transporting organic material impregnation. Consequently, the tuning of TiO_2-films mesoporosity parameters such as pore size and connectivity, wall thickness, and crystallinity appears to be an attractive approach to better understand the physical and chemical parameters that control the efficiency of all solid-state dye-sensitized solar cells. Recently, dye-sensitized solar cells based on liquid electrolytes have been fabricated from mesostructured titania films. [145] These experiments performed with liquid-state electrolyte are very promising. Indeed, the results indicate that these solar cells exhibited energy conversion efficiency (up to 5.31%) comparable to that of nanocrystalline colloidal anatase-TiO_2 films with the same thickness. The mesoporous electrodes are very much different compared to their compact analogs because (i) the inherent conductivity of the film is very low; (ii) the small size of the nanocrystalline particles does not support a built-in electrical field; and (iii) the electrolyte penetrates the porous film all the way to the back-contact making the semiconductor/electrolyte interface essentially three-dimensional. Charge transport in mesoporous systems is under keen debate today and several interpretations based on the Montrol Scher model for random displacement of charge carriers in disordered solids [146] have been advanced. Charge carriers are transported in the conduction band of the semiconductor to the charge collector. The use of sensitizers having a broad absorption band in conjunction with oxide films of nanocrstalline morphology permits to harvest a large fraction of sunlight. [140] When the dye-sensitized nanocrystalline solar cell was first presented perhaps the most puzzling phenomenon was the highly efficient charge transport through the nanocrystalline TiO_2 layer. The researchers have directed towards synthesizing structures with a higher degree of order than the random fractal-like assembly of nanoparticles. A desirable morphology of the films would have the mesoporous channels or nanorods aligned in parallel to each other and vertically with respect to the TCO glass current collector. This would facilitate pore diffusion, give easier access to the film surface avid grain boundaries and allow the junction to be formed under better control. One approach to fabricate such oxide structures is based on surfactant templates assisted prepapration of TiO2 nanotubes as described in recent paper by Adachi et al. [147]. These and the hybrid nanorod-polymer composite cells developed by Huynh et al. [148] have confirmed the superior photovoltaic performance of such films with regards to random particle networks. Ito et al. [149] have applied a conductive transparent thin films made of multilayer coatings consisting of three alternative layers (TiO_2/Ag/TiO_2, TAT) to dye-sensitized solar cells (DSCs). Mesoporous TiO_2 electrodes for DSSCs were coated on TAT by a spincoating and low-temperature sintering method. They were compared to well-known transparent ITO.The resulting mesoporous TiO_2 electrodes suppressed reflection losses of incident light energy, thus improving the performance of DSSCs prepared using the low-temperature sintering method. These DSCs on the TAT coatings yielded a short-circuit photocurrent density of 9 mA/cm^2, a photocurrent of 700 mV, and an overall cell efficiency of 3.9% at one sun light intensity. Also, the Pluronic P123 templated mesoporous TiO$_2$ film was grown via layer-by-layer deposition and characterized by a novel methodology based on the adsorption of n-pentane. Multiple-layer depositions did not perturb the mesoporous structure significantly. TiO$_2$ film prepared was sensitized by a newly developed Ru-bipyridine dye (N945) and was applied as a photoanode in dye-sensitized solar cell (Figure 15). The 1-μm-thick mesoporous film, made by the superposition of three layers, showed enhanced solar conversion efficiency by about 50% compared to that of traditional films of the same thickness made from randomly oriented anatase nanocrystals. [150,151].

SUMMARY AND OUTLOOK

Chapter strategy has been focused on synthesis of mesoporous TiO$_2$, doped transition metals, precious metals and nonmetals/TiO$_2$ including synthetic methods, architecture concepts, and fundamental principles that govern the rational design and synthesis. In this chapter, synthesis mechanisms and the corresponding pathways are first demonstrated for the synthesis of mesoporous TiO$_2$ from the surfactant templating approach. The continuing breakthroughs in the synthesis and modifications of TiO$_2$ nanoparticles have brought new properties with improved photocatalysts and solar cell performance. The chapter covered the applications of mesoporous TiO$_2$ as efficienct photocatalysts and solar cell. This steady progress has demonstrated that mesoporous TiO$_2$ nanoparticles are playing and will continue to achieve an important role in the protections of the environment and in the search for renewable and clean energy technologies.

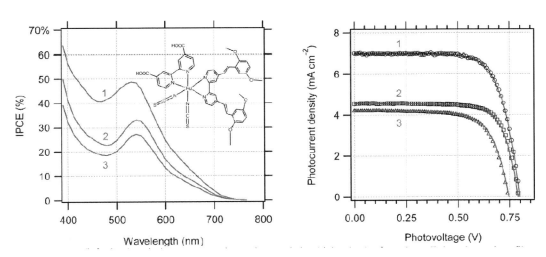

Figure 15. IPCE (left chart) and photocurrent-voltage characteristics (right chart) of a solar cell, based on TiO2 films sensitized by N945. Pluronic-templated three-layer film; 1.0-im-thick (1), nonorganized anatase treated by TiCl4; 0.95-μm-thick (2), nonorganized anatase nontreated by TiCl4; 0.95-μm-thick (3). Inset shows the chemical formula of N945 dye. Reprinted with permission from ref. 151, Copyright 2005 ACS.

REFERENCES

[1] Kresge, CT; ELeonowicz, M; Roth, WJ; Vartuli, JC; Beck, JS. *Nature*, 359, 710(1992).
[2] Ismail, A. A. and Bahnemann, D.W. *Journal of Materials Chemistry,* 21, 11686-11707 (2011).
[3] Chae, W-S; Lee, S-W,Kim, Y-R. *Chem. Mater.*, 17, 3072(2005).
[4] Yang, P; Zhao, D; Margolese, DI; Chmelka, BF; Stucky, GD. *Nature*, 396, 152(1998).
[5] Antonelli, DM; Ying, JY. *Angew. Chem., Int. Ed.*, 34, 2014(1995).
[6] Li, D Zhou; Honma, HI. *Nat. Mater.*, 3, 65(2004).
[7] Lee, J; Orilall, MC; Warrwn, SC; Kamperman, M; Disalvo, F; Wiesner, JU. *Nat. Mater.*, 7, 222(2008).
[8] Niederberger, M; Bartl, MH; Stucky, GD. *Chem. Mater.*, 14, 4364(2002).
[9] Shibata, H; Ogura, T; Mukai, T; Ohkubo, T; Sakai, H; Abe, M. *J. Am. Chem. Soc.*, 127, 16396(2005).
[10] Lee, DW; Ihm, SK; Lee, KH. *Chem. Mater.*, 17, 4461(2005).
[11] Atitar, M. F.; Ismail,A. A.; Bahneman, D.; Afanasev,D.; Emeline, A.V. *Chemical Engineering Journal*, 264, 417-424 (2015).
[12] Chappel, S; Chen, S; Zaban, A. *Langmuir*, 18, 3336(2002).
[13] Ito, S, Katayama, T; Sugiyama, T; Matsuda, M; Kitamura, M; Wada, T; Yanagida, YS. *Chem. Mater.*, 15, 2824 (2003).
[14] Li, BWang, Li, XM; Yan, L. *Mater. Chem. Phys.*, 78, 184(2003).
[15] Nagaveni, K; Sivalingam, G; Hegde, MS; Madras, G. *Appl. Catal., B*, 48, 83(2004).
[16] Huang, Y; Ho, W; Lee, S; Li, L; Zhang, G; Yu, JC. *Langmuir*, 24, 3510(2008).
[17] Choi, H; Aantoniou, MG; Pelaez, M; Delacruz, AA; Shoemaker, OA; Dionysiou, DD. *Environ. Sci. Technol.*, 41, 7530(2007).
[18] Yu, JC; Yu, JG; Ho, WK; Jiang, ZT; Zhang, LZ. *Chem. Mater.*, 14, 3808(2002).
[19] Tojo, S; Tachikawa, T; Fujitsuka, M; Majima, T. *J. Phys. Chem. C*, 112, 14948(2008).
[20] Sinha, AK; Suzuki, K. *J. Phys. Chem. B*, 109, 1708(2005).
[21] Schattka, JH; Shchukin, D; Jia, G; Antonietti, JM; Caruso, RA. *Chem. Mater.*, 14, 5103(2002).
[22] Tayade, RJ; Kulkarni, RG; Jasra, RV. *Ind. Eng. Chem. Res.*, 45, 5231(2006).
[23] O'Regan, B; Gratzel, M. *Nature*, 353, 737(1991).
[24] Nazzerruddin, MK; Kay, A; Podicio, I; Humphy-Baker, R, Muller, E; Liska, P; Vlachopoulos, N; Gratzel, M. *J. Am. Chem. Soc.*, 115, 6382(1993).
[25] Pichot, F; Pitts, JR; Gregg, BA. *Langmuir*, 16, 5626(2000).
[26] Lindstrom, H; Holmberg, A; Magnusson, E; Lindquist, S-E; Malmqvist, L; Hagfeldt, A. *Nano Lett.*, 1, 97(2001).
[27] Chen, L; Yao, B; Cao, Y; Fan, K. *J. Phys. Chem. C*, 111, 11849(2007).
[28] Kalousek, V; Tschirch, J; Bahnemann, D; Rathousky, J. *Superlattices and Microstructures*, 44, 506(2008).
[29] Beyers, E; Cool, P; Vansant, EF. *J. Phys. Chem. B*, 109, 10081(2005).
[30] Wang, H; Miao, JJ; Zhu, JM; Ma, HM; Zhu, JJ; Chen, HY. *Langmuir*, 20, 11738(2004).
[31] Lee, AC; Lin, RH; Yang, CY; Lin, MH; Wang, WY. *Mater. Chem. Phys.*, 109, 275(2008).

[32] Wark, M; Tschirch, J; Bartels, O; Bahnemann, D; Rathousky, J. *Microporous and Mesoporous Materials*, 84, 247(2005).
[33] Shiraishi, Y; Saito, N; Hirai, T. *J. Am. Chem. Soc.*, 127, 12820(2005).
[34] Tian, B; Yang, H; Liu, X; Xie, S; Yu, C; Fan, J; Tu, B; Zhao, D. *Chem. Commun.*, 1824(2002).
[35] Liu, Y; Li, J; Wang, M; Li, Z; Liu, H; He, P; Yang, X; Li, J. *Crystal Growth & Design*, 5, 1643(2005).
[36] Lu, Y; Hoffmann, MR; Yelamanchili, S; Terrenoire, A; Schrinner, M; Drechsler, M; Moller, MW; Breu, J; Ballauff, M. *Macromol. Chem. Phys.*, 210, 377(2009).
[37] Zhan, S; Chen, D; Jiao, X; Tao, C. *J. Phys. Chem. B*, 110, 11199(2006).
[38] Madhugiri, S; Sun, B; Smirniotis, PG; Feraris, JP; Jr. Balkus, KJ. *Microporous Mesoporous Mater.*, 69, 77(2004).
[39] Yu, J; Liu, WH; Yu, A. *Crystal Growth & Design*, 8, 930(2008).
[40] Sugimoto, T; Zhou, XP. *J. Colloid Interface Sci.*, 252, 347(2002).
[41] Liu, J; An, T; Li, G; Bao, N; Fu, G; Sheng, J. *Microporous and Mesoporous Materials*, 124, 197(2009).
[42] Soler-Illia, GJ; de Louis, AA; Sanchez, AC. *Chem. Mater.*, 14, 750(2002).
[43] Patarin, J; Lebeau, B; Zana, R. *Current Opinion in Colloid & Interface Science*, 7, 107(2002).
[44] Yu, J; Zhang, L; Cheng, B; Su, Y. *J. Phys. Chem. C*, 111, 10582(2007).
[45] Shibata, H; Ogura, T; Mukai, T; Ohkubo, T; Sakai, H; Abe, M. *J. Am. Chem. Soc.*, 127, 16396(2005).
[46] Xu, H; Zhang, L. *J. Phys. Chem. C*, 113, 1785(2009).
[47] Tian, G; Fu, H; Jing, L; Xin, B; Pan, K. *J. Phys. Chem. C*, 112, 3083(2008).
[48] Peng, T; Zhao, D; Dai, K; Shi, W; Hirao, K. *J. Phys. Chem. B*, 109, 4947(2005).
[49] Wang, XC; Yu, JC; Ho, CM; Hou, YD; Fu, XZ. *Langmuir*, 21, 2552(2005).
[50] Liu, Z; Sun, DD; Guo, P; Leckie, JO. *Chem. Eur. J.*, 13, 1851(2007).
[51] Kim, DS; Kwak, S-Y. *Applied Catalysis A: General*, 323, 110(2007).
[52] Wang, HW; Kuo, CH; Lin, HC; Kuo, IT; Cheng, CF. *J. Am. Ceram. Soc.*, 89, 3388(2006).
[53] Yu, JC; Zhang, L; Yu, J. *Chem. Mater.*, 14, 4647(2002).
[54] Yu, J; Zhou, M; Cheng, B; Yu, H; Zhao, X. *J. Mol. Catal. A*, 227, 75(2005).
[55] Wessels, K; Minnermann, M; Rathousky, J; Wark, M; Oekermann, T. *J. Phys. Chem. C*, 112, 15122(2008).
[56] Matsumoto, Y; Ishikawa, Y; Nishida, M; Ii, S. *J. Phys. Chem. B*, 104, 4204(2000).
[57] Choi, H; Sofranko, AC; Dionysiou, DD. *Adv. Funct. Mater.*, 16, 1067(2006).
[58] Wang, X; Yu, JC; Hou, Y; Fu, X. *Advanced Materials*, 17, 1, 99(2005).
[59] Wang, K; Yao, B; Morris, MA; Holmes, JD. *Chem. Mater.*, 17, 4825(2005).
[60] Kim, DS; Kwak, SY. *Environ. Sci. Technol.*, 43, 148(2009).
[61] Hoffmann, MR; Martin, ST; Choi, WY; Bahnemann, DW. *Chem. Rev.*, 95, 69(1995).
[62] Vidal, H; Kapar, J; Pijolat, M; Colon, G; Bernal, S; Cordon, A; Perrichon, V; Fally, F. *Appl. Catal., B*, 30, 75(2001).
[63] Georgieva, J; Armyanov, S; Valova, E; Poulios, I; Sotiropoulos, S. *Electrochem. Commun.*, 9, 365(2007).
[64] Xuan, S; Jiang, W; Gong, X; Hu, Y; Chen, Z. *J. Phys. Chem. C*, 113, 553(2009).

[65] Kim, TW; Ha, H-W; Paek, M-J; Hyun, S-H; Baek, I-H; Choy, J-H; Hwang, S-J. *J. Phys. Chem. C*, 112, 14853(2008).
[66] Zhu, J; Ren, J; Huo, Y; Bian, Z; Li, H. *J. Phys. Chem. C*, 111, 18965(2007).
[67] Arnal, P; Corriu, RJP; Leclercq, D; Mutin, PH; Vioux, A. *J. Mater. Chem.*, 6, 1925(1996).
[68] Mori, K; Kondo, Y; Morimoto, S; Yamashita, H. *J. Phys. Chem. C*, 112, 397(2008).
[69] Zhou, W; Fu, H; Pan, K; Tian, C; Qu, Y; Lu, P; Sun, C-C. *J. Phys. Chem. C*, 112, 19584(2008).
[70] Bian, Z; Zhu, J; Wang, S; Cao, Y; Qian, X; Li, H. *J. Phys. Chem. C*, 112, 6258(2008).
[71] Kong, L; Chen, H; Hua, W; Zhang, S; Chen, J. *Chem. Commun.*, 4977(2008).
[72] Zhang, X; Zhang, L; Xie, T; Wang, D. *J. Phys. Chem. C*, 113, 7371(2009).
[73] Yu, JC; Li, G; Wang, X; Hu, X; Leung, CW; Zhang, Z. *Chem. Commun.*, 2717(2006).
[74] Fan, X; Chen, X; Zhu, S; Li, Z; Yu, T; Ye, J; Zou, Z. *J. Mol. Catal. A*, 284, 155(2008).
[75] Rolison, DR. *Science*, 299, 1698(2003).
[76] Bell, AT. *Science*, 299, 1688(2003).
[77] Yu, JC; Wang, XC; Fu, XZ. *Chem. Mater.*, 16, 1523(2004).
[78] Rodrigues, S; Ranjit, KT; Uma, S; Martyanov, IN; Klabunde, KJ. *Adv. Mater.*, 17, 2467(2005).
[79] Pan, JH; Lee, WI. *Chem. Mater.* 18, 847(2006).
[80] Liu, S; Yu, J; Mann, S. *J. Phys. Chem. C*, 113, 10712(2009).
[81] Nakamura, R; Okamoto, A; Osawa, H; Irie, H; Hashimoto, K. *J. Am. Chem. Soc.*, 129, 9596(2007).
[82] Kim, TW; Hwang, SJ; Park, Y; Choi, W; Choy, JH. *J. Phys. Chem. C*, 111, 1658(2007).
[83] Stone, VF; Davis, RJ. *Chem. Mater.*, 10, 1468(1998).
[84] Yu, JC; Zhang, L; Zheng, Z; Zhao, J. *Chem. Mater.*, 15, 2280(2003).
[85] Dai, K; Peng, T; Chen, H; Liu, J; Zan, L. *Environ. Sci. Technol.*, 43, 1540(2009).
[86] Peng, TY; Zhao, D; Song, HB; Yan, CH. *J. Mol. Catal., A*, 238, 119(2005).
[87] Yang, X; Wang, Y; Xu, L; Yu, X; Guo, Y. *J. Phys. Chem. C* 112, 11481(2008).
[88] Stathatos, E; Petrova, T; Lianos, P. *Langmuir*, 17, 5025(2001).
[89] Chu, S-Z; Inoue, S; Li, K; Wada, D; Haneda, H; Awatsu, S. *J. Phys. Chem. B*, 107, 6586(2003).
[90] Dong, W; Sun, Y; Lee, CW; Hua, W; Lu, X; Shi, Y; Zhang, S; Chen, J; Zhao, D. *J. Am. Chem. Soc.*, 129, 13894(2007).
[91] Wang, Z; Zhang, F; Xue, Y; Cui, B; Guan, JN. *Chem. Mater.*, 19, 3286(2007).
[92] Li, G; Zhao, XS. *Ind. Eng. Chem. Res.*, 45, 3569(2006).
[93] Li, Y; Kim, S-J. *J. Phys. Chem. B*, 109, 12309(2005).
[94] Xuzhuang, Y; Yang, D; Huaiyong, Z; Jiangwen, L; Martins, WN; Frost, R; Daniel, L; Yuenian, S. *J. Phys. Chem. C*, 113, 8243(2009).
[95] Li, G; Kang, ET; Neoh, KG; Yang, X. *Langmuir*, 25, 4361(2009).
[96] Wang, XH; Li, JG; Kamiyama, H; Moriyoshi, Y; Ishigaki, T. *J. Phys. Chem. B*, 110, 6804(2006).
[97] Aronson, BJ; Blanford, CF; Stein, A. *Chem. Mater.*, 9, 2842(1997).
[98] Alvaro, M; Aprile, C; Benitez, M; Carbonell, E; Garcıa, H. *J. Phys. Chem. B*, 110, 6661(2006).
[99] Maldotti, A; Molinari, A; Amadelli, R; Carbonell, E; Garcia, H. *Photochem. Photobiol. Sci.*, 7, 819(2008).

[100] Li, G; Bai, R; Zhao, XS. *Ind. Eng. Chem. Res.*, 47, 8228(2008).
[101] Kang, C; Jing, L; Guo, T; Cui, H; Zhou, J; Fu, H. *J. Phys. Chem. C*, 113, 1006(2009).
[102] Shi, KY; Chi, YJ; HTYu, Fu, HG. *J. Phys. Chem. B*, 109, 2546(2005).
[103] Morishita, M; Shiraishi, Y; Hirai, T. *J. Phys. Chem. B*, 110, 17898(2006).
[104] Hu, Y; Martra, G; Zhang, J; Higashimoto, S; Coluccia, S; Anpo, M. *J. Phys. Chem. B*, 110, 1680(2006).
[105] Zhang, W; Froba, M; Wang, J; Tanev, PT; Wong, J; Pinnavaia, TJ. *J. Am. Chem. Soc.*, 118, 9164(1996).
[106] Thangaraj, A; Kumar, R; Mirajkar, SP; Ratnasamy, P. *J. Catal.*, 130, 1(1991).
[107] Rohlfing, DF; Szeifert, J; Yu, MQ; Kalousek, V; Rathousky, J; Bein, T. *Chem. Mater.*, 21, 2410(2009).
[108] Allain, E; Besson, S; Durand, C; Moreau, M; Gacoin, T; Boilot, JP. *Adv. Funct. Mater.*, 17, 549(2007).
[109] Besson, S; Ricolleau, C; Gacoin, T; Jacquiod, C; Boilot, J-P. *Microporous Mesoporous Mater.*, 60, 43(2003).
[110] Ogawa, M; Ikeue, K; Anpo, M. *Chem. Mater.*, 13, 2900(2001).
[111] Ismail, AA; Bahnemann, DW; Robben, L; Wark, M. *Chem. Mater.*, 22, 108(2010).
[112] Srinvasn, M; White, T. *Environ. Sci. Technol.*, 41, 4405(2007).
[113] Bannat, I; Wessels, K; Oekermann, T; Rathousky, J; Bahnemann, D; Wark, M. *Chem. Mater.*, 21, 1645(2009).
[114] Li, X; Peng, J; Kang, J-H; Choy, J-H; Steinhart, M; Knoll, W; Kim, DH. *Soft Matter*, 4, 515(2008).
[115] Alberius, P. C. A., Frindell, K. L., Hayward, R. C., Kramer, E. J., Stucky, G. D., Chmelka, B. F. *Chem. Mater.*, 14, 3284(2002).
[116] Lakshminarasimhan, N; Bae, E; Choi, W. *J. Phys. Chem. C*, 111, 15244(2007).
[117] Asahi, R; Morikawa, T; Ohwaki, T; Aoki, K; Taga, Y. *Science*, 293, 269(2001).
[118] Irie, H; Wanatabe, Y; Hashimoto, K. *J. Phys. Chem. B*, 107, 5483(2003).
[119] Burda, C; Lou, Y; Chen, X; Samia, ACS; Stout, J; Gole, JL. *Nano Lett.*, 3, 1049(2003).
[120] Fang, J; Wang, F; Qian, K; Bao, H; Jiang, Z; Huang, W. *J. Phys. Chem. C* 112, 18150–18156 (2008).
[121] Chi, B; Zhao, L; Jin, T. *J. Phys. Chem. C* 111, 6189-6193(2007).
[122] Cong, Y; Zhang, J; Chen, F; Anpo, M. *J. Phys. Chem. C*, 111, 6976-6982 (2007).
[123] Wang, X; Yu, JC; Chen, Y; Wu, L; Fu, X. *Environ. Sci. Technol.*, 40, 2369-2374 (2006).
[124] Soni, SS; Henderson, MJ; Bardeau, JF; Gibaud, A. *Adv. Mater.*, 20, 1493–1498 (2008).
[125] Martínez-Ferrero, E; Sakatani, Y; Boissière, C; Grosso, D; Fuertes, A; Fraxedas, J; Sanchez, C. *Adv. Funct. Mater.*, 17, 3348–3354 (2007).
[126] Grosso, D; Cagnol, FG; Soler-Illia, AA; Crepaldi, EL; Amenitsch, H; Brunet-Bruneau, A; Bourgeois, A; Sanchez, C. *Adv. Funct. Mater.*, 14, 309 (2004).
[127] Grosso, DG; Soler-Illia, AA; Crepaldi, ELF; Cagnol, F; Bourgeois, A; Brunet-Bruneau, A; Amenitsch, H; Albouy, PA; Sanchez, C. *Chem. Mater.*, 15, 4562(2003).
[128] Di Valentin, C; Pacchioni, G; Selloni, A. *Chem. Mater.*, 17, 6656 (2005).
[129] Liu, R; Ren, Y; Shi, Y; Zhang, F; Zhang, L; Tu, B; Zhao, D. *Chem. Mater.*, 20, 1140–11462008,.
[130] Meng, Y; Gu, D; Zhang, FQ; Shi, YF; Yang, H; Li, F; Yu, Z; Tu, CZ; Zhao, BDY. *Angew. Chem., Int. Ed.*, 44, 7053 (2005).

[131] Zhang, D; Yang, D; Zhang, H; Lu, C; Qi, L. *Chem. Mater.*, 18, 3477-3485(2006).
[132] Lei, Z; Xiao, Y; Dang, L; Hu, G; Zhang, J. *Chem. Mater.*, 19, 477-484 (2007).
[133] Pan, JH; Zhang, X; Du, AJ; Sun, DD; Leckie, JO. *J. Am. Chem. Soc.*, 130, 11256–11257 (2008).
[134] Zhou, JK; Lv, L; Yu, J; Li, HL; Guo, P-Z; Sun, H; Zhao, XS. *J. Phys. Chem. C*, 112, 5316-5321(2008).
[135] Yang, HG; Zeng, HC. *J. Phys. Chem. B*, 108, 3492-3495 (2004).
[136] Yu, J; Wang, W; Cheng, B; Su, B-L. *J. Phys. Chem. C*, 113, 6743–6750 (2009).
[137] Yu, JC; Ho, W; Yu, J; Hark, SK; Iu, K. *Langmuir*, 19, 3889-3896 (2003).
[138] Nazeeruddin, MK; Pechy, P; Renouard, T. *J. American Chemical Society*, 123, 1613(2001).
[139] Gratzel, M. "Photoelectrochemical cells," *Nature*, 414, 338(2001).
[140] Gratzel, M. *Journal of Photochemistry and Photobiology A*, 164, 3(2004).
[141] Nazeeruddin, MK; De Angelis, F; Fantacci, S. *J. American Chemical Society*, 127, 16835(2005).
[142] Green, MA; Emery, K; King, DL; Igari, S; and Warta, W. *Progress in Photovoltaics: Research and Applications*, 10, 355(2002).
[143] Dai, SY; Wang, KJ; Weng, J; et al. *Solar Energy Materials and Solar Cells*, 85, 447(2005).
[144] Hu, L; Dai, S-Y; Weng, J; et al. *J. Phys. Chem. B*, 111, 358(2007).
[145] Hou, K; Tian, B; Li, F; Bian, Z; Zhao, D; Huang, C. *J. Mater. Chem.*, 15, 2414(2005).
[146] Nelson, J. *Phys. Rev.* B 59, 15374(1999).
[147] Adachi, M; Murata, Y; Okada, I; Yoshikawa, SJ. *Electrochem. Soc.* 150, G488(2003).
[148] Huynh, WU; Dittmer, JJ; Alivisatos, AP. *Science* 295, 242 (2002).
[149] Ito, S; Takeuchi, T; Katayama, T; Sugiyama, M; Matsuda, M; Kitamura, T; Wada, Y; Yanagida, S. *Chem. Mater.*, 15, 2824(2003)
[150] Zukalova, M; Prochazka, J; Zukal, A; Yum, JH; Kavan, L. *Inorganica Chimica Acta*, 361, 656(2008)
[151] Zukalova, M; Zukal, A; Kavan, L; Nazeeruddin, MK; Liska, P; Gratzel, M. *Nano Lett.*, 5, 1789(2005).

In: Comprehensive Guide for Mesoporous Materials. Volume 3   ISBN: 978-1-63463-318-5
Editor: Mahmood Aliofkhazraei   © 2015 Nova Science Publishers, Inc.

Chapter 6

POROUS SI STRUCTURES FOR GAS, VAPOR AND LIQUID SENSING

V. A. Skryshevsky[*]

Institute of High Technologies,
Taras Shevchenko National University of Kyiv,
Kyiv, Ukraine

ABSTRACT

In the present chapter, we pay attention both to technical and physico-chemical aspects of porous silicon (PS) sensors operation. The transducers and sensors of resistivity, capacitance, Schottky barrier, MIS, FET, EIS, ISFET, LAPS and optical sensors with PS are considered. The set of gas sensor`s parameters for detection of humidity, CO, NO_2, different volatile organic compounds, alcohols, H_2S, H_2 and other gases is described.

1. INTRODUCTION

At present, the detection of the toxic, inflammable or explosive gases is very important due to their harmful influence on human health. Reliable detection of the low level pollution for air quality monitoring is an important motivation for developing the systems of recognition like electronic nose. In the workplace, it includes the detection of hazardous materials such as SO_2 and H_2S near petro-chemical plants, volatile organic compounds (VOCs) in coating operations, NO_x in heating, ventilation, and air conditioning equipment and others.

Nowadays, the development of the improved sensors of high sensitivity and selectivity is based on the achievements of modern semiconductor technique and the using of new materials. Application of PS, prepared by electrochemical etching of Si substrate, for sensors

---

[*] E-mail:skrysh@univ.kiev.ua

is considered in several reviews [1-7] focusing on the analysis of effects that occur in PS at gas adsorption.

In the present chapter, we pay attention both to technical and physico-chemical aspects of PS sensor operation.

2. RESISTIVITY (CONDUCTIVITY) SENSORS

Conductance changes in PS layers have been observed upon adsorption of molecules, which presumably arise from dielectric-induced changes in the carrier concentration in the Si. Adsorbate-induced conductance changes may also arise from carrier concentration variation due to interaction with dangling bond states on PS surface [7, 8].

Usually, *resistivity* (or *conductivity*) sensor is the simplest electrical transducer consisted of PS layer on Si or isolator substrate with, at least, 2 metal terminals (Figure 1). To reduce the high PS resistivity, sometimes, the metal terminals are formed as interdigitated electrodes [9, 10]. Supplementary discontinuous layer of catalyst that covers porous surface can be deposited.

Series resistivity sensors were proposed to detect low concentration of nitrogen dioxide (NO_2) [9, 11-18]. NO_2 is a toxic air pollutant emitted by combustion engines with attention level at 106 ppb and alarm level at 212 ppb [17]. The detectable level of NO_2 in air by PS resistivity sensor is well below 100 ppb.

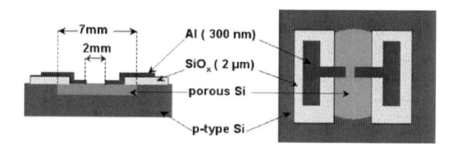

Figure 1. Sketch of the PS resistivity sensor. Reprinted with permission from [19].

Figure 2 shows the dynamic response of the sensor at T_{room} versus the concentration of NO_2 in dry air from 240 till 12 ppb. The sensor shows a very high sensitivity of conductivity: $\Delta G/G=0.26$ at 12 ppb, response and recovery time is about 10 min [17]. The adsorbed NO_2 molecules act as acceptor centers, at its adsorption on PS they inject free holes into PS film increasing the conductivity [12, 20]. The drawback of NO_2 sensors is the strong dependence of sensor current from humidity (Figure 2, b). An increase of RH leads to a decrease in conductivity due to donor-like character of water molecules adsorbed as PS that lowering of the free holes concentration [20]. At that time the relative response of meso-PS (with and without gold film) to interfering species as CO (up to 1000 ppm), CH_4 (up to 5000 ppm) is negligible and anyhow low for alcohol such as methanol at concentrations up to 800 ppm [12]. The sensitivity of resistivity sensors is determined by the porous microstructure that depended on the composition of the electrochemical solution used for anodization, however, there is no direct correlation between porosity and sensitivity [18].

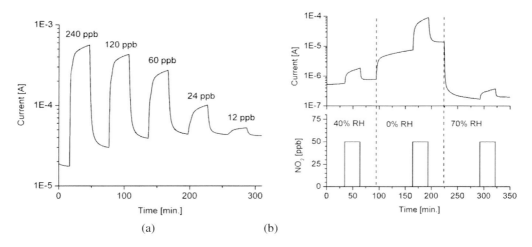

Figure 2. Dynamic response of the sensor to different concentration of NO_2 in dry air (a) and in the presence of different humidity levels and NO_2 concentrations (b). The graph above represents the sensor response; the graph below represents NO_2 concentrations as a function of time. Reprinted with permission from [17].

Table 1. Resistivity (conductivity) PS sensors

N	Si substrate	PS thickness, porosity	Metal contact, coating	Detected gas	Minimal concentration	Sensitivity $\Delta G/G$ or $\Delta R/R$	Ref.
1	p-Si, 3-13 Ω.cm	free standing, 15 µm, 45%	interdigital Pt	NO_2 O_3 CO benzene	100 ppb 200 ppb 1000 ppm 20 ppm	13,8	[9]
2	p-Si, 10 Ω.cm	free standing	planar Au	NO_2	70 ppb	0,39	[15]
3	p$^+$-Si, 6-15 mΩ.cm	32 µm, 80%	planar Au	NO_2	12 ppb in dry air, 50 ppb in humidity	11,7 1,2	[17], [14]
4	n-Si, 1 Ω.cm	40 µm, 45%	planar, nano Au	NO_2	5 ppm	1,6	[11]
5	p$^+$-Si, 5-15 mΩ.cm	20-30 µm, 30-75%	Ti-W-Pt	NO_2	3 ppm		[12]
6	p-Si, 12 Ω.cm	8-10 µm	metallophtalo-cyanine	NO_2	100 ppm	29,3%	[13]
7	p$^+$-Si, 5-15 mΩ.cm		planar Au	NO_2	100 ppb		[18]

Table 1. (Continued)

N	Si substrate	PS thickness, porosity	Metal contact, coating	Detected gas	Minimal concentration	Sensitivity ΔG/G or ΔR/R	Ref.
8	p-Si, 1-20 Ω.cm		planar Au, nano Au/SnO_2 coating	CO SO_2 NH_3 NO_x	5 ppm 1 ppm 500 ppb 1 ppm		[16]
9	p-Si, 2-20 Ω.cm		interdigital Au	HCl HN_3 NO	100 ppm 100 ppm 100 ppm	0,03%/ppm 0,1%/ppm 0,1%/ppm	[21]
10	n-Si, 1 Ω.cm	80 μm, 60%	planar Au	Methanol Ethanol Chloroform Toluene	500 ppm 500 ppm 5100 ppm 3000 ppm	1,75 1,6 0,6 2,8	[22]
11	p-polySi		planar Au	ethanol	2000 ppm	19	[8]

The aged PS sensors show improved reversibility compared with fresh samples. The stabilization of the current is also faster in aged rather than in fresh samples. The electrical response dynamics and hysteresis can also be improved by means of a pre-treatment via prolonged exposure to high concentrations of NO_2 [15]. Selectivity and sensitivity of PS resistivity sensors can be enhanced using catalyst materials, for example, by electroless metal deposition to form a gold or tin oxide nanostructured framework interacting with the nanopore-coated microporous surface [16]. The application of PS resistivity sensors to detect several gases is summarized in table 1.

3. CAPACITANCE SENSORS

The operating principle of PS *capacitance* sensors is based on the increasing of the dielectric constant of porous materials (and, as results, on the increasing of capacity) after pore filling by solution or vapor of water or organic molecules [23-25].

The sketch of typical PS capacitance sensor is presented in Figure 3, a. Silicon wafer (100) with 0.02 Ω.cm resistivity is used as substrate [24, 25]. This device was applied as humidity and ethanol vapor sensor in the range of 0-0.5% alcohol concentration. Alcohol concentration determining in the blood is important in order to prevent traffic accidents (if the alcohol concentration in a driver`s blood exceeds 0.05%, the driver may be punished for violation of drinking and driving regulations). The sensitivity of the capacitance curve measured at 120 Hz indicates an increase of 2.5% capacitance per 0.1% alcohol solution concentration, while the sensitivity at larger frequency was observed to be less.

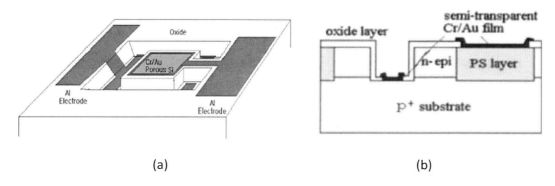

Figure 3. The top view (a) and cross-section (b) of PS capacity sensor. Reprinted with permission from [24].

Under the UV light illumination (λ=254 nm) of sensor surface the increase in sensitivity and initial capacitance is observed. The sensor was examined in human exhaling gases composed typically of N_2, CO_2, water vapor without alcohol gas. N_2 and CO_2 gas concentrations included in exhaling breath had little effect on capacitance response (because the static dielectrical constant of N_2 and CO_2 gases is close to 1, similar to air in unfilled pores in PS).

Capacitance-type humidity sensors in which PS layer is used as a humidity-sensing material were developed also in [26-31] (Table 2). The capacitance in the PS layer will strongly depend on the relative humidity RH because the static permittivity of pure H_2O is 80, which is much more than the 12 of the Si. The better humidity sensitivity was obtained for capacitance sensors with hydrothermally-etched PS pillar array [28]. With the RH ranged from 11 to 95%, the variations of the capacitance reach 1500% at a signal frequency of 100 Hz and 800% at 1000 Hz (Figure 4). About 15 and 5 s, respectively, are needed for the capacitance to reach 90% of its final/initial values during a RH-increasing process and a RH-decreasing process. The faster response to humidity of PS pillar array sensor might be due to the regular morphology and suitable thickness of the sensing layer.

The capacitance sensor with improved humidity sensitivity based on ordered macro-PS with a Ta_2O_5 thin film coating is proposed [31]. The sensor's capacity versus RH shows perfect linearity in two regions respectively over the whole RH range. The sensor shows the fastest response time at 200 kHz: 18–40 s to small RH changes, and 300 s to 100% RH changes, the hysteresis at 200 kHz is 0.3–3.3%.

Another approach that allows for development of high sensitive, stable and reproducible capacitance humidity sensor is thermal oxidation [30] and carbonizations of the PS surface [29]. These treatments change the originally hydrophobic PS surface to hydrophilic, thus improving its humidity sensing properties.

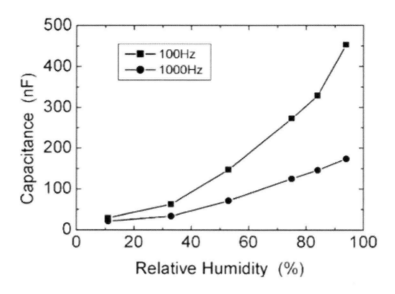

Figure 4. Variation of capacitance with RH at two different electrical signal frequencies at T_{room}. Reprinted with permission from [28].

Table 2. Capacity PS sensors for liquids and humidity

N	Si substrate	PS thickness	Metal contact, coating	Detected gas	Detection range	Sensitivity $\Delta C/C$	Ref
1	p^+-Si, 20mΩ.cm + n-epi layer	4 μm	Cr/Au	ethanol/ water mixture	0-0,5% ethanol	2,5% per 0,1% with UV light 4% per 0,1%	[25]
2	p^+-Si, 20mΩ.cm + n-epi layer	4 μm	Cr/Au	methanol/water mixture humidity	0-0,5% methanol 20-100%	1,6 300%	[24]
3	n^+-Si, 6 mΩ.cm		Au	humidity	10-90%		[26]
4	p-Si, 0,15-0,2 Ω.cm		Al	humidity	10-95%	1500%	[28]
5	n-Si	97 μm	Al, Ta_2O_5 coating	humidity	5-100%	300%	[31]
5	p-Si, 10-20 Ω.cm	30 μm, oxidated	interdigital Al	acetone, methanol, ethanol, i-propanol			[30]

4. SENSORS BASED ON SCHOTTKY BARRIER, MIS STRUCTURE AND HETEROJUCTION

The built- in potentials in the interfaces define the working principle of the sensors with *Schottky barrier, MIS structure or heterojunction* (in contrast to the resistive sensor type in which metal contacts are of Ohmic type). In such structures the electrical characteristics (I-V, C-V, impedance) are changed due to the influence of molecule adsorption on potential barrier or/and the parameters of PS (dielectric function and charge of surface states) [32-36]. The theoretical models of different metal/PS/Si structures were considered elsewhere [37-43].

It is known that the flammability limits of hydrogen in air is 4.0-75 vol%. For this reason it is necessary to develop highly sensitive hydrogen sensors to prevent accidents due to hydrogen - containing gases leakage. Catalyst metal-semiconductor devises utilise the effect of the change in work function of the metal at the interface, because the adsorbed hydrogen atoms in catalyst metal form a dipole layer at the interface [44]. However, operating temperature of such sensors achieves 200-500°C, it corresponds to maximal efficiency of metal catalyst [45] that complicates the measuring instruments.

MIS type structures with PS and Pd as catalyst were applied to detect the H_2 and hydrogen - containing gases [46-51]. Application of MIS type structures with thin Pd and PS layers is a promised way to reduce the operating temperature and improve the hydrogen sensing since: i) H_2 is highly soluble in Pd, the solubility depends on temperature and H_2 partial pressure, ii) deposition of thin Pd film on PS forms the island-type catalyst that reduces the activation energy to dissociate the hydrogen-containing molecules [52], iii) the Pd is semitransparent for penetration of hydrogen atoms that allows to accelerate the sensor response and recovery times [44], iv) observed hydrogen-induced drift in PS sensors response is much less than in Pd/SiO$_2$/Si devices [46].

A typical result of contact potential difference response (CPD) of the Pd/thin PS/n-Si structure in synthetic air (20% O$_2$+ 80%N$_2$) at H$_2$ concentrations from 200 up to 4000 ppm is given in Figure 5,a [46]. Sensors are characterised by a considerable sensitivity, time of response and recovery time is in minutes range when operated at T_{room} and reveal the best sensing comparing with Pd/p-Si and Pd/SiO$_2$/p-Si structures. Experimental values of CPD response are described by the square root dependence of the partial H$_2$ pressure as predicted by theory [44]. To detect H$_2$ it is proposed to use Pd/CH$_x$/PS/Si structures. The role of the CH$_x$ layer is to protect PS against oxidation [51]. Such structures demonstrate a record response and recovery time (2 and 15 s, respectively) at large 0.2 bar hydrogen concentrations.

Schottky-type and MIS structures were applied to analysis of other gases too. The deposition of thin films of catalytic metal which creates island - type covering on the PS surface, or special sensor design that facilitates diffusion of molecules through the pores towards metal layer is applied. The example of Schottky and MIS- type sensors of several gases are presented in Table 3.

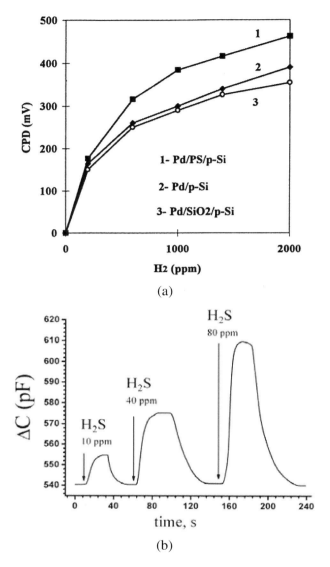

Figure 5. a) Steady-state CPD response of Pd/thin PS/p-Si, Pd/p-Si and Pd/SiO$_2$/p-Si structures versus H$_2$ concentration in synthetic air. Reprinted with permission from [46], b) Kinetics of the signal response of Pd/PS/p-Si under action various concentrations of H$_2$S.

Table 3. Schottky and MIS- type sensors with PS interface

N	Potential barrier structure	PS thickness	Measured value	Detected gas	Detection range	Ref.
1	Pd/PS/n,p-Si	15-75 nm	I-V	H$_2$	200ppm-10%	[49]
2	Pd/PS/n-Si		CPD	H$_2$	200ppm-4%	[46]
3	Pd/PS/p-Si		resistivity	H$_2$	0-1,5%	[48]

N	Potential barrier structure	PS thickness	Measured value	Detected gas	Detection range	Ref.
4	Pd/PS/p-Si		I-V	H_2 O_2	500 ppm 500 ppm	[47]
5	Al-Pd/PS/p-Si	1 µm	C-V	H_2 H_2S	10-1000 ppm 5-80 ppm	[50]
6	Pd/CH_x/PS/p-Si	1-10 µm	I-V	H_2	0,2 bar	[51]
7	Pt-Pd/PS/n-Si	5-10 µm	I-V	NO_2	5 ppm	[53]
8	Au-Pd/SnO_2/PS/p-Si	0,6-19 µm	resistivity	NO_2 LPG		[57]
9	Al/CH_x/PS/p-Si	1-10 µm	I-V, C-V	CO_2	500-1500 ppm	[59]
10	Al/CH_x/PS/p-Si	1-15 µm	I-V, C-V	ethylene, ethanol, propane	>115 ppm	[58]
11	Au/PS/n-Si	41 µm	I-V	ethanol, methanol	0,2% 0,2%	[54]
12	Ag/a-Si:H/PS/p-Si	1µm	I-V	water-ethanol	0-50%	[61]
13	Ti/PS/p-Si	100-300 nm	I-V, C-V	humidity	0-50%	[55]
14	Au/PS/p-Si		C-V	humidity	10-100%	[56]

MIS Pd/PS/Si structure is very sensitive to low (5-80 ppm) concentrations of hydrogen sulphide toxic gas [50]. The kinetic dependence of capacitance at gas inlet and outlet is shown in Figure 5,b. At low gas concentrations the MIS structures had a high speed of performance and short times of response (<10 s). Heterojunctions based on CH_x/PS structure show a good rectifying behavior, and current sensitivity $\Delta I/I_o$ to ethylene (26.8%), ethane (13.7%) and propane (9.4%) at 575 ppm concentration of these gases, response and recovery times at 115 ppm ethylene are of about 3 and 7 min, respectively [58]. It has been applied also as CO_2 sensors in measured range of 500- 1800 ppm [59]. Amorphous Si/PS heterojunction sensors of organic vapors and molecular oxygen are considered in [60, 61].

5. FET AND ISFET SENSORS

The working principle of a gas sensor that is based on the field effect derived from the metal oxide semiconductor field effect transistor (*MOSFET*). In n-channel MOSFET, a p-Si wafer contains two n-type diffusion regions (source and drain), the structure is covered with a thin SiO_2 insulating layer on top of which a metal gate electrode is deposited. In FET gas sensor the space charge layer near SiO_2 interface is induced by the adsorbed gases. The application of PS layer instead of metal gate in FET, seems, very attractive since developed internal porous surface enhances the molecule adsorption and technology of such sensors should be compatible with silicon IC technology. However, the technical realization of declared compatibility of PS fabrication process with the IC technology is one of the bottleneck of PS sensors, since nonstandard processes (like electrochemical etching) are generally used for PS technology. An alternative to this approach is the modification of a standard IC technology with the addition to specific steps for PS fabrication. However, performing the anodization step at PS formation between standard IC processing can result in contamination from PS itself to other wafer parts and PS degradation because of thermal and etching steps. To avoid this problem it is desirable to perform PS fabrication at the end of IC processing [62].

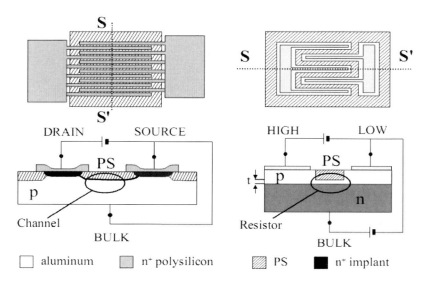

Figure 6. Schematic top view (top) and cross-section (bottom) of fabricated gas sensing devices: the APSFET on the left; the resistor on the right. Reprinted with permission from [66].

Adsorption porous Si-based FET (APSFET) gas sensor was proposed [62, 63]. It is a gas sensor based on a standard bipolar +CMOS+DMOS silicon technology in which PS is fabricated at the end of process (Figure 6,a). APSFET works like open gate FET structure with PS layer directly above the conduction channel in crystalline Si. Electrical conduction does not take place in the PS layer, which only plays the role of sensing material. In this way the electrical properties of the PS sensing film are not involved in the measurement, with advantages in terms of sensors sensitivity and reproducibility.

Figure 7,a shows a typical *I-V* source-drain characteristics of a device for several isopropanol vapor concentrations.

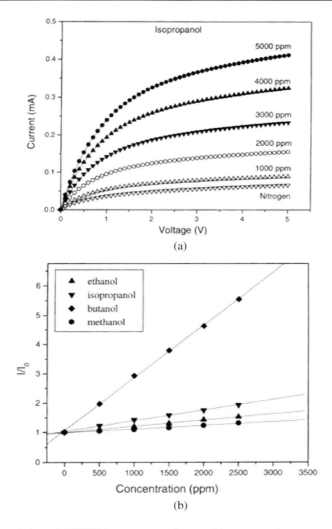

Figure 7. I-V characteristics of APSET in presence of several isopropanol concentrations (a) and sensor response to the several alcohols at V= 5V (b). Reprinted with permission from [62].

As typical for FET structures they have linear region for low voltages and a saturation current at high voltages. The remarkable growth of sensitivity is observed at the increase of PS thickness from 0.15 to 0.5 μm. In Figure 7,b the effect on the sensor current of alcohols vapor at constant voltage bias (5 V) is shown. For every species the sensor response linearly depends on the concentration of the species in the environment, and the sensitivity of the sensor increases with the number of carbon atoms in the molecule or the molecular weight of the molecule. An opposite effect with respect to alcohol vapors, a reduction of the conduction current is observed at exposition in organic acids (formic, acetic, propanoic and butanoic acid).

NO_2 concentration as low as 100 ppb was detected using the APSFET [63]. Nonoxidized sensors show a high sensitivity only for fresh devices reducing with the aging of the sample. Oxidation of the PS layer improves the electrical performance of sensors, in terms of stability, recovery time, and interference with the relative humidity level, keeping the high sensitivity to nitrogen dioxide. For instance, 100 ppb of NO_2 produce a current variation of about one order of magnitude. The drawback of such sensor - RH level affects the percentage variation of the sensor current when exposed to NO_2 and the response time. FET sensor with PS layer under gate layer shows the rapid response (250s) and recovery (240s) time at hydrogen detection (1.2 mbar), however, drain current does not recover completely to the initial value [64] due to hydrogen-induced drift (HID) effect, when hydrogen trapping sites exist in PS [49].

The analog of FET structure with p-n junction gate and adsorbing top layer of PS (that acts as a floating sensing gate) is proposed in [65, 66] (Figure 6,b). This device differs in the mechanisms of conduction, if in the APSFET the current flows in the inversion layer of the FET (electrons); in this structure the transport is due to the majority carriers (holes). The adsorption of polar molecules into the PS layer could change the free carriers concentration in the p- resistor or modulate the resistance value through modifying the localized charge on the interfacial states [32-36] that leads to the change of thickness of the underlying space charge layer.

Other modification of MOSFET structure that can be applied to electrolyte solution measurements is an *ion sensitive field effect transistor (ISFET)*. In this device, the metal gate of MOSFET is replaced by an ion-sensitive membrane, the measured solution, and a reference electrode. The source-drain conductance varies as the function of the *pH value* or ion concentration of the solution in contact with the sensing surface. Input *I-V* characteristics of the ISFET at pH change are shifted along the voltage axis. By measuring this voltage shift, the pH can be determined. ISFET is very sensitive to any kind of potential generation at or near the gate insulator/ electrolyte interface. Thus, each biological or chemical reaction, leading to chemical or electrical changes (for instance, a pH or ion-concentration change) [67-69] at this interface, can be measured by means of these devices coupled with the respective chemical or biological recognition element. The first realization of ultra high sensitivity ISFET devices in which PS structures are placed on the gate region like ion-sensitive membrane is reported [70].

An *electrolyte-insulator-semiconductor (EIS)* is MIS-type structure in which the metallic gate electrode (like in ISFET) is replaced by a sensor layer, which is in direct contact with the analyte (electrolyte) and a reference electrode. EIS is considered to be an alternative to ISFET structures since for the last the poor adhesion and fast leaching-out of the sensitive materials as well as electrochemical corrosion of the passivation layer sometimes is observed. In contrast, EIS exhibits higher long-term stabilities and possess planar surfaces where no passivation of the electronic circuits is necessary [71,72]. As the ISFET, the EIS sensor is very sensitive for any kind of potential generation at or near the gate insulator/electrolyte interface. The analyte concentration or composition to be detected can be measured in capacitance/voltage (*C-V*) or constant capacitance modes. The charge carrier distribution at the interface insulator–semiconductor is controlled by an external *dc* voltage applied to reference electrode, a superimposed *ac* voltage is used to measure the space-charge layer capacity. According to Nernstian law a theoretical value of *C-V* curve shift is 59.1 mV per pH decade under standard conditions.

For pH sensors due to better pH response, hysteresis and drift characteristics compared with SiO_2, the gate materials such as Al_2O_3, Si_3N_4 and Ta_2O_5 are used instead of SiO_2 [73]. Si_3N_4 is known to be resistant to many chemicals. However, miniaturization of these EIS structures down to the micrometer scale, as for ISFETs, is very difficult since the capacitance values become too small for accurate measurements. To overcome this drawback it was proposed to use PS layers [71]. The scheme of the different capacitive planar and two structured EIS sensors Si$_3$N$_4$ (30-70 nm)/SiO$_2$ (30-50 nm)/Si for the pH determination is given in Figure 8. The structured EIS sensor consists of the trenches with the size varied between 2 µm and 8 µm and a depth of about 2 µm or macro-PS (pore diameter and pore depth is 1 and 2 µm, respectively). For all sensors an average pH sensitivity of 54 mV/pH in the concentration range from pH 4 to pH 8 exists. The drift, i.e., the time-dependent shift of the calibration curve varied 4-6 mV/day. However, for the porous EIS sensor, due to the surface enlargement, C_{max} is increased to about 800 nF in comparison to about 40 nF for the structured sensor and to about 25 nF for the planar one.

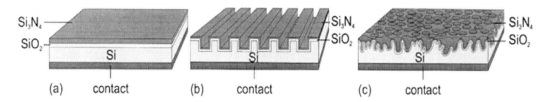

Figure 8. The scheme of the semiconductor insulator capacitors: (a) planar, (b) structured and (c) porous. Reprinted with permission from [71].

Table 4. FET, ISFET, EIS and LAPS sensors

N	Type of PS structure	Detected gas	Detection range	Ref
1	FET	H$_2$	1,2 mbar	[64]
2	FET	NO$_2$ CO$_2$ ethanol	100 -500 ppb 40-100 ppm 250-500 ppm	[63]
3	FET	isopropanol, ethanol, methanol, acetic acid	<10000 ppm <2500 ppm <1500 ppm	[62]
4	FET	isopropanol	<15000 ppm	[65, 66]
5	ISFET	pH-meter	pH=2-4	[70]
6	EIS	pH-meter	pH=4-8	[71]
7	EIS	pH-meter	pH=2-9	[68]
8	EIS	pH-meter	pH=5-10	[67]
9	EIS	pH-meter	pH=4-9	[69]
10	LAPS	pH=3-10	[76]	

The *light-addressable potentiometric sensor (LAPS)* is a semiconductor-based chemical sensor with an EIS. The principle of the LAPS is similar to that of the EIS capacitance sensor, in which the capacitance of the EIS system is measured to determine the ion concentration of the solution [72,74]. In the case of LAPS, a *dc* bias voltage is applied to the EIS structure so that a depletion layer appears at the insulator-semiconductor interface. The width, and therefore, the capacitance of the depletion layer vary with the surface potential. This variation of the capacitance is read out in the form of the photocurrent induced by the modulated light. The light-addressability of the LAPS allows an application of the LAPS as a chemical imaging sensor [75]. By using a focused laser beam as the light source, the local values of the surface potential can be measured. A two-dimensional map of the distribution of the pH value or the ion concentration can be obtained by scanning the sensing area with the focused laser beam.

Porous Si LAPS is developed in [76]. LAPS reveals satisfactory high pH sensitivity of 57.3 mV/pH. As an application of the PS LAPS, a penicillin LAPS was fabricated by the immobilization of the penicillinase on the surface. The pH change due to enzymatic reaction was detected for the penicillin concentration between 250 µM and 10 mM. Several field effect sensors are shown in Table 4.

6. MEMBRANE SENSORS

PS can be used in gas sensor design as micromachined membranes. Miniaturized hotplates are important parts of pellistor and resistor-type gas sensors [2]. The incorporation of thermally isolated micro- hotplates in the device construction allows a considerable reduction in both the power consumption and thermal transient time. For hotplate fabrication, one can use technologies of Si bulk and surface micromachining [7]. In bulk micromachining technology, the Si wafer is etched from the backside of the Si wafer in defined regions. Surface micromachining technology for free-standing membrane fabrication uses sacrificial layers. An easily etchable sacrificial layer is deposited onto the substrate surface followed by a second layer deposition. This layer will form the free-standing membrane after removing of the sacrificial layer by etching. PS is an ideal approach to combine both technologies and can be fabricated with thicknesses up to several tens of micrometers [77, 78]. PS as a sacrificial layer, due to its thickness, offers the possibility to create a large air gap between the membrane and substrate. Due to very high surface-to-volume ratio the sacrificial PS layer can be perfectly etched away.

The resistive sensor based on free–standing PS membrane displays very high sensitivity to detect NO_2 at T_{room} for concentrations as low as 200 ppb, $\Delta G/G= 111$ [79]. This sensor showed low interference from other gases like ethanol, methanol and ozone and no interference from carbon monoxide. On the contrary the humidity influencing the resistivity is significant.

To overcome some drawbacks of PS resistive sensors (poor selectivity to adsorbed gases, low log-term stubility during ageing) it is proposed to use PS as a physical support for another sensing film [80,81]. To monitor aromatic hydrocarbon molecules the mixed semiconducting oxides Sn–V–O have been used as a sensing layer. These oxides have shown peculiar catalytic properties, promoting selective oxidation of specific hydrocarbon compounds such as alkylaromatics and alcohols. The main feature of the device is represented by a permeated suspended macro- PS membrane few tens of microns thick on top of which a heater resistor and a temperature sensor are integrated (Figure 9).

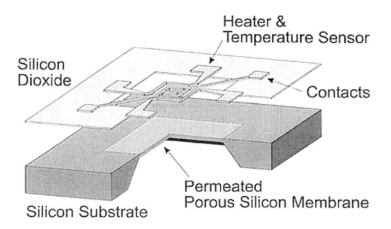

Figure 9. Sensor microstructure based on a suspended PS membrane. Reprinted with permission from [80].

Characteristics of the permeated PS suspended membrane sensor at 10% RH: sensitivity $\Delta G/G = 1.0$ (at 0.5 ppm C_6H_6), response time (90%) is 80 s, recovery time (70%) is 120 s, measurement cycle duration is 15 min.

Analog to EIS working principle membrane sensor is proposed [82]. The sensor design consists of an macro-PS region connected to a KOH etched fluidic channel. The KOH etched fluidic channels are monolithically integrated with the sensing membrane causing the analyte delivery sites and fluidic channels to be self-aligned. The frontside macro-PS region serves as the sensing membrane for solvent detection.

7. OPTICAL SENSORS

Optical properties of PS have been proposed for gas, vapor and liquid sensing since they appear to be extremely sensitive to the presence of dielectric substances inside of the pores. As the refractive index of the void space increases, the effective refractive index of the PS layer increases, causing the optical spectrum of the layer to shift to longer wavelengths. Thus, by monitoring the reflectance or transmission spectrum, one can detect the binding of molecules inside the pores since the capture of targets inside the void space increases the refractive index. In order to monitor the refractive index of PS samples, a convenient way is to fabricate the samples as suitably thin single layers or stacks of thin layers such as a *Bragg mirror, rugate filter, Fabry–Perot filter, luminescent PS microcavity (PSM)* [3,83-86].

Bragg mirrors are periodic stacks of two quarter-wavelength optical thickness layers of different refractive indices. The periodicity gives rise to a photonic bandgap, in which light propagation is forbidden and incident light is reflected. A microcavity consists of two Bragg mirrors and a layer that breaks the periodicity of the refractive index profile. The reflectance spectrum of a microcavity with a half-wavelength optical thickness layer is characterized by a resonant dip in the stop band.

The sensitivity of optical sensors is $\Delta\lambda/\Delta n$, where $\Delta\lambda$ is the wavelength shift and Δn is the change of refractive index. For a system able to detect a wavelength shift of 0.1 nm, the minimum detectable refractive index change is 2.10^{-4} [85]. The problem of PS optic sensors is the instability due to the aging of PS itself. One way to prevent this effect is the PS oxidation [84] or PS chemical modification [87].

Figure 10 shows the reflectivity at normal incidence for the Bragg mirror (it consists of ten layer pairs, of thickness 0.36 and 0.19 μm having porosities of 71% and 54%, respectively, electrochemically etched in the p^+ doped 0.01 Ω.cm Si wafer, yielding indices of 1.56 and 2.09 for the layers, the reflectivity peak at 957 corresponds to the second Bragg condition peaks). The 70-120 nm shift in the Bragg wavelength of the mirror arises from refractive index changes, induced by capillary condensation of the acetone and chlorobenzene vapors in the meso-PS, in the layers of the mirrors [88].

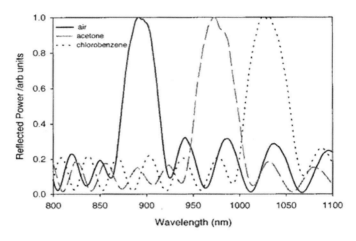

Figure 10. Normalized experimental reflectivity spectra showing peak wavelength shift of Bragg mirror after exposure to acetone and chlorobenzene. Reprinted with permission from [88].

Fabry–Perot fringes measured from the PS layer in ethanol solution are changed as the concentration of the solution varied (Figure 11). Standard deviation of the difference between the fringe patterns obtained in ethanol solutions and deionized water showed an almost linear relationship to the logarithm of ethanol molar concentration in the range between 1.10^{-5} and 1.10^{-14} M [89]. PS vapor sensor based on shift of Fabry–Perot fringes was demonstrated with the ethanol detection limit of 500 ppb and a dynamic range of nearly five orders of magnitude [90].

Optical reflectivity spectra of PS Fabry–Perot filters were also applied to detect low concentration of CO_2 (in ppth range) [87], HF gas with detection limit of 30 ppm [86], hydrocarbons [91].

As it was mentioned before, PSM exhibits well resolved Fabry–Perot interference fringes. In the luminescent PSM a narrow peak of photoluminescence (PL) of 5-10 nm width is observed whose spectral position depends on the optical thickness, hence on the refractive index of the cavity (PL peak position $\lambda_c = nd$, where d is the thickness of the central layer), while the PL intensity depends on substrate doping and on the etching and ambient conditions. Marked red shift of the cavity peak, as well as large changes of the integrated PL intensity, were observed for PSM in different organic solvents [92].

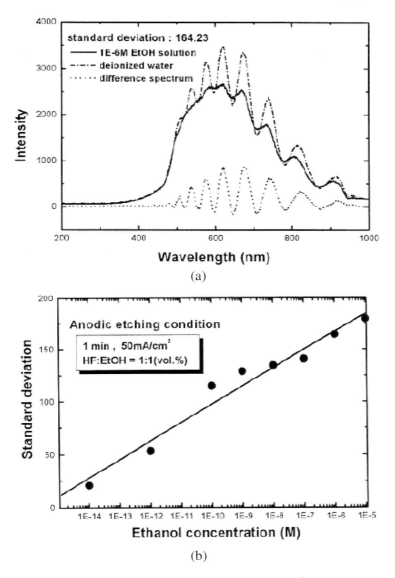

Figure 11. a) Fabry–Perot fringe patterns and difference spectra at 1.10^{-6} M ethanol concentrations and b) Standard deviation of difference spectra versus ethanol concentrations. Reprinted with permission from [89].

If the refractive index of the environmental gas changes, as, for example, when organic vapor is added to the environment, the PL peak as all Fabry–Perot fringes shift in red region up to 100 of nanometers (Figure 12, a). In the case of total filling of the pores the shift can be easily calculated with the Bruggeman effective medium approximation [86, 92]. The results of this calculation are shown together with the experimental data in Figure 12,b. Measurements performed in saturated vapors gave the same results obtained for the liquid solvents, confirming that the red shift is due to capillary condensation into pores. As opposed to the peak shift, the calculation of PL intensities in different organic solvents is much more complicated [92]. PL is strongly quenched by high values of dielectric constant.

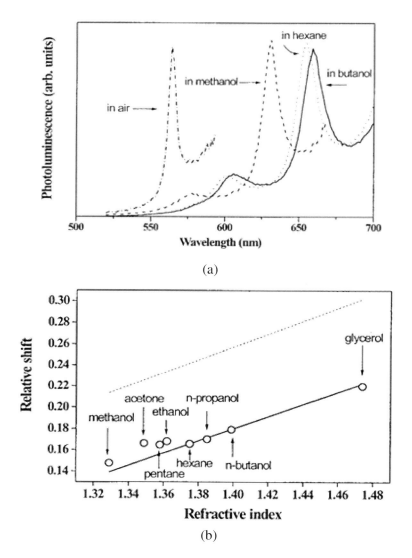

Figure 12. a) PL spectra of a PSM initially centered at 570 nm after immersion in different organic solvents. Intensities are not in scale; b) Relative peak shift for total filling versus refractive index for PSM initially centered at 570 nm. Dashed and solid lines: calculated shift without and with inclusion of variation of PS refractive index with wavelength. Reprinted with permission from [92].

The effect of ethanol vapor concentration on PL of PSM has been studied [94,93]. During the filling of ethanol, a larger ethanol concentration implies a larger effective refractive index and the position of the peak of the cavity resonance redshifts monotonically. The narrowing of the PL peak in a PSM allows the measuring of small variations of n, i.e., the detection of low gas concentrations.

Conjugated polymers entrapped in PSM have been studied as optical sensors for low volatility explosives such as trinitrotoluene [94]. The resonance peak of the microcavity reflectance modulates the fluorescence spectra of entrapped polymers in this device. The exposure of the PSM containing entrapped polymer to explosives vapor results in a red shift of the resonance peak, accompanied by the quenching of the fluorescence. The observed redshift in reflectivity is 2–3 nm larger than 1 nm of redshift in fluorescence.

The response time of optic sensors can be very small that is promising to make artificial electronic nose. Both PL quenching and shifts of reflectivity interference fringes of PS Fabry-Perot sample occurs during a couple of seconds.

Surfaces of PS chips can be stabilized by an anodic oxidation treatment, leading to long shelf life > 1 month, good baseline stability, and acceptable reversibility [95].

Different improved designs of photonic crystals based on PS for sensing application are discussed, among them the methods of PS/polymer micropatterning [96,97]. Micrometer-sized photonic crystals consisting of PS/polymer composite can be prepared by spray coating a fine mist of polymer solution onto PS photonic crystal layer. The uncoated PS is removed by treatment with aqueous base, resulting in composites that retain the spectral interference fringes of the PS photonic crystals but with improved chemical and mechanical stability [96]. The photonic device as planar waveguide in PS with submicrometer lateral patterning was proposed in [98].

The method of ellipsometry can also be applied to measure the change in refractive index of film. The measured quantity is the complex reflectance ratio $\rho = R_p/R_s = tan(\psi) \, exp(i\Delta)$, where R_p and R_s are the complex reflection coefficients of light parallel and perpendicular to the plane of incidence, respectively. The actual quantities measured are $tan(\psi)$ and $cos(\Delta)$ from which the ellipsometric parameters ψ and Δ are obtained. From these instrumental quantities the information on the optical properties in terms the complex refractive index, $n + ik$, or the dielectric function $\varepsilon = (n + ik)^2 = \varepsilon_1 + i\varepsilon_2$ can be obtained. Monitoring over a period of time allows the recording of changes in these parameters caused by adsorption of molecules into pores.

Applicability of ellipsometry for gas sensing was analyzed in [99-102]. For the different types of PS layer, the sensitivity for several gases has been demonstrated including water, ethanol, acetone, toluene. The detection limit threshold for acetone vapors by the layers was 12 ppm. However, as well as other optical methods, ellipsometry of PS is not specific for a certain gas.

CONCLUSION

The advantages of PS application for gas sensor are proved by the several reasons: PS has a huge internal surface area (up to 500-1000 $m^2 cm^{-2}$) and high surface chemical reactivity that can enhance the adsorption efficiency to the gas of interests; it allows to use a number of transduser principles (electrical, optical); simplicity and cheapness of the PS preparation that is compatible with silicon IC technology; there is a possibility to make 3D device and multisensors. It allows to detect very low concentration of CO, NO_2, different volatile organic compounds, alcohols, H_2S, H_2 and organic solutions. The PS gas sensors parameters are among the best gas sensors of different types. However, the PS sensors have some drawbacks. The irreversible PS surface deviation during aging, contact with humidity, UV ilumination leads to the instability of optical and electrical properties of PS and sensors. Thefore, the using of diferent passivation treatments of porous surface is necessary. Selectivity and sensitivity of PS sensors can be enhanced using catalyst materials and using the complex structure as ISFET, Bragg –mirrors, etc.

REFERENCES

[1] Sailor, M. J. (2012). Porous silicon in Practice. Preparation, Characterisation and Application, Wiley-VCH, Weinheim.
[2] Parkhutik, V. (1999). *Solid State Electron.*, 43, pp.1121-1141.
[3] Bisi, O., Ossicini, S., Pavesi, L. (2000), *Surf. Sci. Rep*, 38, pp. 1-126.
[4] Marsh, G. (2002). *Materials Today,* 5, pp. 36-41.
[5] Ozdemir, S. and Gole, J. L. (2007).*Current Opinion in Solid State and Materials Science*, 11, pp.92-100.
[6] Saha, H. (2008)., *Int. J. Smart Sens. Intell. System*, 1, pp. 36-56.
[7] Korotcenkov, G. and Cho, B. K. (2010). *Critic. Rev. Solid State & Mater. Sci.* 35, pp.1-37.
[8] Han, P. G., Wong, H. and Poon, M.C. (2001). *Colloids and Surfaces*, A179, pp.171–175.
[9] Baratto, C., Faglia, G., Comini, E., Sberveglieri, G., Taroni, A., La Ferrara, V., Quercia, L. and Di Francia, G. (2001), *Sensors & Actuators*, B 77, pp. 62–66.
[10] Valera, E., Casals, O., Vetter, M. and Rodriguez, A. (2007). *Spanish Conf. Electron Dev.*, Madrid, pp. 197 – 200.
[11] Baratto, C., Sberveglieri, G., Comini, E., Faglia, G., Benussi, G., La Ferrara, V., Quercia, L., Di Francia, G., Guidi, V., Vincenzi, D. , Boscarino, D. and Rigato V. (2000) *Sensors & Actuators,* B 68, pp. 74–80.
[12] Boarino, L., Baratto, C., Geobaldo, F., Amato, G., Comini, E. , Rossi, A.M., Faglia, G., Le'rondel, G. and Sberveglieri G. (2000), *Mat. Sci. Eng.*, B 69–70, pp.210–214.
[13] Chakane, S., Gokarna, A. and Bhoraskar, S.V. (2003). *Sensors & Actuators*, B 92, pp. 1–5.
[14] Pancheri, L., Oton, C.J., Gaburro, Z., Soncini, G. and Pavesi, L. (2004). *Sensors & Actuators*, B97, pp. 45–48.
[15] Massera, E., Nasti, I., Quercia, L., Rea, I. and Di Francia, G. (2004). *Sensors & Actuators*, B102, pp.195-197.
[16] Lewis, S. E., DeBoer, J. R., Gole, J. L. and Hesketh, P. J. (2005). *Sensors & Actuators*, B 110, pp. 54–65.
[17] Pancheri, L., Oton, C.J., Gaburro, Z., Soncini, G. and Pavesi, L. (2003). *Sensors & Actuators*, B89, pp. 237-239.
[18] Gaburro, Z., Bettotti, P., Saiani, M., Pavesi, L., Pancheri, L., Oton, C.J. and Capuj, N. (2004). *Appl. Phys. Lett.*, 85, pp. 555-557.
[19] Iraji zad, A., Rahimi, F., Chavoshi, M. and Ahadian, M.M. (2004). *Sensors & Actuators,* B 100, pp. 341–346.
[20] Timoshenko, V.Yu., Dittrich, Th., Lysenko, V., Lisachenko, M.G. and Koch, F. (2001). *Phys. Rev,* B 64, p.085314.
[21] Seals, L., Gole, J. L., Tse, L. A. and. Hesketh, P. J. (2002)., *J. Appl. Phys.*, 91, pp.2519-2523.

[22] Baratto, C., Comini, E., Faglia, G. , Sberveglieri, G., Di Francia, G., De Filippo, F., La Ferrara, V., Quercia, L. and Lancellotti, L. (2000), *Sensors & Actuators*, B 65, pp. 257–259.

[23] Anderson, R. C., Muller, R. S. and Tobias, C. W. (1990), *Sensors &Actuators*, A 21–23, pp. 835–839.

[24] Kim, S.-J., Park, J.-Y., Lee, S.-H. and Yi, S.-H. (2000), *J. Phys. D: Appl. Phys.*, 33, pp. 1781–1784.

[25] Kim, S.-J., Jeon, B.-H., Choi, K.-S. and Min, N.-K. (2000). *J. Solid State Electrochem*, 4, pp.363-366.

[26] Foucaran, A., Pascal-Delannoy, F., Giani, A., Sackda, A., Combette, P. and Boyer, A. (1997). *Thin Solid Films*, 297, pp. 317–320.

[27] Foucaran, A., Sorli, B., Garcia, M., Pascal-Delannoy, F., Giani, A. and Boyer, A. (2000). *Sensors & Actuators*, A79, pp.189–193.

[28] Xu, Y. Y., Li, X. J., He, J.T., Hu, X. and Wang, H. Y. (2005). *Sensors & Actuators*, B 105, pp.219-222.

[29] Bjorkqvist, M., Paski, J., Salonen, J. and Lehto, V.-P. (2006), *IEEE Sensor. J.*, 6, pp. 542 – 547.

[30] RoyChaudhuri, C., Gangopadhyay, S., DevDas, R., Datta, S.K. and Saha, H. (2008). *Int. J. Smart Sens. Intell. System*, 1, pp.638-658.

[31] Wang, Y., Park, S., Yeow, J.T.W., Langner. A., and F. Muller. (2010). *Sensors & Actuators*, B 149, pp. 136–142.

[32] Manilov A. I., Skryshevsky, V. A. (2013). *Mat. Sci. Eng.B*, 178, pp.942-955.

[33] Stievenard, D. and Deresmes, D. (1995). *Appl. Phys. Lett*. 67, pp. 1570-1572.

[34] Matsumoto, T., Mimura, H., Koshida, N. and Masumoto., Y. (1999). *Jpn.J.Appl.Phys*, 38, pp.539-541.

[35] Lee, W.H., Lee, C., Kwon, Y.H., Hong, C.Y. and Cho, H.Y. (2000). *Solid State Commun*, 113, pp.519-522.

[36] Skryshevsky, V.A., Zinchuk, V.M., Benilov, A.I., Milovanov, Yu. S., Tretyak, O.V. (2006). *Semicond. Sci. Technol.*, 21, pp. 1605-1608.

[37] Hao, P. H., Hou, X. Y., Zhang, F. L. and Wang, X. (1994). *Appl.Phys.Lett*, 64, pp. 3602-3604.

[38] Deresmes, D., Marissael, V., Stievenard, D. and Ortega, C. (1995). *Thin Solid Films*, 255, pp.258-261.

[39] Dimitrov, D. B. (1995). *Phys.Rev*, B 51, pp. 1562-1566.

[40] Ben-Chorin, M., Moller, F. and Koch, F. (1995), *J. Appl. Phys.*, 77, pp.4482-4488.

[41] Martin-Palma, R.J., Perez-Rigueiro, J., Guerrero-Lemus, R., Moreno, J.D. and Martinez-Duart, J.M. (1999). *J. Appl. Phys.*, 85, pp.583-586.

[42] Vikulov, V. A., Strikha, V. I., Skryshevsky, V. A., Kilchitskaya, S. S., Souteyrand, E. and. Martin, J.-R. (2000). *J. Phys. D: Appl. Phys.*, 33, pp. 1957-1964.

[43] Tretyak, O. V., Skryshevsky, V. A., Vikulov, V. A., Boyko, Yu. V. and Zinchuk, V. M. (2003). *Thin Solid Films*, 445, pp. 144–150.

[44] Lundstrom, I. and Soderberg, D. (1981/1982). *Sensors & Actuators*, 2, pp.105-138.

[45] Kohl, D. (2001). *J. Phys.D: Appl. Phys.*, 34, pp. R125-R149.

[46] Polishchuk, V., Souteyrand, E., Martin, J.-R., Strikha, V. and Skryshevsky, V. (1998). *Anal. Chem. Acta*, 375, pp. 205-210.

[47] Litovchenko, V.G., Gorbanyuk, T.I., Solntsev, V.S. and Evtukh, A.A. (2004). *Appl. Surf. Sci.*, 234, pp. 262-267.
[48] Luongo, K., Sine, A. and Bhansal, S. (2005), *Sensors &Actuators*, B 111–112, pp. 125–129.
[49] Skryshevsky, V. A., Polischuk, V., Manilov, A. I., Gavrilchenko, I. V. and Skryshevsky, R. V. (2008). *Sensor Electron. Microsys. Technol.*, 2, pp.21-27.
[50] Solntsev, V.S., Litovchenko, V.G., Gorbanyuk, T.I. and Evtukh, A.A. (2008). *Semicond. Phys. Quant. Electron. & Optoelectron.*, 11, pp. 381-384.
[51] Lachenani, H., Cheraga, H., Menari, H. and Gabouze, N. (2009). *Mat. Sci. Forum*, 609, pp. 225-230.
[52] Joshi, R. K., Krishnan, S., Yoshimura, M. and Kumar, A. (2009). *Nanoscale Res. Lett.*, 4, pp. 1191–1196.
[53] Zhang, W., de Vasconcelos, E. A., Uchida, H., Katsube, T., Nakatsubo, T. and Nishioka, Y. (2000), *Sensors & Actuators*, B 65, pp.154–156.
[54] Alwan, A. M. (2007). *Eng. & Technol.*, 25, pp.1023-1027.
[55] Strikha, V., Skryshevsky, V.A, Polishchuk, V., Souteyrand, E. and Martin, J.-R. (2000). *J. Porous Mat.*, 7, pp. 111–114.
[56] O'Halloran, G. M., Kuhl, M., Trimp, P. J. and French P. J. (1997). *Sensors & Actuators*, A 61, pp. 415 -420.
[57] Subramanian, N. S., Sabaapathy, R. V., Vickraman, P., Kumar, G. V., Sriram, R. and Santhi, B. (2007). *Ionics*, 13, pp.323–328.
[58] Belhousse, S., Cheraga, H., Gabouze, N. and Outamzabet, R. (2004). *Sensors & Actuators*, B 100, pp. 250–255.
[59] Chiali, A., Ghellai, N., Gabouze, N. and Sari, N.E.C. (2010). *Adv.Mater. Sci.*, 10, pp.4-10.
[60] De Rosa, R., Di Francia, G., La Ferrara, V., Quercia, L., Roca, F. and Tucci, M. (2000). *Phys.Stat.Sol*, A 182, pp. 489-493.
[61] Tucci, M., La Ferrara, V., Della Noce, M., Massera, E., Quercia, L. (2004). *J. Non-Cryst. Solids*, 338–340, pp. 776–779.
[62] Barillaro, G., Nannini, A. and Pieri, F. (2003). *Sensors & Actuators*, B 93, pp. 263–270.
[63] Barillaro, G., Diligenti, A., Nannini, A., Strambini, L.M., Comini, E. and Sberveglieri, G. (2006). *IEEE Sensors J.*, 6, pp. 19 – 23.
[64] Zouadi, N., Gabouze, N., Bradai, D. and Dahmane, D. (2009). *Mater.Sci.Forum*, 609, pp. 261-264.
[65] Barillaro, G., Diligenti, A., Marola, G. and Strambini, L.M. (2005), *Sensors & Actuators*, B 105, pp.278-282.
[66] Barillaro, G., Diligenti, A., Nannini, A., and Strambini L. M. (2005)., *Phys. Stat. Solidi.*, C2, pp. 3424–3428.
[67] Mizsei, J., Shrair, J.A. and Zolomy, I. (2004). *Appl. Surf. Sci.*, 235, pp.376-388.
[68] Reddy, R. R. K., Basu, I., Bhattacharya, E. and Chadha, A. (2003). *Current Appl. Phys.*, 3, pp. 155–161.
[69] Schoning, M.J., Simonis, A., Ruge, C., Ecken, H, Muller-Veggian, M. and Luth, H. (2002). *Sensors*, 2, pp. 11-22.
[70] Zehfroosh, N., Shahmohammdi, M., Mohajerzadeh, S. (2010). *In Nanotechnology 2010: Electronics, Devices, Fabrication, MEMS, Fluidics and Computational*, pp. 390 – 393.

[71] Schoning, M. J., Malkoc, U., Thust, M., Steffen, A., Kordos, P. and Luth, H. (2000). *Sensors & Actuators*, B 65, pp. 288–290.
[72] Schoning, M. J. (2005). *Sensors*, 5, pp. 126-138.
[73] Chung, D.W.-Y., Yang, C.-H. and Wang, M.-G. (2004). *Proc.SPIE*, 5586, SPIE, Bellingham, WA, pp.18-25.
[74] Yoshinobu, T., Schoning, M.J., Finger, F., Moritz, W. and Iwasaki, H. (2004). *Sensors*, 4, pp.163-169.
[75] Lundstrom, I., Erlandsson, R., Frykman, U., Hedborg, E., Spetz, A., Sundgren, H., Welin, S. and Winquist, F. (1991), *Nature*, 352, pp. 47-50.
[76] Yoshinobu, T., Ecken, H., Poghossian, A., Luth, H., Iwasaki, H. and Schoning, M.J. (2001). *Sensors & Actuators,* B 76, pp.388-392.
[77] Steiner, P. and Lang, W. (1995). *Thin Solid Films,* 255, pp.52-58.
[78] Splinter, A., Bartels, O., Benecke, W. (2001). *Sensors & Actuators*, B76, pp.354-360.
[79] Baratto, C., Faglia, G., Sberveglieri, G., Boarino, L., Rossi, A.M. and Amato, G. (2001), *Thin Solid Films*, 391, pp. 261-264.
[80] Angelucci, R., Poggi, A., Dori, L., Cardinali, G.C., Parisini, A., Tagliani, A., Mariasaldi, M., Cavani, F. (1999). *Sensors & Actuators,* A74, pp.95–99.
[81] Parisini, A., Angelucci, R., Dori, L., Poggi, A., Maccagnani, P., Cardinali, G.C., Amato, G., Lerondel, G. and Midellino, D. (2000). *Micron*, 31, pp. 223–230.
[82] Clarkson, J.P., Fauchet, P.M., Rajalingam, V. and Hirschman, K. D. (2007). *IEEE Sensors J.*, 7, pp. 329-335.
[83] Chan, S., Fauchet, P.M., Li, Y., Rothberg, L.J. and Miller, B.L. (2000). *Phys. Stat Solidi*, A 182, pp. 541-546.
[84] Bettotti, P., Cazzanelli, M., Dal Negro, L., Danese, B., Gaburro, Z., Oton, C. J., Prakash, G. V. and Pavesi, L. (2002). *J. Phys. Condens. Matter.,* 14, pp. 8253–8281.
[85] Ouyang, H. and P. M. Fauchet, *In Photonic Crystals and Photonic Crystal Fibers for Sensing Applications,* H. H. Du editor (SPIE, Bellingham, WA, 2005), pp 600508 1-15.
[86] I.I. Ivanov, V.A. Skryshevsky, T. Serdiuk, V. Lysenko, *Sensors and Actuators B*, 174, p.521-526, 2012.
[87] Rocchia, M., Garrone, E., Geobaldo, F., Boarino, L. and Sailor, M. J. (2003). *Phys. Stat. Solidi*, A 197, pp. 365–369.
[88] Snow, P.A., Squire, E.K., Russek, P.St.J. and Canham, L.T. (1999). *J. Appl.Phys.*, 86, pp.1781-1784.
[89] Min, H.K., Yang, H.-S. and Cho, S. M. (2000). *Sensors & Actuators*, B 67, pp. 199–202.
[90] Gao, J., Gao, T. and Sailor, M. J. (2000). *Appl. Phys. Lett.*, 77, pp.901-903.
[91] De Stefano, L., Moretti, L., Rendina, I. and Rossi, A.M. (2003). *Sensors & Actuators*, A 104, pp.179-182.
[92] Mulloni, V., Gaburro, Z. and Pavesi, L. (2000). *Phys. Stat. Solidi*, A 182, pp. 479-484.
[93] Gaburro, Z., Daldosso, N. and Pavesi, L. (2001). *Appl. Phys. Lett.*, 78, pp.3744-3747.
[94] Levitsky, I. A., Euler, W. B., Tokranova, N. and. Rose, A (2007). *Appl. Phys. Lett.*, 90, pp.041904.
[95] Letant, S.E., Content, S., Tan, T. T., Zenhaausern, F. and Sailor, M. J. (2000). *Sensors & Actuators*, B 69, pp. 193-198.
[96] Li, Y. Y., Kollengode, V. S. and Sailor, M. (2005). *Adv. Mater*, 17, pp.1249-1251.

[97] Gargas, D. J., Muresan, O., Sirbuly, D. J. and Buratto, S. K. (2006). *Adv. Mater.*, 18, pp.3164–3168.
[98] Jamois, C., Li, C., Skryshevskyi, R., Orobtchouk, R., Chevolot, Y., Monnier, V., Souteyrand, E. and Benyattou, T. (2010). *Proc. 7th Int. Conf. Porous semicond. - Science and technol*, Valencia, Spain, pp. 126-127.
[99] Zangooie, S., Bjorklund, R. and Arwin, H. (1997). *Sensors & Actuators*, B 43, pp. 168–174.
[100] Zangooie, S., Jansson, R. and Arwin, H. (1999). *J. Appl. Phys.*, 86, pp. 850-858.
[101] Arwin H. (2001). *Sensors & Actuators*, A 92, pp. 43-51.
[102] Wang, G. and Arwin, H. (2002). *Sensors & Actuators*, B 85, pp. 95-103.

Chapter 7

HIERARCHICAL ZEOLITES: PREPARATION, PROPERTIES AND CATALYTIC APPLICATIONS

Ana P. Carvalho[1], Nelson Nunes[1,2] and Angela Martins[1,2]

[1] Centro de Química e Bioquímica, Faculdade de Ciências, Universidade de Lisboa, Lisboa, Portugal
[2] Área Departamental de Engenharia Química, Instituto Superior de Engenharia de Lisboa, Rua Conselheiro Emídio Navarro, Lisboa, Portugal

ABSTRACT

Zeolites are crystalline aluminosilicates widely used in catalysis as well as in separation and purification fields. The unique combination of properties, such as, high surface area, well-defined microporosity, high thermal stability and intrinsic acidity underlie the successful performance of these materials for a great number of applications. However, when bulky reaction intermediate species are involved the purely microporous character of zeolites is a drawback, because it often imposes diffusion limitations due to restricted access to the active sites. This is the case of some applications in the petroleum, petrochemical and fine chemical industries.

This chapter aims to provide a comprehensive examination of the different approaches that can be adopted to enhance the accessibility and molecular transport in the zeolite framework, illustrated with examples from the literature.

Dealumination and, more recently, desilication are post-synthesis treatments that have been extensively used over numerous structures with academic and industrial interest. These treatments lead to Si/Al ratio changes and, simultaneously, hierarchical structures combining micro and mesopores are commonly obtained. As a consequence of the treatments, alterations of some important properties, such as, hydrothermal resistance and acidity are also observed. The catalytic performance is enhanced and generally the structures become less sensitive to deactivation due to coke deposition. There are also several examples pointing out the great advantage of consecutive treatments (e. g. dealumination by acid treatment followed by desilication, or vice-versa) to tune the samples catalytic properties.

In recent years, template based preparation strategies have been increasingly explored to obtain hierarchical structures with two or even three levels of porosity. This will result in a great increase in the accessibility and mass transport of the guest molecules towards the active sites within the zeolite framework.

In spite of the well documented benefits of hierarchical network for catalytic applications, the main industrial application is the steamed dealuminated Y zeolite in fluid catalytic cracking. The production of hierarchical porous materials, through the several methods before mentioned, are still at an exploratory phase, although some attempts to scale-up and supply them to industrial units were recently reported, which can be the start to a new era of hierarchical zeolites in industrial catalysis.

1. INTRODUCTION

Zeolites are crystalline microporous aluminosilicates known since 1756 when the stilbite structure was identified by the geologist Cronsted, but became industrial important only after the studies of Barrer and Milton, reporting the successful synthesis of various zeolite structures [1, 2]. Since these pioneer studies zeolites started to be widely used as catalysts, adsorbents and ion exchangers in many application fields as it is schematized in Figure 1.

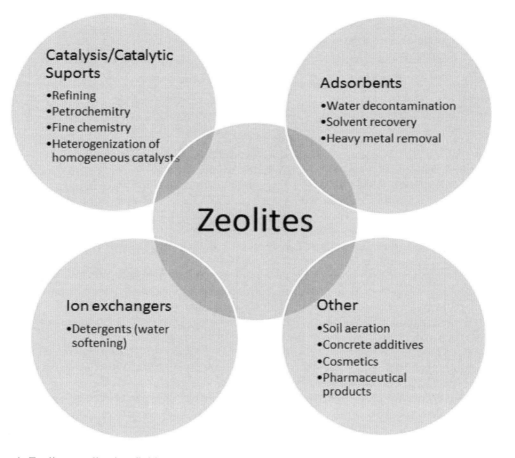

Figure 1. Zeolites application fields.

The incorporation of zeolites in the formulation of detergents is the most important application of these materials as ion exchangers, consuming a very large amount of the annual production of zeolites. Linde Type A is the structure used in this application since an almost complete exchange of sodium cations by calcium and magnesium occurs after a very short contact time, even at room temperature [3]. The use of zeolites, usually natural clinoptilolite, as materials for treatment of radioactive fallout from nuclear tests and accidents is another example of the use of zeolites as ion exchangers, due to their superior selectivity for a great number of radionuclides [4].

The use of zeolites as adsorbents for gas and liquid phase processes is reported in numerous studies. Adsorption of volatile organic compounds [5] is an example of a gas phase process, and the removal of heavy metals [6] or pharmaceutical compounds from aqueous solutions [7, 8] are examples of liquid phase processes.

The review of F. A. Mumpton [9] summarizes various others applications of natural zeolites in somewhat unexpected fields as, for example, components of dimension stones, cement and concrete, as supplements for animal diet, or as soil amendments. Applications of zeolites as deodorizing agents and for medical purposes are also listed.

In spite the great number of possible applications of zeolites, the importance of these materials is undoubtedly related to their use as catalysts in refining and petrochemistry. These high added value applications have been the main reason for the intensive research effort focused on the synthesis and modification of these solids, aiming the preparation of the most appropriate catalyst for a given process.

The use of zeolites as catalysts is a consequence of their unique combination of properties (see Figure 2). Actually, the optimization of a catalytic process implies the search of the experimental conditions that allow, not only high activity values but also high selectivity for a certain product. If the activity is mostly determined by the zeolite intrinsic acidity, due to the presence of Brønsted acid sites, the selectivity is linked with the microporous nature of the zeolite structure (pore width typically between 0.25 nm and 1.5 nm). The strictly microporous framework induces various types of shape selectivity that can be fundamental to increase the yield of a desire product, but can also impose diffusion constrains that will limit the catalytic performance of the zeolites. Restricted molecular transport of the species inside the zeolite crystal occurs when the dimensions of the molecules and the pore size are similar, i.e., when the diffusional regime is molecular and not of Knudsen type, which occurs when wider pores are present [2]. So, especially when bulky molecules are considered, the most probable is that the diffusional constrains may adversely affect the catalytic performance; however it must be mentioned that in other cases they can be a benefit to the catalytic process. The typical example of this last case is xylene isomerization over ZSM-5. The diffusional limitations for the molecular transport of *o*-xylene and *m*-xylene lead to an effective trapping of these two unwanted isomers inside the zeolite structure, which can only leave the ZSM-5 framework after being converted into the most valuable isomer, *p*-xylene. The percentage of *p*-xylene in the products is then higher than that expected by the thermodynamic equilibrium, and a direct dependence of the selectivity with the ZSM-5 crystals dimensions is observed [10].

In the majority of the cases a low diffusivity in the channels of the zeolite framework represents a drawback for catalytic process (Figure 2). In this sense, several strategies have been explored to enhance the accessibility and molecular transport towards the active sites. Altering the synthesis protocol to obtain nanosized crystals will shorter the micropore path length and is then one of the possible options to overcome the problem [11]. Other possible approach is to increase the diffusion inside the pores, which can be achieved through different procedures: synthesis of wide-pore and large cavity zeolites [12], delamination of zeolites [13] and the use of composites of zeolites and mesoporous materials [14].

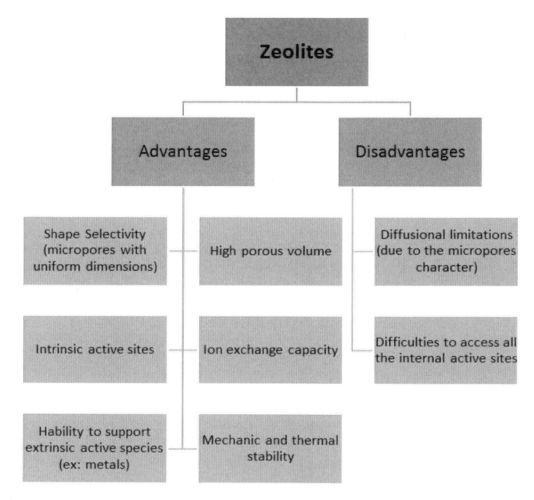

Figure 2. Main advantages and disadvantages of zeolite catalysts.

The preparation of hierarchical structures, i.e., structures presenting multimodal pore systems is an alternative route to obtain materials with much less diffusion constrains than that present in the characteristic micropore network of zeolites. The pore structure of these solids is a complex system of micro and mesopores (and in some cases, also macropores) that can be obtained following various synthesis and post-synthesis procedures as it is schematically presented in Figure 3.

The purpose of this chapter is to provide a global view on the various methods to prepare hierarchical zeolite structures, reporting the impact of each one on the materials` properties and catalytic performance. The designation of zeolite structures that will appear along the text are given according to the framework type codes attributed by the International Zeolite Association (IZA) and can be consulted at www.iza-structure.org, although in many cases the common names are also used.

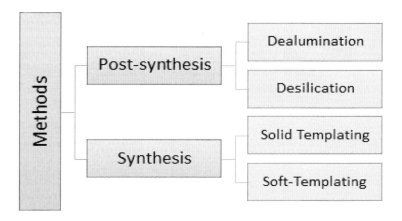

Figure 3. Different approaches to prepare hierarchical zeolite structures.

2. DEALUMINATION

Dealumination consists on the selective removal of Al from the zeolite framework and was originally developed aiming to control the concentration and strength of the acid sites increasing the Si/Al ratio of low-silica zeolites. Simultaneously, hydrothermal stabilization of the structure and mesoporosity development are also achieved. This post-synthesis treatment was first explored in the 60's by Barrer and Makki [15] in a study focused on the modification of clinoptilolite structure through hydrochloric acid treatment. Since then a great number of studies where, not only acid treatments but also other procedures have been employed to dealuminate several zeolite structures.

Considering the mechanism, the various dealumination processes reported in the literature can be divided in two major groups:

– Processes that promote the removal of framework Al atoms generating a vacancy which depending on the treatment conditions, can be reconstructed or lead to a developed mesopore structure:

- Hydrothermal treatments – self-steaming and steaming [16-25]
- Acid treatments [15, 23, 24, 26-32]
- Treatments with EDTA and other chelating species [33-36]
- Treatments with chlorinated compounds [37, 38]
- Treatments with F_2 and fluorinated compounds [39, 40]

– Processes where the isomorphic substitution of framework Al usually occurs in a large extension leading generally to a much less mesoporosity development:

- Treatment with $SiCl_4$ [22, 41-46]
- Treatment with $(NH_4)_2SiF_6$ [22, 47-51]
- Treatments with BCl_3 [52]

To optimize the characteristics of the samples, successive and different types of treatments can be conjugated. An example is the pioneer study developed by Scherzer [53] who reported the preparation of faujasite structures with Si/Al ratio higher than 200 obtained after successive hydrothermal and acid treatments.

Despite the considerable number of different options that can be followed to obtain a dealuminated structure it must be mentioned that hydrothermal treatments, and especially steaming, are by far the most explored methodology. Considering that ultrastable Y zeolite, the base of the catalyst for the cracking process, is prepared through steaming, the extensive research effort made over this dealumination process is not surprising. Among all the other processes mentioned above, the acid treatment is currently employed, in most of the cases as a second step treatment to enable the removal of extra-framework species formed during the hydrothermal process [17].

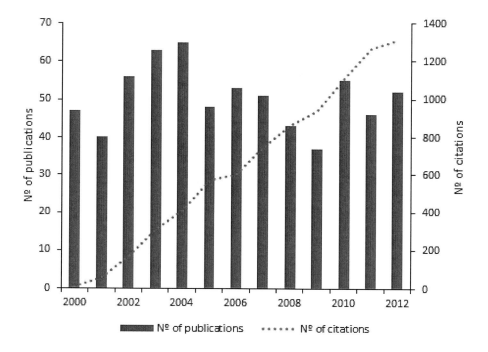

Figure 4. Approximate annual number of publications and citations on zeolite dealumination since 2000. Source: ISI Web of Knowledge, 23rd October 2013. Search terms: "dealumination" AND "zeolite".

Another aspect that is interesting to consider is that, although dealumination process was applied to a great number of structures, and extensively studied in the last decades of the 20th century, the use of this post-synthesis process to modify the characteristic microporous network of the zeolites continues to attract the interest of the scientific community. Since 2000 the annual number of publications on indexed journals about this subject is between 35 and 67, and their citations has been growing steadily, as can be observed in Figure 4. However it must be noted that in many cases the dealumination process is used in association with other post-synthesis treatment. The study developed by Verboekend et al. [54] is illustrative of this fact since authors treated desilicated ZSM-5 samples with HCl solutions to promote the leaching of the amorphous Al rich debris present in the alkaline treated samples, uncovering the micro and mesopore network. The authors tested the modified samples as catalysts in the alkylation of toluene with benzyl alcohol verifying that the alkaline treated solid showed an improved catalytic performance when compared to the parent material. However an even better result was obtained when the acid leached sample was used, pointing out the relevance of the two-steps protocol to enhance the catalytic properties of the samples. On the other hand, there are also examples showing that the acid treatment, besides leaching the extra-framework species created in a previous treatment, also promotes a further modification of the structure [55].

A brief review of the various dealumination processes will be presented, more detailed in the case of hydrothermal and acid treatments. Some examples of catalytic applications will also be presented to illustrate the effectiveness of the treatments to modify the samples properties.

2.1. Hydrothermal Treatments

As previously mentioned, hydrothermal treatments have been widely explored to modified zeolite structures by dealumination. Some examples of the studies developed in recent years are listed in Table 1.

The most studied zeolite for hydrothermal dealumination is, undoubtedly, zeolite Y [8, 22, 62, 63]. However other structures have also been considered as is the case of mazzite [21], mordenite [45, 64], ferrierite [45], ZSM-5 [18], BEA [23, 45, 60, 65], offretite [66], MCM-22 [24, 67, 68], and ZSM-5 [19, 20, 45, 61, 69].

Experimentally hydrothermal treatments consist in a calcination that can be made without submitting the sample to a continuous flux of steam, that is, the water molecules involved in the process are those present in the zeolite framework which has to be previously saturated – self-steaming process - or under N_2 and steam flux – steaming process. The calcination temperature is usually in the range of 500 - 850⁰C, and the zeolites can be modified either in the ammonium or protonic form.

In any case the mechanism involves the hydrolysis of Al-O-Si bonds leading to the removal of aluminum that provokes the formation of a vacancy in the framework, designated as "hydroxyl nest", the appearance of extra-framework Al species [70], and partial amorphization of the framework. The presence of amorphous material is of fundamental importance for the process since, after the study of Maher et al. [71], it is consensually admitted that the amorphous material is the source of mobile Si species that will heal the "hydroxyl nests". The need of an initial partial amorphization of the structure to obtain high dealumination rates is experimentally demonstrated, for example, by the results reported by Wang et al. [72]. These authors studied the steaming over Y zeolite and proved that for short treatment times there is only a decrease of the crystallinity; however when the duration of the treatment increases, the dealumination becomes progressively more important and no further amorphization is detected.

Table 1. Examples of recent studies where hydrothermal treatments were applied to modify zeolite structures

Zeolite	Experimental conditions	Results	Reference
NH$_4$NaY (Si/Al = 2.48)	Steaming at 600 and 750°C, for 6 h, under a flow of N$_2$-steam of 30 cm^3 min^{-1} and a water partial pressure of 69.9 kPa. The samples were further treated with an EDTA solution (7.5 wt%) at 100°C.	Sample submitted to the more severe treatment: • Dealumination rate - 73%; crystallinity - 47%. • V$_{micro}$ decreased from 0.327 to 0.152 cm^3 g^{-1}; V$_{meso}$ increased from 0.028 to 0.171 cm^3 g^{-1}. • EDTA treatment did not lead to changes in the dealumination rate and crystallinity; V$_{micro}$ increased to 0.189 cm^3 g^{-1} and V$_{meso}$ to 0.310 cm^3 g^{-1} • Acidity - the ratio of *strong/total* acid sites increased from 0.48 to 0.55. • Based on the NMR data and acidity characterization it was demonstrated that the steaming treatment at 600°C produces charged monomeric EFAL species, and at 750°C monomeric Al species + oligomeric aluminosilicate species. The extra-framework phase of the sample steamed at 750 + EDTA consists mainly of polymeric Si, Al species.	[22] (2000)
Beta (Si/Al= 12-15) MOR (Si/Al = 5-10) ZSM-5 (Si/Al = 9-47) FER (Si/Al = 17)	Calcination of the NH$_4^+$ forms at 550°C for 5 h with a heating rate of 60°C h^{-1} under shallow bed conditions.	• Beta: dealumination rate ~ 75%; large amount of EFAL was formed; XRD pattern and textural parameters are not significantly changed. • MOR: higher dealumination rate achieved ~ 75%, a large part of the crystalline structure was destroyed and consequently an accentuated decrease of the specific surface area was observed. • ZSM-5 and FER: the conditions used did not promote an effective dealumination of the zeolites.	[45] (2000)
MCM-22 (Si/Al = 15)	Steaming of the NH$_4^+$ form with 37.1 kPa steam at 500, 600 and 700°C, for 2 h. Treated samples were leached with HNO$_3$, 2 M at 100°C.	• The increase of the treatment severity allowed the increase of the Si/Al ratio up to 31.7 without any significant change in the micropore structure, but resulting in the loss of ≈ 84% of the initial Brønsted acid sites. Acid sites in the external surface are preferentially removed. • Disproportionation of ethylbenzene: dealuminated samples showed improved *p*-diethylbenzene selectivity due to the selective removal of acid sites on the external surface.	[56] (2001)
ZSM-5 (Si/Al = 27.2)	Steaming at 570, 660, 730 and 790°C, for 6 h, under a flow of N$_2$-steam of 30 cm^3 min^{-1} and a water partial pressure of 90.4 kPa. The sample treated at 730°C was further treated with an HCl, 0.2 M at 80°C for 24 h.	• Si/Al ≈ 27 (steaming) reaching 35 after acid leaching; crystallinity ≥ 80%. • ^{27}Al MAS-NMR showed the presence EFAL that became more important as the severity of the treatment increased. The removal of EFAL by the acid treatment did not result in the decrease of the intensity of the peak assigned to octahedral coordinated Al which is explained considering that in steamed ZSM-5 sample there are extra-framework oligomeric alumina "invisible" to NMR. • V$_{micro}$ and V$_{meso}$ remained practically unchanged. • Severely steamed samples showed no strong acidity.	[57] (2001)

Zeolite	Experimental conditions	Results	Reference
HBeta (Si/Al = 33)	Steaming at 500 or 700°C under water-air mixture flow (water partial pressure - 93.3 kPa) for 3 or 6 h.	• An important increase of Si/Al is observed only upon steaming at 700°C for 3 h (Si/Al=100, estimated using an empirical equation $N_{Al}=f(v_1)$[58]). • All the treatments led to the decrease of V_{micro} associated to pore blockage by EFAL; V_{meso} also slightly decreased. • Both types of acid sites considerably decreased upon steaming. • IR spectra of the OH region demonstrated that the treatments had only a limited effect on the band at 3782 cm^{-1} ascribed to structure defects.	[23] (2003)
MCM-22 (Si/Al = 22.9)	Steaming at 600°C (heating rate 10°C min^{-1}) for 4 h under flow of 12 cm^3 min^{-1} g^{-1} of water and 100 cm^3 min^{-1} g^{-1} of air.	• Slight increase of Si/Al ratio to 23.1; crystallinity was retained. • Important increase of the amount of five-coordinated EFAL species. • Steaming resulted in the decrease of the total number of acid sites (10^{-4} mol g^{-1}): 9.44 (parent) vs 7.13 (modified sample), and of the number of strong Brønsted acid sites. • Steamed sample mixed with a commercial methanol synthesis catalyst was tested in the conversion of syngas conversion to dimethyl ether showing enhanced selectivity to dimethylether along with least generation of byproducts (e. g. hydrocarbons and CO$_2$).	[24] (2006)
ZSM-5 (Si/Al = 41.5)	Steaming at 400, 550, and 600°C for 5 h with a 100% water vapor flux fed at a rate of 1 g H$_2$O/g. Treated samples were leached with HCl, 2 M, at 80°C for 2h.	Sample steamed at 550°C + acid leaching: • Si/Al=65.5; crystalline structure was retained. • Micropore area decreased from 398 (parent) to 321 m^2 g^{-1}, and mesopore area increased from 63 (parent) to 97 m^2 g^{-1}. • The number and strength of the acid sites decreased upon the treatments. • Methylation of 2-methylnaphthalene: modified sample showed enhanced stability and selectivity to 2,6-dimethylnaphthalene (59.1% after 8 h t.o.s.).	[59] (2008)
ZSM-5	Steaming at 540 °C, for 2 h, under water partial pressure between 12 and 154 Torr.	• No significant change in the micro and mesopore structure was observed. • The number of Brønsted acid sites decreased (≈50%) but their strength was enhanced.	[18] (2012)
ZSM-5 (Si/Al = 90)	Steaming at 550 °C, under N$_2$ flow caring 0.33 mol h^{-1} water, for 2 -48 h.	• No structural of textural modifications were observed after 48 h of treatment. • NMR results proved that crystallographic different Al sites are differently susceptive to dealumination. Those correspond to T-O-T bond angle of 155° being very stable under the experimental condition used.	[19] (2012)
ZSM-5 (Si/Al = 12.5)	Steaming at 400-550 °C, for 6 h, in pure steam flow.	Sample submitted to the more severe treatment: • Si/Al = 14.5; crystallinity -57%. • V_{micro} - decreased from 0.112 to 0.067 cm^3 g^{-1}; V_{meso} increased from 0.069 to 0.121 cm^3 g^{-1}. • Acidity - loss of 35% of total acid sites; for strong acid sites B/L ratio decreased from 3.3 to 1.4. • Ethanol dehydration to ethylene: catalyst steamed at 550°C presented excellent catalytic stability (for a t.o.s. of 350 h conversion reached 95%) and good selectivity for ethylene.	[20] (2013)

Table 1. (Continued)

Zeolite	Experimental conditions	Results	Reference
Beta (Si/Al = 4.5)	Hydrated sample in NH_4^+ form was heated in a covered vessel and steamed at 600°C (1 °C min^{-1}) for 2h. The treated sample was further leached with HNO_3 (0.1, 0.5 and 6 M) for 3 h at 80°C, using a solution zeolite ratio of 50 cm^3 g^{-1}	• Si/Al = 12.4 (after steaming); 55 (after leaching with HNO_3, 6 M). • Steaming resulted in the decrease of V_{micro} from 0.24 to 0.19 cm^3 g^{-1} and the development of mesoporosity (0.05 cm^3 g^{-1}). • Alkylation of benzene with ethane or propene: benzene conversion increases after dealumination due to the mesoporosity created which increased the accessibility of the acid sites; however the progressive increase of the acid concentration resulted in loss of activity attributed to the decrease of the acidity. • Sample treated with 6 M solution is almost inactive for the alkylation of benzene.	[60] (2013)
ZSM-5	Steaming at 550 – 750°C, for 2 h, under He flow (6 cm^3 s^{-1}) with 50% H_2O; heating rate of 5°C min^{-1}.	Sample submitted to the more severe treatment: • Dealumination rate – 77%, crystallinity - 92%. • V_{meso} increased from 0.09 to 0.15 cm^3 g^{-1} • Steaming reduced the concentration of Brønsted acid sites • Oxidation of phenol to dihydroxybenzenes with N_2O: steamed samples allowed a 70% N_2O conversion and dihydrobenzenes selectivity ≥ 90%; coke formation is reduced due to the decrease of acidity.	[61] (2014)

B/L – ratio between the number of Brønsted and Lewis acid sites; EDTA - ethylenediamine tetraacetic acid; EFAL – extra-framework Al species; MAS-NMR – magic angle spinning-nuclear magnetic resonance spectroscopy; N_{Al} – number of framework Al in the unit cell; v_1 – wavenumber of the T-O-T asymmetric stretching vibration; t.o.s. – time on stream; V_{micro} – micropore volume; V_{meso} – mesopore volume; XRD – X-ray diffraction.

Considering the mechanism described above, it becomes clear that the mobility of the Al and Si species determines the dealumination rate and consequently samples characteristics. The presence of steam greatly enhances the mobility of both these species, and so steaming treatments allow better compromises between dealumination rate, textural modification and preservation of crystalline structure. On the contrary, self-steaming process, due to lack of water molecules, lead to more modest dealumination rates and, as it will be discussed below, can also lead to different textural modifications. The study developed by Carvalho et al. [62] is one of the very few works reporting the effect of the two types of hydrothermal treatments over the same sample (Y zeolite). The results presented in Figure 5 clearly show the importance of steaming to extract high percentages of Al from the framework. The results of this study also reveal that with steaming treatments higher dealumination rates are achieved without great loss of crystallinity. For example, the samples submitted to two consecutive calcinations at, respectively, 540 and 760°C under self-steaming conditions, lost 16% of the pristine crystalline structure while the solid steamed lost 28%. However, as can be seen in Figure 5, the dealumination rate attained with steaming is almost 90% while using self-steaming conditions the dealumination rate reached only 40%.

The healing of the framework by the insertion of a Si atom on the framework position previously occupied by an Al results in the shrinkage of the unit cell dimensions [22, 62, 63, 66] since the length of the Si-O bond (1.61 Å) is smaller than that of the Al-O bond (1.71 Å). The experimental results show that the decrease of the unit cell parameters is directly related with the unit cell composition. So, especially in the case of zeolite Y several linear relations between the value of a_0 (cubic unit cell parameter) and N_{Al}, the number of framework Al atoms per unit cell, can be found in the literature. Most probably, the first of these relations was proposed in 1968 by Breck and Flaningen [73] but since then other equations have also been reported [62, 74-76]. The difference between the various relations $N_{Al} = f(a_0)$ are mainly related with the methodologies and techniques used by the authors to evaluate the values of N_{Al} of the samples obtained after post-synthesis treatments. In the case of the more recent ones the values of N_{Al} are obtained from the results of solid-state ^{27}Al and ^{29}Si NMR, i.e., with a technique that allows the distinction between framework and extra-framework Al.

Since the bonds Si-O and Al-O are also energetically different, the dealumination process also leads to changes in infrared spectra in the framework vibration region. The maxima of the bands assigned to the asymmetric (v_1) and symmetric (v_2) stretching vibrations shift to smaller wavenumbers as the number of framework Al decreases. So, similarly to the relations based on the unit cell parameter obtained from X-ray diffraction (XRD) pattern mentioned above, various relations $N_{Al} = f(v_1)$ and $N_{Al} = f(v_2)$ have been proposed in the literature for different zeolite structures [58, 62, 74].

The framework vacancies that are not rebuilt may grow to form larger pores, usually in the mesopore range (i.e., pores with widths between 2 and 50 nm). The creation of hierarchical structures (micro+meso) upon hydrothermal treatments is then a commonly reported result in the literature [20, 22, 63-65]. An example is the study recently developed by H. Xue et al. [64] over commercial mordenite with Si/Al ratio of 7.4. As a consequence of the

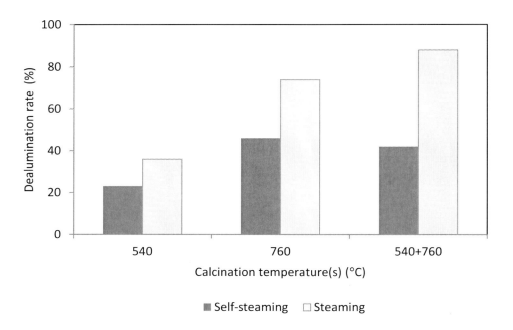

Figure 5. Dealumination rate of Y zeolites modified by self-steaming and steaming at different temperatures for 3 hours. Sample 540+760 was obtained after two consecutive calcinations Adapted from ref [62].

steaming treatment conducted at 750°C for 5 h in a fixed-bed reactor using a gas stream containing 30% H_2O in N_2, the mesopore volume increased from 0.072 cm^3 g^{-1} to 0.170 cm^3 g^{-1}. In another study Janssen et al. [63] modified a commercial Y zeolite by a protocol consisting of two steaming treatments followed by acid leaching. The sample obtained showed a mesopore volume of 0.21 cm^3 g^{-1}, and a broad mesopore size distribution, ranging from 4-40 nm. In the same study the authors characterized a commercial dealuminated Y sample which was prepared following a patented steaming protocol [77]. This sample exhibited a mesopore volume of 0.47 cm^3 g^{-1} and a mesopore size distribution centered at 10 nm. Combining the results of the textural analysis with 3D-TEM (three-dimensional transmission electron microscopy) the authors conclude that a large fraction of the mesopores created by steaming are cavities formed inside the zeolite structure that are connected to the external surface through micropores. Only a small fraction are cylindrical mesopores directly connected to the crystallites surface, that is, the dealumination treatments promoted the formation of intracrystalline mesoporosity.

There are also studies reporting that the textural modifications introduced by the hydrothermal treatment lead to the broadening of the micropore size distribution with practically no impact in the mesopore structure. The results of the textural characterization by nitrogen, n-hexane, and 3-methylpentane adsorption reported by Brotas de Carvalho et al. [66] for self-steamed offretite samples allowed the authors to conclude that, under the experimental conditions used at least a fraction of the microporous system characteristic of offretite network was converted into larger micropores, i.e., supermicropores (pores with width between 0.7 and 2 nm).

After the hydrothermal treatment, the new porosity created is, in a large extension, filled with the debris of the partial amorphization of the structure generated by the extraction of Al from the framework. This implies that hydrothermal treatments are usually followed by an acid leaching under mild conditions [63, 65, 66] to remove the amorphous material which, when blocking the porosity, seriously decreases the number of effective active sites.

Considering the mechanism of the hydrothermal treatments it is evident that, as it was already mentioned, the creation of hierarchical zeolite structures is always associated to a decrease of crystallinity, directly related with the severity of the treatments [62, 63, 66]. The values of crystallinity reported by the group of M. Guisnet [62] for Y zeolite steamed at progressively higher temperatures clearly show the influence of this experimental parameter in the preservation of the zeolite framework, and on the dealumination rate. The sample steamed at 540^{0}C presents 36% of dealumination rate, and practically no destruction of the crystalline structure was observed. However, the increase of the temperature to 760^{0}C has a great impact in the Al removal (dealumination rate = 74%), leading also to a significant decrease in the crystallinity (only around 40% of the initial structure remains after treatment). A further increase of the steaming temperature to 820^{0}C results in a total amorphization of the structure.

In addition to the loss of crystalline structure, and consequent decrease of catalytic active sites, hydrothermal treatments have other two important disadvantages. Since the Brønsted acid sites of the zeolites - determinants for the materials' performance as catalysts - are mainly associated to the framework Al, upon the dealumination process the number and the nature of the acid sites is changed [18-20, 66, 78]. This transformation does not necessarily have a negative impact in the performance of the samples as catalysts. In fact, it has long been recognized that steaming can lead to the formation of "superacid" sites associated to extra-framework species [79] what results in an increase of the catalytic activity [18, 78, 79]. The data recently obtained by Niwa et al. [18] with HZSM-5 steamed samples show that, increasing the partial pressure of water up to around 30 Torr the dealumination process lead to the decrease of the number of acid sites, but their strength increases. This evolution was reflected in a higher hexane cracking activity.

The decrease of the acid sites density, especially when allied to a considerable mesoporosity development has a positive effect on the catalytic stability of the samples, because both aspects inhibit the deactivation of active sites by coke (general name for the carbonaceous deposits formed due to side reactions) deposition. This can be exemplified by the results reported by Sheng et al. [20] with a ZSM-5 steamed at 550°C. The material exhibited excellent catalytic activity and good selectivity on the ethanol dehydration to ethylene transformation, and the coke content determined on spent catalyst is less than that present in the parent ZSM-5 essayed in the same conditions. These findings were interpreted considering that the mesopores created after steaming may accommodate part of coke deposition, suppressing at some extent, the formation of coke deposits in the zeolite intrinsic microporosity.

Other side effect of the hydrothermal treatments is the surface enrichment in Al, especially important in the case of steaming treatments [63], and less significant under self-steaming conditions [17].

2.2. Acid treatments

Experimentally, acid treatments are the dealumination process that demand less equipment. So, being easy to perform it is not surprising that many studies found in the literature are focused on zeolite structure modifications by this type of treatment. Some examples of recent published studies are listed in Table 2.

Dealumination through acid treatment is restricted to zeolites that are stable in acid media. This is not the case of zeolite Y, thus the number of studies focused on acid dealumination of this zeolite structure is not very high. An example is the work developed by Zou et al. [80] where ultrastable Y zeolite was further dealuminated by treatments with oxalic acid, leading to hierarchical zeolites with enhance catalytic properties for hydrolysis of hemicellulose. On the contrary, a great number of examples of acid modified mordenite [81-87], MCM-22 [28, 29, 68, 88, 89] clinoptilolite [29, 31, 90-94], ZSM-5 [7, 95-97], ferrierite [27] and Beta [23, 98, 99] are reported in the literature.

The mechanism of acid dealumination was elucidated by Barrer and Makki [15] and it is generally accepted that the process involves, as in the case of hydrothermal treatments, the formation of a "hydroxyl nest" for each Al removed from the framework. However fifty years after the first mechanism proposal there are still studies searching for data to unambiguously confirm the existence of the hydroxyl nests. An example is the recent work of Senderov et al. [103], focused on acid dealuminated Y zeolite where, from the analysis of the TPD (temperature programmed desorption) data, the authors concluded that hydroxyl nests cannot be present in samples calcined above 200 °C.

It is generally accepted that the acid treatment process occurs into three consecutive steps: (i) the original cations are exchanged by H^+; (ii) dealumination of the framework leading to the formation of "hydroxyl nests", without visible structural changes; and in severe conditions (iii) disintegration of the framework may be observed resulting in an amorphous phase. Thus beside acid concentration, temperature and contact time also the nature of the exchangeable cation is an important factor that determines the extension of dealumination by acid treatments, as it is demonstrated by, for example, the results obtained by Rivera et al. [31] in a study focused on clinoptilolite dealumination through HCl treatments. The authors proved that according with the fact that the ion exchange reaction Ca^{2+}/H^+ is less favored than Na^+/H^+, the acid attack to the Al−O bonds occurs at lower acid concentration when the zeolite is in the sodium form.

In acid treated samples no extra-framework species will remain blocking the porosity since they will be leached to the solution. Therefore mesopores are created but since the vacancies are not reconstructed, to obtain high dealumination ratios high crystallinity loss is necessarily observed, with the consequent loss of active sites. However, in controlled and mild conditions this type of dealumination treatments allows the preparation of solids with very interesting catalytic properties.

Table 2. Examples of recent studies where acid treatments were applied to modify zeolite structures

Zeolite	Experimental conditions	Results	Reference
MOR (Si/Al = 5.8)	HNO_3 0.5, 1 and 2 M or oxalic acid 1, 2 and 6 M; at 100°C, for up to 20 h; acid solution to zeolite ratio of 12 $cm^3 g^{-1}$.	• Oxalic acid was the more effective, leading to Si/Al ratio up to 12. • Samples treated with oxalic acid presented a considerable amount of EFAL. Conversely, HNO_3 dissolved almost all the EFAL and removed silica units leading to the formation of an intracrystalline secondary pore structure. External surface area increased from 6 $m^2 g^{-1}$ (parent) to 14 $m^2 g^{-1}$ (sample treated with HNO_3 6 M for 20 h). • Oxalic acid resulted in the increase of V_{micro} from 0.208 (parent) to 0.257 $cm^3 g^{-1}$ (sample treated with oxalic acid 2 M for 8h) but no change in the mesoporosity was observed.	[86] (2000)
ZSM-5 (Si/Al = 29.5)	HCl, 0.85 M.	• Si/Al = 35.9. • Micropore area decreased from 346 (parent) to 310 $m^2 g^{-1}$, mesopore area increased from 10.7 to 70.6 $m^2 g^{-1}$. • m-Xylene conversion: dealuminated samples presented higher activity than the starting material due to the increase of the overall acidity; the para/orto ratio was similar to the thermodynamic value due to the presence of a developed mesopore structure.	[100] (2000)
Beta (Si/Al = 12-15) MOR (Si/Al = 5-10) ZSM-5 (Si/Al = 9-47) FER (Si/Al = 17)	Oxalic acid, 0.5, 1 or 2 M, at room temperature for Beta, and under reflux for other samples; acid solution to zeolite ratio of 20 $cm^3 g^{-1}$.	• Beta: Si/Al attained 100; XRD patterns showed that a part of the long-range order is lost upon the treatment although the porosity was maintained. • ZSM-5: the highest framework Si/Al ratio obtained was 58; even with the less concentrated acid solution some decrease of the specific surface area was observed. • MOR: low initial Si/Al allowed to acheive Si/Al of 59 with a significant impact on the porosity and crystallinity; high initial Si/Al led to samples with Si/Al≈28, preserving the porosity. • FER: with 2 M solution Si/Al framework ratio increased to 27, but most of the Al atoms remained in the pores in the form of EFAL.	[45] (2000)
ZSM-5 (Si/Al = 27.2)	HCl, 1.5 M; at 90°C for 24 h.	• Si/Al = 29.3; crystallinity - 94%. • No significant change in the micro and mesopore structure was observed.	[59] (2001)
HBeta (Si/Al = 33)	HCl, 1 M, at 30 and 100°C for 2 h then washed with warm distilled water; acid solution to zeolite ratio of 10 $cm^3 g^{-1}$.	• The more severe condition allowed to obtain a Si/Al = 82, estimated using an empirical equation $N_{Al}=f(v_1)$[58]. • The treatments had no effect on micropore and mesopore volumes. • Acidity was greatly affected by the treatments: the number of Bronsted and Lewis acid sites reached, respectively, 84 and 8 $\mu mol\ g^{-1}$ upon the more severe conditions. • The disappearance of the infrared bands at 3662 and 3782 cm^{-1} prove the extraction of EFAL and tricoordinated Al species present in the parent material.	[23] (2003)
Clinoptilolite (Turkish natural zeolite)	HCl, 0.032, 0.16, 0.32, 1.6 and 5 M, at 25, 40, 75 and 100°C for 3 h; acid solution to zeolite ratio of 20 $cm^3 g^{-1}$.	• The dealumination rate of the sample treated with the more concentrated solution reached 10% for treatments at 25 or 40 °C, and 80% for treatments at 100°C. • In the case of the treatment at 100 °C: V_{micro} decreased from 0.122 to 0.089 $cm^3 g^{-1}$, and V_{meso} increased from 0.008 to 0.051 $cm^3 g^{-1}$.	[91] (2005)

Table 2. (Continued)

Zeolite	Experimental conditions	Results	Reference
MCM-22 (Si/Al = 22.9)	Oxalic or citric acid, 0.5 M, at 85°C for 24 h (reflux); acid solution to zeolite ratio of 20 cm^3 g^{-1}.	• Si/Al ratio increased to 31.2 (citric acid) and 37.9 (oxalic acid); crystal structure was retained regardless the acid used. • Sample treated with oxalic acid presented the higher decrease of the total number of acid sites (10^{-4} mol g^{-1}): 9.44 (parent) vs 6.54 (modified sample); but retained a great number of strong Brønsted acid sites. • The treatment with citric acid led to the formation of a large amount of five-coordinated EFAL species. • The modified samples were mechanically mixed with a commercial methanol synthesis catalyst and tested in the process of syngas conversion to dimethyl ether: compared to the parent zeolite, modified samples possess higher selectivity to dimethylether along with least generation of byproducts (e. g. hydrocarbons and CO$_2$).	[24] (2006)
MOR (Si/Al = 5)	Acetic acid, 6 M, reflux for 5 to 12 h.	• Si/Al ratio increased to 35 (5 h) and 110 (12 h); crystallinity was retained. • The treatments resulted in the enlargement of the micropores and formation of a mesopore structure; V_{micro} reached 0.18 cm^3 g^{-1} and V_{meso} 0.13 cm^3 g^{-1}. • Alkylation of cumene with 2-propanol: the acid treatment enhanced the catalyst lifetime due to the reduction of acid site concentration; high para-selectivity (≈45%) is retaining albeit the enlargement of micropores.	[85] (2008)
MOR (Si/Al = 6.5)	HCl, 6 M, heated at 100°C under microwave refluxed or autoclaved for 15 min and 2 h; acid solution to zeolite ratio of 30 cm^3 g^{-1}. Another set of samples were prepared under the same conditions but using a conventional oven.	• Microwave led to faster dealumination than conventional heating regardless the conditions used. • Microwave irradiation in autoclave allowed the preparation of a sample with Si/Al = 23.6, crystallinity = 68%, and high mesoporosity development (micropore area/non-micropore area decreased from 8.9 (parent) to 4.9 (treated sample); total pore volume decreased from 0.059 to 0.120 cm^3 g^{-1}. • The use of microwave radiation led to lower acidity. • Isomerization of styrene oxide: treated sample presented enhanced conversion and higher selectivity for β-phenylacetaldehyde.	[101] (2009)
MCM-22 (Si/Al = 14.5)	HNO$_3$, 4 M at 100°C for 3, 30, 60 and 480 min.	Sample submitted to the more severe treatment: • Dealumination rate = 20%; crystallinity = 100%. • No change on V_{micro} was observed; V_{meso} increased from 0.34 to 0.37 cm^3 g^{-1} • Acidity -loss of 48% of the Brønsted acid sites; Lewis acid sites increased 9% due to EFAL or structural defects formed during dealumination. • Methylcyclohexane transformation: initial activity decreased with dealumination; cracking products increased at expense of isomers which was attributed to diffusion limitations of the isomers induced by EFAL deposition in the inner pore system.	[28] (2009)

Zeolite	Experimental conditions	Results	Reference
Clinoptilolite (Cuban natural zeolite) (Si/Al = 4.48)	HCl, 0.6 M at 100°C for 2 h (1, 2 or 3 times); afterwards washed 3 times in an ultrasonic bath with HCl, 0.05 M (acid solution to zeolite ratio 25 cm^3 g^{-1}) at 70°C for 15 min.	Sample submitted to the more severe treatment: • Si/Al = 10.0; the loss of crystallinity was visible by the appearance of the broad low baseline in the XRD pattern likely due to the partial amorphization of the zeolite phase of the sample. • V$_{micro}$ slightly increased from 0.098 cm^3 g^{-1} (parent) to 0.114 cm^3 g^{-1}. • ^{27}Al MAS NMR indicates that low levels of octahedral aluminum species were created during the treatments.	[29] (2010)
ZSM-5 (Si/Al = 13)	Tartaric acid, 1 M, at 60°C for 4 h, acid solution to zeolite ratio of 20 cm^3 g^{-1}.	• Si/Al = 25.3. • Developed mesoporosity (V$_{meso}$ = 0.35 cm^3 g^{-1}). • The number of strong acid sites decreased to about half of the initial value, but the number of the external acid sites (i.e., acid sites at the mesopore walls) decreased from 0.23 to a very low value (0.05 mmol g^{-1}). • Methanol-to-hydrocarbon conversion: dealumination enhanced the catalyst lifetime due to the fact that coke is formed mainly on mesopore walls of the modified sample.	[49] (2010)
MOR (Si/Al = 6.5) Beta (Si/Al = 10) ZSM-5 (Si/Al = 20)	HCl, 6 M under microwave irradiation at 100°C for 15 min, and for Beta also 5 min. For comparison zeolites were treated in the same conditions in a conventional oven.	• The extent of dealumination was function of the zeolite structure and of the heating method used. • Microwave assisted treatment led to faster and higher dealumination. • Microwave treated samples presented higher relative amounts of EFAL, and a more extensive development of mesoporosity. • Dealumination rate follow the order Beta> MOR> ZSM-5; crystallinity did not significantly decrease. • Styrene oxide isomerization: MOR dealuminated using microwave radiation presented enhanced yield to β-phenylacetaldehyde, due to the strong Brønsted sites formed.	[98] (2011)
Clinoptilolite (Si/Al = 4.8)	HCl, 0.001, 0.05, 0.1, 0.3, 0.5, and 1 M, for 1 h at room temperature, acid solution to zeolite ratio of 250 cm^3 g^{-1}. Samples exchanged with NaCl were treated under the same conditions.	• The more dealuminated sample (Si/Al of the framework = 9.0) presents 67% of crystallinity. • Nitrogen adsorption isotherms indicated an increase of V$_{micro}$ upon treatment. V$_{micro}$ of the sample treated with 1 M HCl is almost double than that of starting material. This evolution is mainly due to the creation of new narrow micropores. • The importance of the exchangeable cation to the resistance of the acid attack was demonstrated since acid treatments made over sodium form led always to smaller dealumination rate than that observed with natural sample which is mainly a calcium form.	[94] (2013)
HUSY (Si/Al = 3.2)	Oxalic acid, 0.1, 0.2 and 0.3 M, at 80°C, for 8h, acid solution to zeolite ratio of 25 cm^3 g^{-1}.	• Si/Al = 18.7; crystallinity = 76%. • V$_{micro}$ decreased from 0.19 to 0.17 cm^3 g^{-1}; V$_{meso}$ increased from 0.12 to 0.18 cm^3 g^{-1}. • Acidity decreased after acid treatment. • Hydrolysis of hemicellulose: dealuminated samples showed to be effective catalysts for this transformation; the yield of total reducing sugars increased from 5.8% (starting material) up to 55.7%. This behavior was attributed to the textural modifications induced by acid treatment which improved the accessibility to the acid sites on the internal surface favoring mass transfer of the reactants and products	[80] (2013)

Table 2. (Continued)

Zeolite	Experimental conditions	Results	Reference
ZSM-5 (Si/Al = 12)	HCl, 1-3 M, at 90°C for 24 h.	• Acid treatment with concentration up to 2 M did not destroy the zeolite framework but removed a small amount of the EFAL present in the parent material, thus the fraction of framework Al and of the Bronsted acid sites increased. • The treatments had no effect on micropore and mesopore volumes. • Xylose dehydration: dealumination led to a slightly decrease of xylose conversion but enhanced furfural selectivity reaching 67%. Under the same conditions the parent material presents a furfural selectivity of 45%.	[95] (2014)
ZSM-5 (Si/Al = 47)	Oxalic acid, 0.5 M, acid solution to zeolite ratio of 50 cm^3 g^{-1}. The mixture was introduced in a Teflon container and heated at 120°C for 2h.	• Si/Al = 51; crystallinity = 70%. • ^{27}Al NMR results reveal that the treatment had no effect on the structural and extra-structural Al, most likely due to the mild conditions used. • No textural modification was observed from the N$_2$ adsorption isotherms. • Dehydration of ethanol: ethanol conversion was not affected by the dealumination and did not decrease with t.o.s.; selectivity of ethylene decreased in comparison to the value obtained with the parent material what was attributed to the decrease of the weak acid sites which seems to be fundamental for ethylene production.	[97] (2014)
MCM-68 (Si/Al = 12)	HNO$_3$, 0.5, 2.0, 4.0, and 6.0 M, acid solution to zeolite ratio of 30 cm^3 g^{-1}, under reflux in an oil bath (130 °C) for 2 - 24 h.	• Si/Al = 51; crystallinity was maintained. • No mesopore structure was formed. • n-Hexane cracking: for high reaction temperature (600 °C) dealumination enhanced n-hexane conversion and catalyst stability. Propylene selectivity reached 45-50%. • The regenerated catalysts after the reaction at 600 °C exhibit almost the same performance as the initial catalyst.	[102] (2014)

EFAL – extra-framework Al species; MAS-NMR – magic angle spinning- nuclear magnetic resonance spectroscopy; N_{Al} – number of framework Al in the unit cell; v_1 – wavenumber of the T-O-T asymmetric stretching vibration; t.o.s. – time on stream; V_{micro} – micropore volume; V_{meso} – mesopore volume; XRD – X-ray diffraction.

This is the case of the study recently published by Inagaki et al. [104] where it was proved that the post-synthesis treatment of ZSM-5 zeolite (synthesized in the absence tetrapropylammonium cation) with HNO_3 selectively removed framework Al from the external surface. The sample was assayed as catalyst for the cracking of *n*-hexane and the results revealed that the very few acid sites remaining in the surface can be readily deactivated in the early moments of the reaction turning more difficult the coke formation. The obtained solid can then be considered a potentially long-life catalyst for cracking of hexane or other paraffin.

The use of acid treatment under microwave irradiation was also recently assayed [98, 99, 101]. The results obtained by González and co-workers [98] concerning acid treatment over Beta, mordenite and ZSM-5 showed that microwaves lead to faster and higher dealumination ratio than conve,ntional heating for all the three zeolite structures. In the particular case of mordenite the HCl treatments under microwave irradiation resulted in a mesoporosity development and Brønsted acid sites with higher strength than the mordenite treated by conventional heating. When tested as catalysts in the styrene oxide isomerization the mordenite treated using microwave radiation showed higher yield to β-phenylacetaldehyde than the pristine commercial mordenite. This result was interpreted as being due to the presence of Brønsted acid sites with higher strength formed during dealumination.

The nature of the acid is an experimental parameter that plays a determinant role in the mesopore development, being HCl [15, 27, 29, 31, 90-95, 98] and HNO_3 [28, 84, 86, 87, 89, 102, 104-106] the most commonly employed. Organic acids as oxalic [24, 45, 80, 86, 88, 97], tartaric [49, 96, 107] or acetic [85] acids are used in a more restricted way. In the particular case of oxalic acid the limited efficiency to promote mesopore formation is justified by the ability of oxalate ions to form complexes with aluminum, inhibiting the mesopore creation [86].

The use of citric acid [24] is also reported although being less effective than, for example oxalic acid, what is attributed to the large size of citric anion.The diffusion of this specie into the zeolitic channels is difficult, leading to a voluminous coordination compound with aluminum which has even more diffusion constrains than the anion to escape from the zeolite framework.

The use of toluene-4-sulfonic acid to promote framework dealumination is also reported in the literature [99, 108, 109]. The advantage of this procedure is that, due to the incorporation of the sulfonic groups, the number and strength of Brønsted acid sites considerably increases.

The number of studies where acid treatments are used to promote the leaching of extra-framework species created during steaming, or other post-synthesis treatments, is considerably higher than the number of studies where acid treatment is used as the sole dealumination process [56, 57, 60, 89]. As it was already mentioned, the leaching of the species that remained occluded in the framework is of fundamental importance because only after the dissolution the porosity becomes available.

To restrict the effect of acid treatment only to the dissolution of the debris deposited in the micro- and mesopores, mild conditions must be used, that is, the treatments must be performed with diluted acid solutions to prevent further dealumination of the structures. The results of several studies show that, in optimized conditions, after the acid leaching treatment the micropore volume of the hydrothermally treated samples considerably increases [83, 110].

Although in the majority of the cases acid treatment is made after a first process that originates extra-framework species, it can also be used previously to modify the zeolite structure to enhance the effectiveness of the post-synthesis treatment. This strategy was followed in the study made by Tian and co-workers [32] where BEA zeolite structure was treated with oxalic acid in different concentrations to obtain samples with different Si/Al ratio that allowed a more effective mesopore formation via NaOH treatments.

2.3. Other Treatments

Apart from hydrothermal and acid treatments all the other dealumination processes listed in the beginning of this section have been much less explored, and some have no more than academic interest, even though some are patented process. The representative example of this situation was the proposal made by Fejes et al. [37] to use phosgene to modify the structures of clinoptilolite, mordenite and Y zeolites. The results reported were very interesting (high dealumination rates with small loss of crystallinity) but not sufficient to overcome the great danger of working with such a poisoning compound, and so, for the best of our knowledge this is the only study where this methodology was used.

The study made by Ghosh and Kydd [111] is another example of a less common dealumination process. The authors used aqueous hydrofluoric acid to modify a ZSM-5 structure and the main effect of the aluminum removal was the decrease of the number of strong Brønsted acid sites. However in optimized conditions an increase in the number of Lewis acid sites was observed.

Dealumination using chelating agents, namely EDTA (ethylenediamine tetraacetic acid) was proposed by Kerr in late 60´s [33] as a possible strategy to obtain ultrastable Y zeolite. Although much less frequently reported, acetylacetone is other chelating agent that can be used for dealumination purposes [100]. However in the majority of the studies reporting the use of chelating agents, these compounds are used to remove the extra-framework species created by hydrothermal treatments [22, 35, 36].

The reaction of the zeolite structures with $(NH_4)_2SiF_6$ or $SiCl_4$ allow isomorphic substitution of the framework Al by Si leading, in principle, to a relatively modest mesoporosity development. Studies illustrating the use of these dealumination processes are presented in Table 3.

The modification of zeolite structures by treatment with $SiCl_4$ was proposed by Beyer and Belenykaja [46]. According to the results presented in this pioneer study, with this method a Y zeolite with Si/Al ≈ 100, could be obtained, keeping the pristine crystallinity. In principle this method can be used to modify zeolites with pore opening larger than the critical dimensions of $SiCl_4$ (6.23 Å [42]), otherwise the Al removal will be restricted to the external surface, as it happens when the procedure is applied to ZSM-5 (5.1×5.5 Å and 5.4×5.5 Å) [45] or ferrierite (4.2×5.4 Å and 3.5×4.8 Å). On the contrary, when the method is applied to Y zeolite (7.4 Å) [8, 22, 44, 46] and mordenite (6.7×7.0 Å and 2.6×5.7 Å) [41, 42] high dealumination ratios are achieved allied to some mesoporosity development, and preservation of the micropore structure.

Table 3. Examples of studies where treatments with $SiCl_4$ and $(NH_4)SiF_6$ (AHFS) were applied to modify zeolite structures

Zeolite	Experimental conditions	Results	Reference
Treatment with $SiCl_4$			
NaY ($Si/Al = 2.48$)	Dehydrated samples were exposed to the $SiCl_4$ vapor for 3 h at 250, 350, or 450°C. The temperature was kept for 3 h under dry N_2 flow to remove the $AlCl_3$ formed after what the samples were thoroughly washed with water.	Sample submitted to the higher temperature • Dealumination rate – 88%; crystallinity – 83%. • V_{micro} decreased from 0.327 to 0.244 $cm^3 g^{-1}$; V_{meso} increased from 0.028 to 0.063 $cm^3 g^{-1}$. • Acidity - the ratio of *strong/total* acid sites increased from 0.48 to 0.61. • Conjugating NMR results and acidity characterization it was concluded that the extra-framework species formed by $SiCl_4$ treatment are aluminosilicate phases, which in the case of the studied samples have $Si/Al = 2-3$.	[22] (2000)
Beta ($Si/Al= 12-15$) MOR ($Si/Al = 5-10$) ZSM-5 ($Si/Al = 9-47$) FER ($Si/Al = 17$)	Treatment according to the procedure described in ref [46]: temperature between 457 and 557°C, for 2 h after what the sample is purged with dry N_2 and then washed until negative chloride test ($AgNO_3$ reaction).	• Beta - Si/Al reaches 28; crystallinity and porosity were retained. • MOR - in optimized conditions framework Si/Al ratio reached 48 with no impact in the texture or structure. • ZSM-5 and FER: due to the relative dimensions of the pores and $SiCl_4$ only Al atoms on the external surface are removed. No significant modifications on the textural properties were observed.	[45] (2000)
Treatment with $(NH_4)SiF_6$ (AHFS)			
NH_4NaY ($Si/Al = 2.48$)	AHFS, 0.1 M was added drop-wised (0.005 mol AHFS/mol of Al/min) to a stirred slurry of zeolite suspension in water buffered at pH 5-6, during up to 3 h at 80°C.	Sample submitted to the longer treatment • Dealumination rate – 50%; crystallinity- 88%. • V_{micro} decreased from 0.327 to 0.283 $cm^3 g^{-1}$; V_{meso} increased from 0.028 to 0.099 $cm^3 g^{-1}$. • Acidity - the ratio of *strong/total* acid sites increased from 0.48 to 0.67. • NMR results demonstrated that a dealumination degree of 50% obtained by AHFS treatment results in a Si-rich amorphous phase with insignificant quantities of EFAL.	[22] (2000)
ZSM-5 ($Si/Al = 27.2$)	Sample 1 - treated twice with AHFS, 0.1 N, using an amount of AHFS equal to N_{Al}. Sample 2 – treated 5 times with an excess of AHFS, 0.5 N	• $Si/Al = 40.6$ (sample 1) and 49.6 (sample 2); crystallinity- 92%. • ^{27}Al MAS-NMR results proved that this delumination procedure does not originate EFAL. • V_{micro} and V_{meso} remained practically unchanged.	[57] (2001)

Table 3. (Continued)

Zeolite	Experimental conditions	Results	Reference
HBeta (Si/Al = 33) Zeolyst	The treatments were made at 25 and 75°C adding the AHFS to obtain 0.10, 1.5, 3.0 and 4.35 times the stoichiometric amount needed to complete dealumination of the sample.	Sample submitted to the severe treatment • Dealumination rate - 50%; crystallinity- 88%. • The decrease of V_{micro} from 0.176 to 0.119 cm^3 g^{-1} was ascribed to SiO$_2$ deposits; V_{meso} decreased from 0.510 to 0.366 cm^3 g^{-1}. • The treatment allowed the complete elimination of the large amount of EFAL and tricoordinated Al species present in the parent structure.	[23] (2003)
ZSM-5 (Si/Al = 13)	1 g of zeolite was treated with 10 cm^3 of 0.1 M AHFS solution at room temperature for 30 min.	• Si/Al = 20.7. • Developed mesoporosity (V_{meso} = 0.29 cm^3 g^{-1}). • The number of strong acid sites decreased for about half of the initial value as well as the amount of acid sites at the mesopore walls that reached 0.14 mmol g^{-1} • The results of the methanol-to-hydrocarbon conversion show that dealumination considerably enhanced the catalyst life-time.	[49] (2010)

V_{micro} – micropore volume; V_{meso} – mesopore volume; MAS-NMR – magic angle spinning- nuclear magnetic resonance spectroscopy; N_{Al} – number of framework Al in the unit cell; v_1 – wavenumber of the T-O-T asymmetric stretching vibration; EFAL – extra-framework Al species.

Müller and co-workers [45] applied this dealumination procedure to BEA, FER, MOR and MFI structures proving that when protonic forms were used one of the reactions products is HCl, which will dealuminate the sample as well. On the other hand, the AlCl$_3$ formed is hydrolyzed during the washing step and, since the resulting aluminum hydroxide is scarcely soluble, extra-framework species will remain within the porosity.

The dealumination process using (NH$_4$)$_2$SiF$_6$ was proposed in 1984 by Skeels and Breck [50]. The method is easily executed but the control of the experimental conditions is of strictly importance. The influence of the pH of the reaction media was mentioned in the work where the method was proposed since when the protonic formed is used one of the products is HF, which will attack the zeolite structure, leading to some degree of amorphization. Garralón et al. [112] demonstrated the importance of controlling other experimental parameters, namely, temperature, time and the ratio *equivalents of SiF$_6^{2-}$/mass of zeolite.*

The method has been used in the dealumination of several structures, particularly FAU [22, 47, 50, 51], MOR [50], MFI [49, 100] and BEA [23, 48]. The treatments originate a preferential dealumination of the outer surface of crystallites and only in the case of Y zeolite a low mesoporosity development is reported [46].

2.4. Catalytic Applications

Concerning the application of dealuminated zeolites in catalysis, in addition to the various examples already briefly described, there is a special case that, due to its industrial importance in the fluid catalytic cracking (FCC) process, has to be mentioned in more detail. The structure is USHY, which, as it was already mentioned, is obtained through steaming. The dealumination treatment will, as previously discussed, decrease the number of acid sites but leads to the formation of very strong sites that will favor the catalytic activity for the cracking of middle distillates (150-340 °C) [113]. However this high activity leads to a very quick deactivation of the catalyst due to coke deposition implying its regeneration. In the industrial units the reactor and the regenerator are side-by-side to allow a continuous regeneration of the spent catalyst by combustion of the coke in air. This process is performed at temperatures around 680-760 °C, during 5 to 10 minutes, leading to the formation of carbon monoxide, carbon dioxide and water. To support these very severe conditions the catalyst has to present a high hydrothermal stability which is greatly improved as a consequence of the changes, namely unit cell shrinkage [25] induced by the dealumination process. This is the reason why the catalyst is designated as ultrastable.

3. DESILICATION

An alternative strategy to create mesoporosity in zeolites consists in the selective removal of Si atoms from the zeolite framework in alkaline medium, through a process called desilication. Although this technique has been known for many years, the first works used this method to modify the Si/Al ratio of the zeolites, without promoting significant changes on the acidity [114, 115], but no textural modifications were reported. Only in 2000 Ogura et al. [116] described the presence of uniform mesopores as consequence of the alkaline treatment performed over ZSM-5 zeolite, without destruction of microporous structure. Table 4 shows representative examples of hierarchical zeolites prepared through several strategies of desilication treatments: conventional heating using NaOH as desilicating agent or other inorganic/organic bases; combinatorial treatments base+acid treatment or steaming; and desilication treatments using microwave as heating source.

3.1. Conventional Treatments

Since the pioneer work of Ogura, systematic studies carried out by Groen et al. over ZSM-5 zeolite [117, 134-136] have shown that, by controlling the experimental parameters such as NaOH concentration, temperature and time it was possible to obtain hierarchical zeolites with intracrystalline mesoporosity, maintaining the original microporosity typical of the zeolite structure, without significant impact on the zeolite acidity. Following this methodology other zeolites structures were submitted to alkaline treatments such as BEA [119], Y [125] FER [121], MOR [122, 133, 137] TON [124, 138] or MCM-22 [55, 139], among others. Soon, it was found that, the conditions typically applied for suitable mesoporosity development by desilication are structure dependent. For instance, Beta and MCM-22 zeolites allow an easy Si extraction, probably because both structures are not very stable frameworks and present relatively large interconnected porous systems [55, 140]. On the other hand, zeolites like mordenite and ferrierite require more severe treatment conditions for the creation of mesoporosity. The application of more severe alkaline treatment leads to a partial dissolution of the outer surface of the crystals, that is, the development of intercrystalline mesoporosity, along with the formation of a wider intracrystalline mesoporosity [140].

The SEM images displayed in Figure 6 illustrate the damage promoted by the alkaline treatment made over FER zeolite [141]. The effect of several experimental parameters of desilication treatment, namely NaOH concentration, heating temperature and duration of the treatment were investigated over a commercial MOR zeolite by Paixão et al. [122]. In this study the authors showed that the increase in the severity of all the tested experimental parameters led to higher mesopore volumes, ranging from 0.05 $cm^3\ g^{-1}$ in the parent zeolite for 0.19 $cm^3\ g^{-1}$ for the optimized conditions of 0.2 M NaOH, 85^{0}C for 2 h, maintaining almost unchanged the microporosity of the samples. The severity of the alkaline treatment also damaged the structure of mordenite, especially in the presence of high NaOH concentration. In fact, the base concentration is the operational parameter with a more strong influence on the zeolite structure, since in the presence of 1 M NaOH solution an almost complete loss of crystallinity (more than 70%) was verified.

Table 4. Examples of recent studies where desilication treatments were applied to modify zeolite structures

Zeolite	Experimental conditions	Results	Reference
Conventional heating – NaOH as desilicating agent			
ZSM-5 (Si/Al = 20)	NaOH, 0.2 M at 80°C for 5 h.	<ul><li>Presence of uniform mesopores without destruction of the microporous structure.</li><li>The mesopores formed through the alkaline treatment have uniform pore sizes such as M41S materials.</li></ul>	[116] (2000)
ZSM-5 (Si/Al = 37)	NaOH, 0.2 M at 35 to 85 °C for 15 to 120 min.	<ul><li>Optimal condition - 0.2 M NaOH at 65°C during 30 min resulting in a significant increase of S_{meso} (from 40 to 225 m^2 g^{-1}) and a relatively small decrease in V_{micro} (25%).</li><li>The mesopore formation is a result of the preferential dissolution of Si from the zeolite framework, keeping the long-range ordering and the micropore size.</li></ul>	[117] (2004)
ZSM-5 (Si/Al = 30)	NaOH, 0.2 M at 65°C for 120 min performed over large ZSM-5 crystals.	<ul><li>Two orders of magnitude of gas transport improvement (neopentane).</li><li>The alkaline treatment of large ZSM-5 crystals originates an accessible interconnected network of intracrystalline mesopores, preserving the crystallinity which decreased the diffusion path, increasing the gas diffusion inside the pores.</li></ul>	[118] (2007)
Beta (Si/Al = 35)	NaOH, 0.2 M, heated between 25 and 65°C during 10 to 60 min.	<ul><li>Extensive silicon extraction at mild treatment conditions originated substantial mesoporosity development. Microporosity, acid properties and crystallinity are negatively affected.</li><li>The low framework stability of the aluminum in the zeolite led to a significant loss of acidity. Catalytic reaction of benzene alkylation with ethylene revealed a significant decrease in activity when compared with the parent zeolite.</li></ul>	[119] (2008)
ZSM-12 (Si/Al = 40)	NaOH, 0.05 - 1 M for 1 h at 85°C at atmospheric pressure. NaOH, 0.1 M under hydro-thermal conditions at 100°C.	<ul><li>ZSM-122 crystalline structure was not affected when changing the treatment from atmospheric pressure to hydrothermal conditions.</li><li>Treatment with 0.5M NaOH solution led to a significant increase in the liquid phase α-pinene conversion.</li></ul>	[120] (2009)
FER (Si/Al = 27)	NaOH, 0.1 to 1 M, heated at 60 to 90°C for 0.5 to 9 h.	<ul><li>FER requires harsher conditions to extract Si when compared to other commercial zeolites.</li><li>Under optimal conditions (0.5 M NaOH, T=80°C for 3 h) S_{meso} increased 3 - 4 times when compared to the parent zeolite, mostly preserving the native acidity and crystallinity.</li><li>In the catalytic pyrolysis of LDPE, desilicated FER changes the product distribution as a consequence of the improved accessibility of the polymer to the active sites.</li></ul>	[121] (2009)
MOR (Si/Al = 10)	NaOH, 0.2 – 1.0 M solution, heated between 50 – 100°C for 0.5 – 10 h.	<ul><li>A linear correlation between the number of framework Si per unit cell and the asymmetric stretching band wavenumber was observed</li><li>Optimal desilication conditions: NaOH, 0.2 M, at 85°C for 2 h.</li><li>Under the optimal desilication conditions, there is no significant modification of the acidic properties.</li></ul>	[122] (2010)

Table 4. (Continued)

Zeolite	Experimental conditions	Results	Reference
MCM-22 (Si/Al = 15)	NaOH, 0.05 to 0.5 M solution, heated at 50°C during 45 min.	• For treatments with NaOH solutions with concentration up to 0.2 M new micropores are formed that gradually grow in size. Above 0.3 M NaOH the interconnections between the two independent channel systems of MCM-22 occur and new porosities are developed. • ^{129}Xe NMR spectroscopy data show the formation of new porosities and interconnections that grow in size after treatment with high NaOH concentration. • The results of toluene disproportionation show that, for low NaOH concentration, the catalytic activity increases due to the formation of additional pores. At high NaOH concentration the catalytic activity decreases due to the partial destruction of the active sites.	[123] (2011)
ZSM-22 (Si/Al = 31)	NaOH, 0.2 M at 80°C for 30, 120 and 300 min. NaOH 0.05, 0.2, 0.4 and 0.8 M at 80 °C during 30 min.	• The alkaline treatment led to the development of mesopores with ~20 nm diameter. • In 1-butene isomerization reaction the desilication treatment had no effect on the catalyst stability but increased the initial activity, which can be correlated with V_{meso}, since the generated mesopores facilitate the access of 1-butene to the active sites.	[124] (2011)
Y (Si/Al = 2.4) USY (Si/Al = 2.6, 15 and 30)	NaOH, 0.10 to 5 M at 65°C for 30 min.	• The treated zeolites maintained the intrinsic microporosity, along with the development of mesoporosity (1.5-fold in total uptake of aromatic probe molecule, when compared with untreated zeolite) • The adsorption of toluene and catalytic properties in the alkylation of benzyl alcohol with toluene and the pyrolysis of LDPE of the hierarchical Y and USY zeolites were superior to the parent zeolites.	[125] (2012)
Conventional heating – Other desilicating agents			
ZSM-5 (Si/Al = 19.5) Y (Si/Al = 2.5)	Na$_2$CO$_3$, 0.8 M at 80 °C for 4 h. NaOH added to control the pH (0.1 - 0.5 M for Y and 0.005 – 0.02 M for ZSM-5).	• NaOH is added to the sodium carbonate solution for a more effective removal of Si by adjusting the initial pH of the suspension. • The higher the Si/Al ratio, the less basic is the solution needed for the treatment.	[126] (1995)
ZSM-5 (Si/Al = 37)	NaOH, KOH or LiOH 0.2M at 65 °C for 15 or 30 min.	• Application of KOH leads to a lower S_{meso} and the use of LiOH resulted in a even more lower S_{meso} when compared to NaOH treated samples. • The kinetics of silica dissolution follows the order: LiOH < NaOH < KOH.	[127] (2007)
ZSM-5 (Si/Al = 42)	NaOH, 0.2 M, TPAOH, 0.2 M or mixtures of both, keeping the OH$^-$ concentration at 0.2 M. Heating at 65 °C for 30 min.	• The presence of organic cations enhances S_{meso} without a severe penalization of V_{micro}. • The quaternary ammonium cations act like pore growth moderators, protecting the zeolite crystal during the desilication process.	[128] (2009)
Beta (Si/Al = 35)	TMAOH, 0.1 M, for 30 min to 6 h. Temperature changing from 65 to 100 °C, during 30 min to 6 h.	• The use of TMAOH promoted the development of mesoporosity in a very sensible zeolite, preserving the crystallinity. • The temperature increase from 50 to 100 °C accelerates the mesopore formation. • For long desilication times and high temperatures a slight decrease in V_{micro} was noticed. • Since the charge compensation TMA$^+$ ions in the zeolite framework decompose during calcinations, giving the protonic form, subsequent ion exchange procedure is avoided.	[129] (2009)

Zeolite	Experimental conditions	Results	Reference
Conventional heating - combinatorial treatments: desilication + acid treatment or steaming			
ZSM-5 (Si/Al = 15, 35 and 200)	Steaming at 600 and 800 °C, for 6 or 50 h under a flow of He-steam of 30 cm^3 min^{-1} and a water partial pressure of 300 mbar. Desilication at 65 °C for 30 min with NaOH, 0.2 M using a ratio of 10 cm^3 NaOH for 330 mg of zeolite.	• When alkaline treatment is performed after steaming extra-framework species created during steaming undergo realumination and subsequently inhibit Si extraction and mesopore formation. • ZSM-5 zeolites are hydrothermally stable up to at least 800 °C, with shape selective effects and improved intracrystalline transport properties.	[130] (2005)
ZSM-5 (Si/Al = 40)	Desilication with NaOH, 0.2 M, at 65 °C for 30 min. Acid treatment with HCl, 0.1 M at 70 °C for 6 h.	• The alkaline treated samples washed with HCl presented modifications on the surface acidity due to the removal of Lewis acid sites, in extra-framework positions, located at the external surface of the mesopores, previously generated by the alkaline treatment. • In m-xylene transformation, the samples submitted to HCl washing presented approximate two-fold increase selectivity in p-xylene and reduced deactivation rate.	[131] (2010)
ZSM-5 (Si/Al = 10-1000))	Alkaline treatment with NaOH, 0.1-1.8 M at 65 °C for 30 min. Acid washing with HCl, 0.02-0.1 M at 65 °C for 6 h	• The highest desilication efficiency was obtained in the Si/Al between 25 and 50. • For samples with Si/Al lower than 20 the acid treatment uncovers the micro-mesopore network enabling the restoration of the acidity. • The acid washing step performed after desilication enhanced the catalytic activity to over 5 times in toluene alkylation with benzyl alcohol reaction, when compared to the parent zeolite.	[54] (2011)
MCM-22 (Si/Al = 14)	Alkaline treatment with NaOH, 0.05 or 0.1 M at 50 °C for 45 min. Acid treatment with HCl, 0.1 M at 70 °C for 6 h.	Sample submitted to the more severe treatment • Combinatorial extraction of both Al and Si from the internal porosity of MCM-22 led to the interconnection of the two internal pore systems – transformation of a 2D to a 3D zeolite structure. • In m-xylene transformation reaction a significant increase in conversion was observed as consequence of easier and faster molecular traffic in the 3D structure.	[55] (2012)
Microwave heating			
ZSM-5 (Si/Al = 37 and 40)	Desilication treatments performed on a CEM Explorer single-mode microwave reactor, using NaOH, 0.2 M at 65 °C with power adjustments to maintain the temperature constant and duration of the treatment from 3 to 30 min.	• Faster development of the intracrystalline mesoporosity for microwave treated samples. The optimum duration of the desilication treatment changed from 30 min (conventional heating) to 3-5 min. (microwave heating). • The microwave heating promoted a more efficient transfer of the thermal heating to the zeolite suspension, resulting on the enhancement of the rate of silicon extraction.	[132] (2009)
MOR (Si/Al = 10)	Desilication treatments performed on a CEM Discover microwave reactor, using a NaOH, 0.2 M at 85 °C with power adjustments to maintain the temperature constant and duration of the treatment of 3, 15 and 30 min.	• The samples heated on microwave reactor presented identical amount of Si extraction in less time. Additionally, they present larger micropores formed by the transformation of the microporosity characteristic of the zeolite structure. • For m-xylene transformation reaction, the acid strength and density of catalytic active sites do not change upon irradiation if the exposure time is less than 15 min.	[133] (2011)

LDPE – low density polyethylene; NMR – nuclear magnetic resonance spectroscopy; TMAOH – tetramethylammonium hydroxide; TPAOH – tetrapropylammonium hydroxide; V$_{micro}$ – microporous volume; S$_{meso}$ – mesopore surface area; V$_{meso}$ – mesoporous volume.

Figure 6. SEM images of (A) parent FER zeolite and (B) sample treated with 0.5 M NaOH at 75 °C for 10 h.

The influence of NaOH concentration was also studied by Machado et al. [55] on a more sensitive structure, that is, MCM-22 zeolite. In this study 0.05 and 0.1 M NaOH solutions were used as desilicating agents, at 50°C during 45 min. For this zeolite the lowest NaOH solution was enough to promote the mesoporosity development. When 0.1 M solution was used no further mesoporosity was achieved but a decrease in microporosity was also observed. In this case Al extraction took place along with Si and consequently, some disaggregation and dissolution of the crystals, along with a more significant loss of crystallinity occurred, as can be observed on Figure 7.

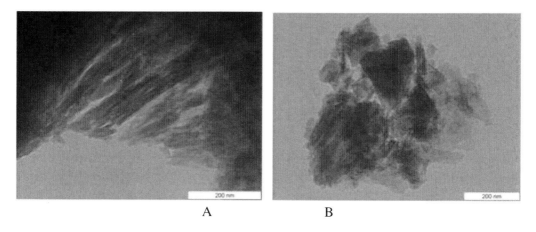

Figure 7. TEM images of MCM-22 samples after treatment with (A) 0.05 M and (B) 0.1 M NaOH solutions at 50°C during 45 min.

The two case studies reported above clearly show that the optimized desilication conditions differ significantly according to the type of zeolite structure. However, even for the same structure the alkaline treatment efficiency is also dependent of the Si/Al ratio. In the case of ZSM-5 zeolite, which is undoubtedly the most studied structure, especially by the Pérez-Ramírez group [54, 117, 134], the optimal Si/Al ratio was proven to be in the range 25-50. In this line, the the studies performed by Cizmek et al. [142] pointed out that the desilication rate constant is a linear function of the Si/Al ratio of the zeolite framework. Accordingly, the presence of a high concentration of Al in ZSM-5 (i.e., small Si/Al ratio) prevents the extraction of Si from the zeolite framework. For samples with Si/Al ratios above 50 no selective extraction occurs and thus, the mesopore formation is random and a broad pore size distribution around 10 nm is obtained [134].

Recently, Verboekend et al. [54] presented a systematic study regarding the desilication of ZSM-5 zeolite with Si/Al ratios ranging from 10 to 1000 using NaOH concentrations from 0.1 to 1.8 M, with additional acid leaching treatments in some cases. The authors interpreted the obtained results using descriptors such as "indexed hierarchy factor" and "desilication efficiency" that allowed the categorization of the studied materials.

The "indexed hierarchy factor" relates the relative mesoporosity S_{meso}/S_{BET} (ratio of the surface area of mesopore and the value obtained applying BET equation [143]), with the relative microporosity V_{micro}/V_{pore} (ratio between the micro and total pore volumes). This descriptor relates the mesoporosity development with the simultaneous reduction in microporosity. The authors found that for ZSM-5 structure, to maximize the mesoporosity development and mimimize the loss of microporosity, the ZSM-5 structure with Si/Al = 40 should be treated with 0.5 M NaOH. When samples with Si/Al ratios lower than 15 are treated with NaOH concentrations between 0.9 and 1.2 M a remarkable reduction in microporosity is verified.

The "desilication efficiency" surveys the degree of zeolite dissolution upon the increase of the external surface area, by relating the increase in mesopore surface area with the mass loss upon desilication [138]. Verboken et al. [54] applied this descriptor in a study with ZSM-5 zeolite and found that the most efficient alkaline treatment (largest external surface area with smallest relative weight loss) was obtained for Si/Al ratio in the range 25-50.

The use of descriptors can be applied to interpret the results of all types of hierarchical materials. A less refined version of the "hierarchy factor" was presented by Pérez-Ramírez and co-workers in 2009 [128]. In this study a vast number of literature data on hierarchical zeolites prepared by several methods (dealumination; desilication; seeding; delamination; composites; nanocrystals; templating) was gathered, aiming to quantitatively compare and correlate the textural properties of various hierarchical materials obtained by different synthetic or post-synthetic methodologies. Compared to all other methods, desilication proves to be the most versatile method, since it allows obtaining materials combining very different values of "hierarchical factor".

Although in the majority of the works concerning desilication of zeolites NaOH is the most used desilicating agent, some studies using different inorganic bases have also been reported, such as, sodium carbonate [126], lithium hydroxide and potassium hydroxide [127]. Recently, alkaline treatments in the presence of organic bases, like tetrapropylammonium hydroxide and tetrabuthylammonium hydroxide were reported [128, 129]. These organic bases are intrinsically less reactive and less selective towards Si removal when compared with strong inorganic bases, as NaOH. So, to attain the same mesoporosity characteristics, the alkaline treatments have to be carried out for longer periods of time, or heated at higher temperatures. However, the use of organic bases can be advantageous since it allows a more accurate control of the Si removal extent [128], which can have important implications for catalytic applications, as will be discussed ahead. On the other hand, by using "softer" desilicating agents it is possible to generate intracrystalline mesoporosity in more fragile zeolite structures without significant dissolution of the crystals and important crystallinity losses. Holm et al. [129] reported the efficient desilication of BEA, a very sensitive structure, using tetraprophylammonium hydroxide, obtaining a hierarchical micro-mesoporous structure with completely retained crystallinity. An additional advantage of this method is the, so called, "one pot" desilication and ion-exchange procedures. In fact, after the common desilication treatments using NaOH a subsequent ion-exchange step has to be carried out to obtain the zeolite in the protonic form. In this case, the charge compensation ion decomposes during calcination step, giving the protonic form, avoiding the ion exchange step.

As it was foreseen, the post-synthesis mesoporosity introduced by desilication treatments lead to a great improvement on the molecular diffusion inside the pore systems of the treated zeolites. These enhancements were attested by diffusion studies using probe molecules such as *n*-heptane, 1,3-dimethylcyclohexane, *n*-undecane in desilicated ZSM-22 [144], or diffusion and adsorption studies of cumene in mesoporous ZSM-5 [123]. From the results of these studies it could be concluded that the hierarchical systems increase the diffusion rates by several orders of magnitude, when compared with conventional zeolites, due to the improved accessibility as a consequence of desilication treatments.

One of the major advantages of desilication over dealumination treatments is the development of mesoporosity without significant impact on the zeolite acidity. Paixão et al. [122] characterized the acidity of desilicated MOR zeolite using the catalytic model reaction of *n*-heptane cracking and through the estimation of acid sites concentration chemically bonded to pyridine followed by infrared spectroscopy. The authors concluded that the acidity properties of desilicated materials remained practically unchanged when optimized desilication treatments were applied. However a decrease in acid site concentration can also be observed when the desilication conditions become too severe, especially in the case of more sensitive structures. An illustrative example is the case of MCM-22 when submitted to alkaline treatment with increasing NaOH concentration [55]. The use of 0.1 M NaOH solution lead to a notorious decrease of acid sites concentration, as it was detected by two distinctive techniques: pyridine adsorption followed by IR spectroscopy and the catalytic model reaction of *m*-xylene transformation. In line with these results ^{29}Si and ^{27}Al MAS-NMR spectra analysis showed that the removal of Si during the alkaline treatment was also accompanied by Al extraction from the zeolite framework.

3.2. Combined Treatments

To optimize the mesoporosity development and the acidity modification induced by the alkaline treatments, in various studies the combination of distinct methodologies has been explored.

Groen et al. [139] evaluated the impact of desilication with 0.2 M NaOH solution at 65 °C and dealumination by steaming at 600°C performed independently and sequentially over a series of ZSM-5 samples with Si/Al ratio between 17 and 137. The results obtained proved that to tailor acidity and mesoporosity the optimal approach is a consecutive desilication-dealumination procedure. This protocol allowed the preparation of a sample with a mesopore volume of 0.45 cm^3 g^{-1} with a pore size distribution centered at 10 nm. When the alkaline treatment was made over the steamed sample the extra-framework species created by steam treatment inhibit the removal of Si and the consequent development of mesoporosity.

Other possible combination - alkaline plus acid treatment - was recently applied by Machado et al. [55] to modify MCM-22 zeolite. Attending to the results already reported on alkaline treatment over MCM-22 [120, 139, 145] the authors prepared a sample with considerable mesoporosity and no significant loss of crystallinity, using 0.05 M NaOH solution at 50°C. The use of 0.1 M NaOH solution led to practically no gain in the mesopore volume but promoted an important decrease in microporosity, and crystallinity. The subsequent acid treatment had different effects on the two alkaline treated samples. No relevant effect on the textural properties was observed in the case of the solid previously treated with the less concentrated alkaline solution. On the contrary, the acid treatment performed over the more fragilized MCM-22 framework, obtained after treatment with 0.1 M NaOH, resulted in the dealumination of the structure, as demonstrated by the ^{27}Al MAS-NMR results. This further structural modification led to a significant decrease of the microporosity that was interpreted as the consequence of Al removal from the two internal pore systems characteristics of this structure, that become interconnected. Thus, the pristine 2D structure characteristic of the MCM-22 turned into a 3D porous system, what had a marked effect on the catalytic behavior on the *m*-xylene transformation. Higher conversion and selectivity for the desired *p*-xylene isomer were observed. A simplified scheme of the porosity evolution upon the various treatments performed is presented in Figure 8.

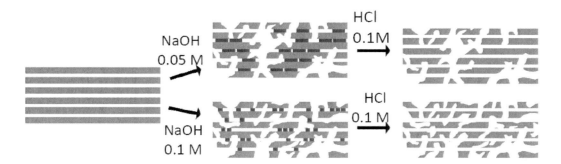

Figure 8. Schematic representation of the influence of the alkaline and acid treatments made over MCM-22 zeolite.

3.3. Microwave Assisted Treatments

As in so many other research areas, also in the case of zeolite modification through desilication treatments, the use of microwave radiation was already considered to accelerate the process. However the number of studies reporting microwave assisted desilication is scarce in comparison with the number of studies where microwave radiation is applied for synthesis as in the case of the synthesis of silicalite-1 reported by Motuzas et al. [146].

The first study reporting microwave assisted desilication was published in 2009 by Abelló and Pérez-Ramírez [132]. The authors found that the treatment of commercial ZSM-5 samples with NaOH after 3 minutes irradiation led to the creation of a mesopore volume identical to that obtained after 30 minutes under conventional heating. In any case the temperature used was 65 °C.

This methodology was also recently applied by Paixão et al. [133], and Machado et al. [55] to modify the structures of, respectively, MOR and MCM-22 zeolites. The results obtained with mordenite demonstrate that microwave radiation led to identical amount of Si extraction from the framework than that obtained with conventional heating, but after considerably shorter treatments. On the other hand, a different textural evolution was observed depending on the heat source used. As it can be seen in Figure 9 conventional heating essentially promotes the increase of the mesopore volume, whereas microwave radiation results in the enlargement of the micropres characteristic of the zeolite structure (values of ultramicropore volume in Figure 9) into supermicropres (values of supermicropore volume in Figure 9). To explain these findings the authors considered the differences in the characteristics of the two modes of heating. Thus, while in the conventional heating the temperature of the reaction media is not uniform, due to convection currents created during the heat transfer from the recipient walls to the solution, when microwave radiation is applied the temperature is more uniform and most important, the zeolite framework will absorb radiation directly promoting a more quick and efficient heating up.

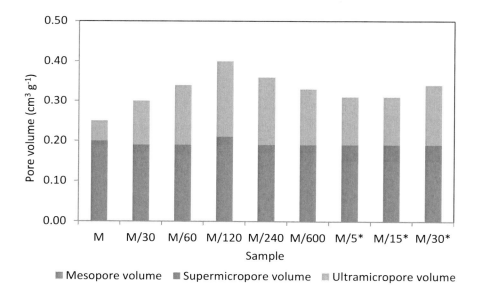

Figure 9. Volume of ultra and supermicropores and volume of mesopores of parent mordenite (M) and desilicated samples under conventional heating (M/duration of the treatment (minutes)) and microwave assisted (M*/duration of the treatment (minutes)). Adapted from ref. [133].

The use of microwave radiation to assist the alkaline treatment of MCM-22 was attempted by Machado et al. [55]. However, under the experimental conditions used (NaOH concentration: 0.1 M; temperature: 50 °C) the irradiation for 5 and 10 minutes led to almost amorphous materials, confirming the susceptibility of MCM-22 structure to basic solution reported in literature [120, 139].

3.4. Catalytic Applications

The first published works about desilication treatments were focused essentially on the optimization of the alkaline treatments, foreseeing high mesoporosity development, as well as the application on several characterization techniques [130, 134, 136]. However, soon the catalytic applications of these materials started to be explored. Table 5 resumes catalytic applications of desilicated zeolites. Some selected examples will be discussed in more detail in the following.

Table 5. Relevant publications concerning catalytic studies using desilicated zeolite structures

Structure	Reaction	Observation	Reference
MFI	Alkylation of toluene with benzyl alcohol	Increase the activity about five times	[54]
MFI	Isomerization of o-xylene	Increase two times the selectivity into p-xylene	[131]
MFI	Methanol to hydrocarbon (MTH) conversion	Increase of catalyst lifetime about three times	[49]
MFI	1-Hexene aromatization	Improve the catalytic stability	[147]
MFI	Oxidation of benzene to phenol	Lower deactivation	[148]
FAU	Liquid phase alkylation of toluene with benzyl alcohol	Improve the catalytic behavior	[125]
BEA	Pyrolysis of low density polyethylene	Enhanced catalytic activity	[149]
BEA	n-Hexane hydroisomerization	Milder reaction conditions; reduced extent of coking	[150]
FER	Polyethylene pyrolysis	Enhanced catalytic performance	[121]
IFR	Low-density polyethylene (LDPE) pyrolysis	Improved catalytic performance	[151]
MOR	Benzene alkylation	Improved catalytic performance; formation of coke suppressed	[137]
MOR	n-Hexane hydroisomerization	Increased selectivity into dibranched isomers	[152]
MTW	Isomerization of α-pinene	Improve the catalytic activity	[120]
MWW	Toluene disproportionation	Increased conversion	[139]
MWW	m-Xylene isomerization	Increased conversion; higher selectivity into p-xylene	[55]
TON	Oligomerization of propene	Increased activity and selectivity; higher resistance to deactivation	[153]

A significant amount of catalytic studies concerns ZSM-5 zeolite since this was the first zeolite structure submitted to intensive research focused on its modification through desilication treatments. The versatility of this zeolite and its wide application on refining and especially in petrochemistry, allowed the exploration of the catalytic behavior in several processes, such as alkylation[54], isomerization [131], and aromatization [147] reactions, among others with promising results.

On the other hand, zeolite Y and especially USHY, due to its high importance in fluid catalytic cracking (FCC), are among the most important zeolites. So, the search of possible improvements on the catalytic behavior of this zeolite structure is highly relevant. Verboekend et al. [125] used a strategic combination of post-synthesis treatments – dealumination and desilication – to design a broad family of hierarchical Y and USHY zeolites with preserved crystallinity and microporosity. The prepared samples were catalytically tested in the liquid phase alkylation of toluene with benzyl alcohol and showed superior catalytic behavior when compared with the untreated materials. These results created some expectations about future applications of desilicated FAU structure on refining process, especially on hydrocracking reactions.

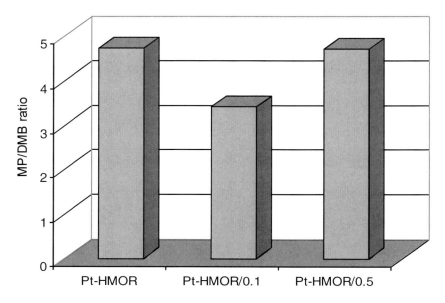

Figure 10. Ratio between the sum of methylpentanes and dimethylbutanes (MP/DM) for the isomerization of n-hexane. Reaction conditions: T=250 °C, $H_2/n-C_6$=9; WHSV = 6.6 h^{-1}, data presented at similar conversion of 15% (mol%). Adapted from [152].

Pt/MOR zeolite is the industrial catalyst for short chain n-alkane hydroisomerization, aiming to transform linear into branched paraffins, improving the octane number of gasoline. Monteiro et al. [152] studied the influence of the alkaline treatment, with increased NaOH concentration, on a commercial mordenite sample. The samples were characterized focusing on the implications of alkaline treatment on acidity, mesoporosity development and metal dispersion. The catalytic behavior of the bifunctional catalysts was studied in n-hexane hydroisomerization reaction. The main reaction products obtained in this reaction were the monobranched isomers, methylpentanes (MP), and the dibranched isomers, dimethylbutanes (DMB), which are the most valuable ones due to higher octane number. The authors observed that under very mild alkaline conditions (0.1 M NaOH) the discrete modification promoted on the zeolite porosity improved the diffusion and the access to the acid sites, resulting in a higher selectivity into dimethylbutanes. Figure 10 shows the ratio between the sum of the isomers methylpentanes and dimethylbutanes (MP/DMB), being the decrease of this ratio notorious for the sample treated with 0.1 M NaOH, Pt-HMOR/0.1.

When a more concentrated base solution was used, although a higher mesoporous volume was reached, the Pt particles tend to agglomerate at the intercrystalline mesoporosity, forming large metal particles. TEM images of Figure 11 clearly show the presence of large metal particles at the intercrystalline mesoporosity of Pt-MOR/0.5 in opposition to Pt-MOR/0.1

where the metal particles are much smaller and seem to be confined to the intracrystalline mesopores.

The results obtained in this study demonstrate that the presence of high mesopore volumes does not necessarily lead to an improved catalytic behavior, thus the desilication conditions need to be tailored for each envisaged application.

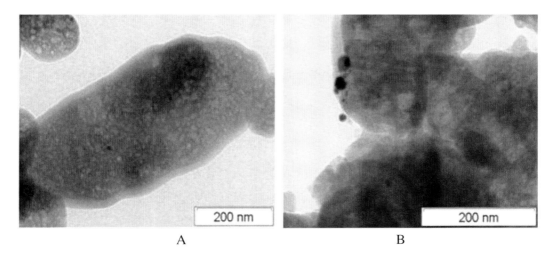

Figure 11. TEM images of Pt dispersed on alkaline treated (A) Pt-MOR/0.1 and (B) Pt-MOR/0.5 samples.

MCM-22 zeolite is an interesting zeolite structure with two internal pore systems, one of them with large supercages and high external surface area due to the presence of hemisupercages. However the access to the inner supercages is restricted due to the narrow accesses. Machado et al. [55] applied alkaline treatments and alkaline followed by acid treatments to MCM-22. The transformation of m-xylene over this structure showed very interesting results, revealing that the combination of alkaline with acid treatments can have a beneficial effect on the catalytic behavior, leading to an increased catalytic conversion as a consequence of easier and faster molecular traffic into a 3D structure created by the combination of alkaline plus acid treatments.

The application of desilicated zeolites for catalytic purposes comprises also the use of hierarchical porous materials as catalysts supports. This is a new and very promising field since it envisages the immobilization of homogeneous catalysts at the mesopores of solid matrices, with economic and environmental advantages. A few examples can be referred, such as the use of desilicated ZSM-5 as support for lipase enzyme [154] or the use of desilicated MOR zeolite as support for organometallic catalysts [155, 156]. In both cases the catalysts showed improved catalytic activity and selectivity, when compared to the homogeneous catalyst.

4. TEMPLATED ZEOLITE STRUCTURES

During the past decade the concept of introducing a supplementary pore system (meso/macropores) in addition to the intrinsic zeolite microporosity during the zeolite synthesis, without any further treatment, has captivated the attention of a large number of research groups. In fact, the number of publications about this subject, as well as the corresponding citations has been growing quite significantly as can be visualized in Figure 12.

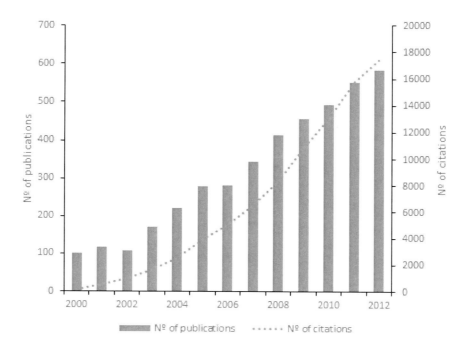

Figure 12. Rising interest on templated zeolite structures from 2000 to 2012. The graph shows the approximate annual number of publications and citations. Source: ISI Web of Knowledge 25th October 2013. Search terms "zeolite" AND solid templating" OR "soft templating" OR "supramolecular templating."

The use of templates to produce hierarchical zeolites is quite appealing strategy since it allows to modulate the supplementary pore system with customized pore dimensions and shapes. According to the nature and size of the template, bi-modal (micro-mesoporous) or tri-modal (micro-meso-macroporous) hierarchical zeolites can be produced, opening a wide set of applications for these engineered materials for catalysis and separations processes.

Figure 13 resumes the main strategies to produce hierarchical porous zeolites, the most common templates used, as well as the type of porosity obtained.

The present section covers the recent developments in templating methods to fabricate hierarchical porous zeolites, reviewing the synthesis strategies and the templates used, along with the advantages and limitations of each method. Literature examples related to catalytic applications of templated hierarchical zeolites are also presented.

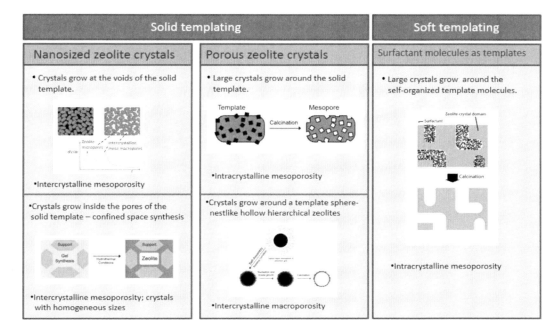

Figure 13. An overview of the various synthesis routes to generate hierarchical zeolites.

4.1. Solid Templating Strategies

The first published works concerning the use of solid templates to obtain hierarchical structured zeolites were reported by Madsen and Jacobsen [157] in 1999, and Jacobsen et al. [158] in 2000. In these pioneer papers the authors described two main strategies that use carbon black as template to develop mesoporosity in zeolites: nanosized zeolite crystals and mesoporous zeolite crystals.

In the first study published in 1999 [157], the fabrication of nanosized ZSM-5 crystals was obtained by incipient wetness impregnation of carbon black BP700 or BP2000, from Cabot Corp., with the zeolite synthesis gel. After aging, the paste was transferred to stainless steel autoclaves containing a few drops of water to produce saturated steam and thus, force the gel to fill the voids of the carbon blacks, where the crystallization took place. The same methodology of confined space synthesis, or nanocasting, can be extended to alternative templates.

An interesting variation from carbon blacks is the use of porous materials with pores large enough to promote the crystals' growth inside the confined space of the pores, for this reason this approach is called confined space synthesis. For example, Kim et al. reported the use of colloid-imprinted carbons (CIC) as matrix to produce nanosized ZSM-5 zeolite with regular dimensions that are exposed upon removal of the matrix by combustion [159]. The uniform size of the produced zeolite crystals can be customized according with the pore dimensions of the matrix, which is a clear advantage of this methodology, when compared with the use of common carbon blacks. However, the high cost involved in the production of these sacrificial matrixes has also to be taken into consideration.

An alternative strategy to produce nanosized zeolite crystals was proposed by Jacobsen and co-workers in 2000 [158], describing the synthesis of mesoporous zeolite crystals. In this study BP2000 carbon black was used as a solid template but in this case, an excess of concentrated ZSM-5 zeolite gel was added. This procedure allows the zeolite to grow around the carbon particles, thus obtaining large zeolite single crystals that are able to encapsulate the solid template. After the aging period, the removal of the carbon matrix by combustion leads to the isolation of large zeolite crystals with additional porosity present in each individual zeolite crystal. Thus, the hierarchical zeolite crystals comprise the typical micropores inherent to the zeolite structure along with an additional intracrystalline mesopore system. Finally, another pore system will exist as a consequence of the packing of zeolite crystals, determined by the size and shape of the crystals [160]. Using the same method other zeolite structures such as ZSM-12 [161], silicalite-2, ZSM-11 [162] and TS-1 [163] were successfully synthesized.

As shown previously, the production of nanosized zeolites or mesoporous zeolite crystals can be achieved by using the same templates, leading to the same zeolite structure but with distinct textural characteristics, depending on the experimental conditions used on the synthesis that led to different nucleation and crystal growth rates. When high nucleation rates occurred the formation of nanosized zeolites takes place. On the other hand, a high growth rate favors the synthesis of zeolite single crystals [160].

The use of carbon black materials as templates usually lead to randomly oriented mesopores, exhibiting a high tortuosity and broad pore size distribution [164], or even cavern-like mesoporosity [165]. A better control over the pore diameters and spatial arrangements can be achieved using carbon nanotubes or nanofibers as solid templates. These materials allow the development of straight mesoporous channels, connecting the inner pore structure to the external surface, increasing the mass transfer and accessibility to the active sites [166, 167]. Ordered mesostructured carbons, with pore sizes of 2-10 nm (CMK-n) were considered promising templates for the synthesis of mesoporous zeolites, reproducing the ordered mesostructured carbon matrix. ZSM-5 zeolite was successfully synthesized using CMK-1 (cubic structure) [168-170] and CMK-3 (2-D hexagonal structure) [170, 171], although the replication was not fully faithful. This result was due to the occurrence of a fragmentation phenomena of the porous carbon during the crystallization process leading to the formation of disordered mesopores [165]. Carbon aerogels (CAs) are more robust materials than CMKs due to the presence of larger mesopores and thicker walls. When using CAs as templates the first step is the introduction of the zeolite gel into the mesopores of CAs. Upon crystallization, zeolite crystals grow within the three dimensional mesopore system. Since CAs usually present monolithic morphologies, mesoporous zeolite monoliths of type ZSM-5[172] and Y [173] were obtained after the complete removal of the carbon matrix by combustion.

An alternative strategy to solid templating using carbons is the so-called *in situ* carbon templating that consists in the decomposition of a carbohydrate (example: a sugar), inside the pores of the silica source, during the crystallization period. This strategy was successfully applied in the synthesis of ZSM-5 [174, 175] and has the advantage of allowing a careful control of the mesopore volume by adjusting the amount of carbon precursor added during the synthesis. However, the poor control of pore size and geometry attained with this method is a disadvantage that must be taken into consideration.

The application of alternative solid templates soon arose to compete with carbon materials. Polymers are among the most used templates that have been used to produce zeolites with intracrystalline mesoporosity. For instance, ZSM-5 zeolites with intracrystalline mesoporosity were successfully synthesized using silane-functionalized polymer as template [176]. Other hierarchical structures were produced following the same method, such as BEA and MEL, synthesized in the presence of polyvinyl butyral gel as the mesopore directing agent [177]. Resorcinol-formaldehyde-based organic aerogels were also used as templates to produce zeolite A with three-dimensional mesopores [178] or hierarchically structured monolithic silicalite-1 [179]. Nanosized $CaCO_3$ is an example of an inorganic compound that was used as hard template for the creation of intracrystalline porosity within silicalite-1 crystals [180].

Another interesting approach for the preparation of hierarchical zeolites using solid template strategies is the use of biological templates that offers a great diversity of shapes, with the advantage of being abundant and relatively inexpensive. So far, only a few examples have been reported in the literature. For instance, organized bacterial superstructures were explored as 3D templates for the production of zeolite fibers with hierarchical structure [181]. Wood cells [182] and starch [183] were also applied as templates to produce hierarchical zeolites tissues. However, one of the most exotic templates that was reported concerns the use of *Equisetu arvense* - a silica-rich plant - to produce hierarchical MFI and BEA structures [184, 185].

Although a significant number of papers concerning hierarchical porous zeolites are centered in the synthesis of micro-mesoporous systems, an increasing number of works is also focusing interest on micro-macropore or even tri-modal systems micro-meso-macroporous hierarchical zeolites. These materials with very large pores or cavities have revealed some curious behavior as catalysts, since it is believed that the presence of macropores could hinder the coke formation, avoiding the severe conditions during the catalyst regeneration [165]. Another application for the large pore materials is their use as supports for the immobilization of catalysts or catalyst precursors on a solid support, combining the advantages of homogeneous and heterogeneous catalysis, with easy separation and recycling [186, 187].

The use of macroscale hard templates to produce hollow zeolite spheres is the main strategy to create a macroporous structure in zeolitic materials. Polystyrene (PS) beads are the most common and versatile solids used as matrixes. With these materials the production of hollow zeolite spheres of LTA, FAU, BEA and MFI structures was reported [188, 189]. The versatility of PS spheres comes from the possibility to easily modify its surface by depositing several layers of polyelectrolytes, that will charge the surface and thus enhance the adhesion strength between the building particles through electrostatic interactions [190].

Carbon is also an important hard template to generate macroporous zeolites in a similar way as described for PS spheres. Chu and co-workers [191] successfully synthesized nestlike hollow hierarchical MCM-22 microspheres using an expedite one-pot hydrothermal method involving self-assembly synthesis with carbon black microspheres - with a diameter of 4-8 µm - as hard templates. After high temperature combustion of the carbon-zeolite composite, a hollow macroporous core is obtained, surrounded by microporous MCM-22 zeolite. Additional post-synthesis treatments can be performed on these micro-macroporous materials like desilication, to tune-in the porosity, allowing to obtain combinations of micro-meso-macroporous hierarchical zeolite materials [192].

The use of solid templates is an appealing strategy to create hierarchical zeolites due to its versatility to tune the size and shape of the additional pore system by choosing the proper hard template. On the other hand, as the matrix is chemically inert, it can be applied to all zeolites independently of their chemical composition [193], and therefore also to zeotype materials that do not contain silicon or aluminum as, for example, SAPO type materials [194]. The chemical composition of the zeolite or zeotypes can be properly optimized in the synthesis gel composition and thus, independent control of acidity from the mesopore characteristics is attained.

4.2. Soft Templating Strategies

Soft templating is a synthesis strategy to produce hierarchical zeolites that uses macromolecules as templates. The name "soft" comes from the fact that the template used is actually another molecule that is added to the synthesis gel.

A significant number of studies concerning the application of soft templates to create mesoporosity in zeolites have emerged over the last ten years and the main synthesis procedures and templates used were the object of recent reviews [165, 195]. Some of the most relevant contributions on soft or supramolecular templating strategies will be mentioned briefly on this section.

One of the most important class of macromolecules applied as templates are surfactants, which are macromolecules with the particularity of possessing an alkylic chain with a hydrophobic tail and a hydrophilic head. When these molecules are added to the zeolite gel they tend to organize themselves to form micelles. Thus, the choice of the adequate surfactant to be used as template is crucial to create the desired mesoporous structure, since characteristics like alkylic chain length, morphology, stability and type of interactions with the zeolite gel have to be carefully considered.

Hydrophilic cationic polymers, such as polydiallyldimethylammonium chloride (PDADMAC), can strongly interact with silica-based materials due to their high positive charge density and high stability under alkaline conditions at high temperature. The variety of available low cost cationic polymers promotes their application as templates in a wide range of commercially important hierarchical zeolites at industrial scale. Several examples of their use in the production of common zeolites, such as BEA, MFI and FAU structures, where intracrystalline mesopores ranging from 5-40 nm were reported [196, 197].

The synthesis of mesoporous siliceous materials inspired the synthesis of zeolites using the same type of templates such as cetyltrimethylammonium bromide (CTAB), which is a conventional surfactant used in the synthesis of MCM-41. Unfortunately, the weak interaction between this type of surfactant and the synthesis gel led to a phase separation between zeolite and mesoporous phases [198].

Organosilanes are another type of soft templates used to prepare hierarchical zeolites since these macromolecules easily interact with silica-based species. Amphiphylic organosilanes of type $[CH_3O)_3SiC_3H_6N(CH_3)_2C_nH_{2n+1}]Cl$, built from a long chain hydrophobic group and silanols, were used to fabricate MFI zeolite structure with 2-20 nm mesopores [199]. By using this template the mesoporosity could be well adjusted by tuning the molecular structure of the organosilane template along with the adjustment of hydrothermal synthesis conditions. This can be attained since the presence of silanol groups on the surfactant molecule favors a strong interaction between the template and the silica-based species on the zeolite synthesis gel. The same methodology was successfully applied to other aluminosilicate and aluminophosphate materials [200, 201]. These few examples pretend to illustrate that the synthesis of hierarchical zeolites using soft templates is only in its first steps and is still a big challenge. The key factor to produce hierarchical zeolites with this method is a good interaction between the organic template and the synthesis gel of the zeolites, which will condition the phase separation of micro and mesoporous materials or as envisaged, the formation of intracrystalline mesoporosity at the zeolite crystals.

Recently, Garcia-Martínez et al. [202] reported the generation of controlled mesoporosity into FAU structure through a surfactant templating approach, using CTAB in basic medium (NH_4OH, NaOH, etc.). Contrary to the common application of surfactant templates during the zeolite synthesis, in this case the CTAB is used during a post-synthesis treatment. The authors used a commercial Y previously submitted to an acid washing with citric acid, which was then put in contact with the CTAB template in basic medium to introduce mesoporosity. This procedure lead to a significant development of intracrystalline mesoporosity, keeping the original zeolite acidity.

4.3. Catalytic Applications

The catalytic behavior of templated zeolite crystals has been tested in several reactions with promising results. The improved catalytic activity of these materials has been attributed to the enhanced accessibility to the active sites and fast diffusion of reactants and reaction intermediates. Representative examples of the catalytic application of templated zeolites are displayed in Table 6. Some selected cases will be presented in this section.

Among the few studies carried out with nanozeolites prepared by confined space synthesis, Derouane et al. [215] reported the catalytic behavior of Beta zeolite structure prepared within the pores of a carbon black matrix in the Friedel-Crafts acetylation of anisole by acetic anhydride. The turnover frequency (TOF) of nanosized Beta increased when compared to conventional one which was attributed to enhanced access and better use of the intracrystalline porosity by the reactants, as well as, an easier desorption of the products, due to the reduced dimensions of the nanosized catalysts.

Nanosized silicoaluminophosfates (SAPO-11), a zeotype catalyst commonly used in the hydroisomerization of long chain *n*-alkanes aiming to obtain a low degree of branching, exhibit a superior performance for *n*-octane isomerization [218]. These promising results open new possibilities for the application of nanosized materials in reactions that typically take place at the pore mouth of the catalysts.

Christensen et al. [203] studied the alkylation of benzene by ethylene over mesoporous ZSM-5 synthesized with nanoparticles of carbon. The results of the catalytic tests showed improved activity and selectivity when compared to conventional ZSM-5. More recently, Musilova et al. [205] tested carbon templated ZSM-5 in several alkylation reactions such as toluene disproportionation, toluene alkylation with isopropyl alcohol and *p*-xylene alkylation with isopropyl alcohol to understand the role of mesopores developed during carbon templating synthesis. The authors reported an increase in toluene and *p*-xylene conversion with increasing mesoporous volume of ZSM-5. The selectivity into xylenes (all isomers), and isopropyltoluenes increased due to the shorter contact time spent in the mesopores of ZSM-5. However, the decrease in *p*-xylene selectivity indicates some loss of product shape selectivity effect, typical of this catalyst.

Table 6. Relevant examples of catalytic studies using templated zeolite structures as catalysts

Structure	Template	Reaction type	Reference
MFI	Carbon	Alkylation	[203]
MFI	Organosilane	Alkylation	[204]
MFI	Carbon	Alkylation	[205]
MFI	Polystyrene colloidal spheres	Alkylation	[206]
MFI	Organosilane	MTH conversion	[207]
MFI	Mesoporous silica	MTH conversion	[49]
MFI	Organosilane	Cracking	[208]
MFI	Carbon	Cracking	[209]
MFI	Macroporous silica gel	Cracking	[210]
MFI	Carbon	Cracking	[211]
MFI	Starch	MTH conversion	[212]
MFI	Organosilane	Cracking	[213]
BEA	Cationic polymers	Alkylation	[214]
BEA	Confined space synthesis	Alkylation	[215]
BEA	CTAB	Hydroisomerization	[177]
TS-1	Seed	Epoxidation	[216]
MTW	Carbon	Hydroisomerization	[217]
FAU	CTAB	FCC	[202]

Cracking reactions for processing larger molecules have also been quoted as good examples for the application of templated zeolite crystals, especially when soft templating method is used to generate intracrystalline mesoporosity. Some examples were reported in the literature. For instance, Park et al. [213] demonstrated the benefit of small intracrystalline mesopores for the transformation of vacuum gas-oil using mesoporous ZSM-5 zeolite templated with organosilane. It was found an improved product selectivity towards gasoline, LPGs and diesel, simultaneously, with a lower amount of coke, when compared with traditional ZSM-5. Garcia-Martinez et al. [219] demonstrated that controlled mesoporosity introduced into zeolite Y crystals, through a surfactant templating method, showed superior hydrothermal stability and improved selectivity into gasoline and light cycle oil (LCO) since the developed mesopores allowed larger molecules present in the feedstock to access the active sites within the zeolite crystals.

Hierarchical templated zeolites are also showing potentialities as catalysts supports in bifunctional catalysis. An illustrative example is given with nestlike hollow hierarchical MCM-22 microspheres impregnated with 6% Mo [191]. This bifunctional catalysts were tested in methane dehydroaromatization reaction showing an exceptional catalytic performance. This performance was attributed to faster diffusion of reactants and improved metal dispersion. Reduced catalyst deactivation was also found which contributed to improve the catalyst lifetime.

5. OTHER METHODOLOGIES

The production of hierarchical structures is not limited to templating strategies. Other methods are currently under intense research, such as mixtures of solids with different porosity, that is, composite materials. An alternative approach deals with the reproduction of traditional protocols but the synthesis takes place in the presence of microwave radiation. In this section the current tendencies on the synthesis methods and materials that are being produced, as well as examples of the applications under exploration, will be reviewed.

5.1. Composite Materials

Composite materials are characterized by the mixture between two materials with different pore sizes: the zeolite with its characteristic microporosity and another porous material that acts like a support. Thus, in opposition to mesoporous zeolite crystals and nanosized zeolite crystals, described above, composite materials are not purely zeolitic materials. Accordingly, the type of porosity and pore sizes depends strongly on the type of solids involved: the zeolite crystals contribute to the intracrystalline microporosity; the support contributes with intracrystalline mesopores or macropores, depending on the type of support used; finally, the relative sizes of the two components are a key parameter to the intercrystalline mesoporosity, resulting from the packing of the two types of materials [160].

Several examples can be found in the literature that illustrate the beneficial effect of composite materials on catalysis. Parton et al. [220] referred the occurrence of a synergistic effect when two materials with different pore structures contact with each other, leading to an improved catalytic behavior. According to Van Mao [221] the beneficial effects of the composite micro-mesopore systems, when compared to zeolites alone, is attributed to the formation of a so called micropore-mesopore continuum within the composite material, which decreases the resistance to the molecular diffusion. This effect was tested by preparing a mechanical mixture of ZSM-5 zeolite and several mesoporous supports, such as, mesoporous silica, silica-aerogel or alumina. The particles of microporous zeolite were embedded in the larger particles of mesoporous material using the technique of extrusion with a binder. The authors found higher conversion and product selectivities in the aromatization of *n*-butane and the upgrading of the propane steam-cracking products when compared to the same reaction occurring on zeolites. Kinger et al. [222] reported the catalytic behavior of composites made of Pt-containing HMCM-22 or HBeta and mesoporous MCM-41 in hydroisomerization and hydrocracking of *n*-heptane. The composite catalysts showed improvements both on the activity and selectivity when compared to solely Pt-zeolites, especially in the case of HMCM-22, which has a pore system with smaller pore openings when compared with HBeta.

The mesoporous component of the composite not only contributes with an additional pore system but also can act as a support for metals in the case of bifunctional catalysts. In this case, possible diffusional constraints caused by the presence of the metal particles inside the zeolite pore systems are supressed, leading to an increase in the catalytic activity. According to this idea, composite catalysts made of ZSM-5 mixed with an amorphous silica-alumina doped with molybdenium and cerium oxides were tested in *n*-hexane cracking and a very significant increase on catalytic activity was observed [223]. The hydroisomerization of

n-hexane was carried out over composite materials based on HBeta zeolite mechanically mixed with mesoporous materials loaded with Pt, such as MCM-41, SBA-15 and a porous clay heterostructure (PCH) [14]. The results obtained from the three composites studied showed that only Pt/PCH+HBea showed increased selectivity in dibranched isomers which are the most valuable products. This effect was attributed to an improved molecular diffusion on bidimensional PCH material, when compared to monodimensional SBA-15 and MCM-41 structures.

5.2. Microwave Synthesis of Nanoporous Materials

Microwave is a form of electromagnetic energy that falls at the lower frequency end of the electromagnetic spectrum and is defined from 300 to about 300 000 MHz frequency range. Within this region of electromagnetic energy, only molecular rotation is affected, not the molecular structure, thus, the effect of microwave absorption is purely kinetic. Microwave heating is commonly used in slow reactions where high activation energies are required to perform transformations. With the high molecular energy generated by the transfer of microwave energy, reactions that required many hours, or even days, have been accomplished in a few minutes, making these processes environmentally friendly since they require less energy than conventional ones [224, 225]. As the synthesis time is decreased the crystal growth period is also shortened and small crystal size materials are obtained, with intercrystalline mesoporosity, resultant from the crystal packing. Accordingly, one of the areas where microwave technology had significant developments is the synthesis of nanoporous materials, with recognized advantages, such as: i) reduction of the time needed for the synthesis, by over an order of magnitude; ii) uniform particles in terms of dimensions and compositions; iii) possibility to obtain products with more diverse compositions.

The first patent regarding the microwave synthesis of zeolites was registered in 1988, attributed to Mobil, for the synthesis of NaA and ZSM-5 [226]. It was claimed that small zeolite crystals with high rates of reproducibility were obtained. Since then, the synthesis of several zeolite structures has been reported. An extensive selection of materials synthesized using microwave radiation was reviewed by Tompsett et al. [224], covering several types of materials such as zeolites, aluminophosphates, silicioaluminophosphates and mesoporous silica-based materials. In the following some representative examples are presented.

A common conclusion for all studies regarding the synthesis under microwave radiation as heat source is that microwave synthesis takes considerably less time than conventional hydrothermal synthesis. For instance, Arafat et al. [227] obtained uniformly sized FAU zeolite structure upon microwave irradiation during 10 min, whereas the conventional heating requires 10-15 h, depending on the Si/Al ratio of the zeolite. Park et al. [228] applied microwave radiation on the synthesis of small and uniform zeolite crystals of ZSM-5 and Beta, and mesoporous materials like MCM-41 and SBA-15. The authors concluded that, besides decreasing the synthesis time, microwave technique provides an effective way to control particle size distribution and the morphology of the crystals.

Microwave radiation was also found useful for the synthesis of metal loaded nanoporous materials. For example, zirconium and titanium supported SBA-15 were successfully synthesized using microwave radiation, leading to the preparation of promising materials for selective oxidation catalysts involving large molecules [229, 230].

Aluminophosphates (AlPOs) and more recently, SAPOs have been efficiently synthesized using microwave radiation as heat source. For instance, AlPO4-5 was successfully synthesized under microwave irradiation after only 60 s of heating; no amorphous phase or by-products were detected [231]. Jhung et al. [232] reported the synthesis of SAPO-5 and SAPO-34 using microwave radiation and found that the structure transformation of SAPO-5 into SAPO-34, that occurred during the synthesis, can be controlled with the heating rate of microwave radiation.

As a final remark, it has to be taken into consideration that distinct results have been presented by different research groups, concerning the same material. The synthesis of FAU zeolite structure is an illustrative example of this situation: according to Arafat et al. [227] the synthesis took 40 min, but Katsuki and co-workers [233] claimed that 1-3 h is needed, and Slangen et al. [234] reported 4 h, with all these syntheses being carried out at the same temperature. These diverse results led to the conclusion that the operational parameters such as stirring speed, temperature ramp rate, radiation frequency, among others, are crucial parameters to perform reproducible synthesis and must be taken into consideration. Some guidelines for the optimization of these parameters were reviewed and summarized by Tompsett et al. [224].

6. COMMERCIAL APPLICATIONS OF HIERARCHICAL ZEOLITES

The commercial applications for hierarchical zeolites are well established in the case of dealumination treatments with great impact in the refining industry, particularly for the FCC catalyst, USHY, obtained through steaming dealumination of Y zeolite [113]. One of the oldest patent reporting the manufacture process of ultrastable Y was claimed by Maher and McDaniel in 1968 [235] and since then a very large number of patents claiming the preparation of USHY were granted. Nevertheless, several patents concerning the dealumination of other zeolite structures, such as FER [236], BEA [237] or MOR [238] have also been claimed. Additionally, other applications besides FCC process are patented and some illustrative examples are listed on Table 7.

The production of hierarchical zeolites prepared by desilication or templating methods transcended already the academic interest and some patents were claimed, some of them assigned by well-known enterprises, envisaging commercial applications. Table 7 summarizes some relevant examples.

Concerning the scale-up production and commercialization of hierarchical zeolites prepared by desilication and templating strategies, the first attempt for large-scale production of desilicated MFI structure was reported in 2007 by Pérez-Ramírez and co-workers [127]. The researchers demonstrted that the synthesis and modification of this zeolite prepared in a 1.5 m^3 reactor can be reliable extrapolated to the plant scale, opening new horizons for the applications of desilicated materials towards industrial applications. On the other hand, Garcia-Martinez and co-workers [257, 259, 260], founded Rive Technology Inc. in 2006 that developed and scaled-up a methodology to produce hierarchical or mesostructured zeolites through a surfactant template process. After the initial studies at laboratory scale, mesostructured Y zeolite production was scaled-up and in 2011 the modified Y zeolite was

Table 7. Examples of patents for fabrication and/or application of hierarchical zeolites

Name	Patent number	Assignee	Publication date	Reference
Dealumination				
Selective surface dealumination of zeolites and use of the surface-dealuminated zeolites	CA 2094947	Mobil Oil Corp.	Feb. 2003	[239]
Method for preparing modified Beta zeolite catalyst used for alkylation reaction of isobutane/butylene	CN 1124891	Fudan University	Oct. 2003	[240]
Using zeolite selectively deluminated to remove nonframework aluminum; making alkyl naphthalenes for example	US 6747182	ExxonMobil Chemical Patents Inc.	Jun. 2004	[241]
Extremely low acidity ultrastable Y zeolite catalyst composition and process	US 6860986	Chevron USA Inc.	Mar. 2005	[242]
Method of preparing zeolite single crystals with straight mesopores	EP 1284237	Haldor Topsoe A/S	May 2006	[243]
Hydrocracking catalyst containing Beta and Y zeolites, and process for its use to make jet fuel or distillate	US 7585405	UOP Llc	Sep. 2009	[244]
Hydrocarbon conversion processes using a catalyst comprising a UZM-8HS composition	US 7638667	UOP Llc	Dec. 2009	[245]
High temperature catalyst and process for selective catalytic reduction of NOx in exhaust gases of fossil fuel combustion	US 7695703	Siemens Energy, Inc.	Apr. 2010	[246]
Hydrocracking catalyst containing Beta and y zeolites, and process for its use to produce naphtha	CA 2627337	UOP Llc	Jun. 2012	[247]
Catalysts for improved cumene production and method of making and using same	US 8524966	UOP Llc	Sep. 2013	[248]
Desilication				
Mesoporous zeolitic material with microporous crystalline mesopore walls	US 6669924	Université Laval	Dec. 2003	[249]
Material with hierarchized porosity, comprising silicium	CN 100522810	IFP	Aug. 2009	[250]
Fabrication of hierarchical zeolites	US 7824657	Haldor Topsoe A/S	Nov. 2010	[251]
Crystallised material with hierarchized porosity and containing silicon	EP 2197794	IFP Energies Nouvelles	Oct. 2011	[252]
Mesostructured material having a high aluminium content and composed of spherical particles of specific size	EP 2274236	IFP Energies Nouvelles	Jun. 2012	[253]
Amorphous silicon-containing material with hierarchical and organized porosity	US 8236419	IFP Energies Nouvelles	Aug. 2012	[254]
Extra mesoporous Y zeolite	US 8361434	ExxonMobil Research and Engineering Company	Jan. 2013	[255]
Three-dimensional interconnected hierarchical-structured zeolite molecular sieve material and preparation method thereof	CN 102390843	Fudan University	May 2013	[256]
Introduction of mesoporosity in low Si/Al zeolites	US 8486369	Rive Technology, Inc.	Jul. 2013	[257]
Crystallized material with hierarchical porosity containing silicon	US8685366	IFP Energies Nouvelles	Apr. 2014	[258]

successfully introduced into a North America refinery showing superior hydrothermal stability and improved catalytic selectivity into gasoline and light cycle oil (LCO), as well as less bottom products and coke [202]. Rive Technology began the ongoing supply of the first commercial FCC catalyst with surfactant-templated mesoporous Y zeolite in April 2013, starting the new era of hierarchical zeolites in industrial catalysis [259].

CONCLUSION

To increase the molecular diffusion inside the zeolite pore systems and improve the access to the active sites different strategies can be followed: post-synthesis methods, such as, dealumination, desilication, or the use of several templates during the synthesis procedures. Chronologically the first method extensively explored was dealumination, namely steaming treatments, which allowed the preparation of the industrial important USHY zeolite, the catalyst used in the FCC process. All types of hydrothermal treatments, as well as the majority of the other dealumination strategies, led to the extraction of aluminum from the zeolite framework with the formation of lattice vacancies that, depending on the experimental conditions, are reconstructed in more or less extension. Hence, the facilitated transport provided by the introduction of mesoporosity may be partially canceled out by the reduction of the acid sites density, with predictable negative consequences for some catalytic applications of zeolite materials.

More recently, the controlled removal of silicon from the zeolite framework through alkaline treatments, in a process called desilication, opened new perspectives for the preparation of hierarchical structures, with low impact on the acidic properties as long as optimized experimental conditions are used. The importance of several experimental conditions, such as temperature, base type and concentration, duration of the treatment and type of heat source, on the characteristics of desilicated samples have been evaluated. Additionally intrinsic properties of the zeolite, namely the Si/Al ratio proved to be determinant for the development of intracrystalline mesoporosity without significant decrease of crystallinity and acid properties.

The use of solid templates is an appealing strategy to create hierarchical zeolites that has been increasingly applied due to its versatility to tune the size and shape of the additional pore system by choosing the proper hard template. This method allows the production of nanosized zeolites with intercrystalline mesoporosity when the zeolite synthesis takes place at the voids of template particles or confined inside the template pores. Alternatively, the zeolite crystals may grow surrounding the template particles originating porous zeolite crystals with intracrystalline mesoporosity that can be customized according to the size and shape of the template used. The use of macroscale hard templates such as polystyrene or carbon spheres is also appealing since it allows the production of nestlike hollow zeolite spheres with intercrystalline meso or macroporosity.

As the solid templates are chemically inert, this strategy can be applied to all zeolites independently of their chemical composition [193], and therefore also to zeotype materials such as SAPO or AlPO [194]. The chemical composition of the zeolite or zeotypes can be properly optimized in the synthesis gel composition and thus, independent control of acidity from the mesopore characteristics is attained.

Even though several solid matrixes have been explored, carbon based materials especially carbon black, carbon nanofibers and nanotubes are the most used ones. These templates present the advantage of being totally removed by calcination, allowing the total recovery of the pure zeolite. However, the negative aspect is the large amount of carbon needed for the appropriate creation of mesoporosity, where values up to 75% have been reported [164]. Additionally, from a safety point of view, it is very difficult to promote the removal of such high amount of carbon by calcination/combustion on a large industrial plant, under safety conditions, which can compromise the application of this strategy to develop commercial hierarchical zeolites.

Soft template synthesis methods have also been recently used to prepare hierarchical zeolite, using surfactant molecules, among other macromolecules. However, this strategy implies a careful selection of the template since characteristics like alkylic chain length, morphology; stability and type of interactions with the zeolite gel are key factors for the success of the synthesis.

Alternative strategies to produce hierarchical zeolites have been reported such as the mixture between two materials with different pore sizes producing composite materials or even the synthesis using microwave radiation as heat source to produce nanosized zeolites crystals.

The interest in preparing hierarchical zeolite is continuously increasing with improved post-synthesis experimental procedures, such as the use of microwave heat source, and innovative templates used during the synthesis. The fact that improved catalytic performances have been observed in several examples using hierarchical catalysts is an important driving force to the continuous development of new hierarchical structures.

Recent works have demonstrated the feasibility of scaling-up the synthesis of hierarchical zeolites, as well as the possibility for providing them with technical shapes and sizes, suitable to be use in industrial reactors [261]. In 2007 the Pérez-Ramírez group [127] reported the first attempt for the large-scale preparation of desilicated zeolites in a 1.5 m^3 reactor that, according to the authors, can be extrapolated to plant scale. More recently Garcia-Martinez and co-workers [259] successfully produced a mesostructured Y zeolite through a surfactant templating approach that revealed a superior catalytic performance in FCC process. The authors founded a company that produces hierarchical zeolites in commercial scale and successfully supplies an industrial refining unit.

The economic aspects can be a drawback for the general commercialization of hierarchical zeolites since most of the preparation procedures developed so far imply extra-cost when compared with the preparation of conventional zeolites. Removal of framework species involves some loss in the zeolite mass, whereas in hard templating approaches the solid template is destroyed after the zeolite crystallization. On the other hand, in soft templating strategies, the use of relatively expensive reagents is needed. To overcome this drawback recent studies show that is possible to recover and reuse the surfactant to further decrease the production costs [262, 263].

As a final conclusion it can be mention that the rapid development of a great number of synthesis strategies, and the good catalytic performances of new hierarchical zeolites allow foreseen their use in an increase number of industrial applications is a near future.

ACKNOWLEDGMENTS

The authors thank the support of FCT (pluriannual program of CQB through strategic project PEst-OE/QUI/UI0612/2013). T. Conceição is acknowledged for the illustrations of Figures 8 and 13.

REFERENCES

[1] Guisnet, M.; Ramôa Ribeiro, F., Les zéolithes, un nanomonde au service de la catalyse. *EDP Science, Les Ullis*, 2006.
[2] Figueiredo, J.L.; Ramôa Ribeiro, F., Catálise Heterogénea. 2ª ed,Fundação Calouste Gulbenkian, *Serviço de Educação e Bolsas*, Lisboa, 2007.
[3] Kurzendörfer, C.P.; Kuhm, P.; Steber, J., Zeolites in the Environment, Detergents in the Environment, Schwuger, M.J., Editor., *Marcel Dekker*, New York, 1997.
[4] Shenber, M.A.; Johanson, K.J., Influence of zeolite on the availability of radiocaesium in soil to plants. *Science of the Total Environment*, 113(3) 1992, 287-295.
[5] Valdés, H.; Solar, V.A.; Cabrera, E.H.; Veloso, A.F.; Zaror, C.A., Control of released volatile organic compounds from industrial facilities using natural and acid-treated mordenites: The role of acidic surface sites on the adsorption mechanism. *Chemical Engineering Journal*, 244 2014, 117-127.
[6] Mihaly-Cozmuta, L.; Mihaly-Cozmuta, A.; Peter, A.; Nicula, C.; Tutu, H.; Silipas, D.; Indrea, E., Adsorption of heavy metal cations by Na-clinoptilolite: Equilibrium and selectivity studies. *Journal of Environmental Management*, 137 2014, 69-80.
[7] Martucci, A.; Cremonini, M.A.; Blasioli, S.; Gigli, L.; Gatti, G.; Marchese, L.; Braschi, I., Adsorption and reaction of sulfachloropyridazine sulfonamide antibiotic on a high silica mordenite: A structural and spectroscopic combined study. *Microporous and Mesoporous Materials*, 170 2013, 274-286.
[8] Kawai, T., Adsorption characteristics of polyvinyl alcohols on modified zeolites. *Colloid and Polymer Science*, 292(2) 2014, 533-538.
[9] Mumpton, F.A., La roca magica: Uses of natural zeolites in agriculture and industry. *Proceedings of the National Academy of Sciences of the United States of America*, 96(7) 1999, 3463-3470.
[10] Chen, N.Y.; Garwood, W.E.; Dwyer, F.G., Shape selective catalysis in industrial applications. 2nd Ed,M. Dekker, New York, 1996.
[11] Larsen, S.C., Nanocrystalline zeolites and zeolite structures: Synthesis, characterization, and applications. *Journal of Physical Chemistry C*, 111(50) 2007, 18464-18474.
[12] Corma, A.; Diaz-Cabanas, M.J.; Jorda, J.L.; Martinez, C.; Moliner, M., High-throughput synthesis and catalytic properties of a molecular sieve with 18-and 10-member rings. *Nature*, 443(7113) 2006, 842-845.
[13] Corma, A.; Fornes, V.; Guil, J.M.; Pergher, S.; Maesen, T.L.M.; Buglass, J.G., Preparation, characterisation and catalytic activity of ITQ-2, a delaminated zeolite. *Microporous and Mesoporous Materials*, 38(2-3) 2000, 301-309.

[14] Paixão, V.; Santos, C.; Nunes, R.; Silva, J.M.; Pires, J.; Carvalho, A.P.; Martins, A., n-Hexane hydroisomerization over composite catalysts based on BEA zeolite and mesoporous materials. *Catalysis Letters,* 129(3-4) 2009, 331-335.

[15] Barrer, R.M.; Makki, M.B., Molecular sieve sorbents from clinoptilolite. *Canadian Journal of Chemistry-Revue Canadienne de Chimie,* 42(6) 1964, 1481-1487.

[16] Kerr, G.T., Chemistry of crystalline aluminosilicates.7. Thermal decomposition products of ammonium zeolite Y. *Journal of Catalysis,* 15(2) 1969, 200-204.

[17] Carvalho, A.P.; de Carvalho, M.B.; Ribeiro, F.R.; Fernandez, C.; Nagy, J.B.; Derouane, E.G.; Guisnet, M., Dealumination of zeolites.4. Dealumination of offretite through hydrothermal treatment. *Zeolites,* 13(6) 1993, 462-469.

[18] Niwa, M.; Sota, S.; Katada, N., Strong Brønsted acid site in HZSM-5 created by mild steaming. *Catalysis Today,* 185(1) 2012, 17-24.

[19] Ong, L.H.; Dömök, M.; Olindo, R.; van Veen, A.C.; Lercher, J.A., Dealumination of HZSM-5 via steam-treatment. *Microporous and Mesoporous Materials,* 164 2012, 9-20.

[20] Sheng, Q.; Ling, K.; Li, Z.; Zhao, L., Effect of steam treatment on catalytic performance of HZSM-5 catalyst for ethanol dehydration to ethylene. *Fuel Processing Technology,* 110 2013, 73-78.

[21] Dutartre, R.; de Menorval, L.C.; DiRenzo, F.; McQueen, D.; Fajula, F.; Schulz, P., Mesopore formation during steam dealumination of zeolites: Influence of initial aluminum content and crystal size. *Microporous Materials,* 6(5-6) 1996, 311-320.

[22] Triantafyllidis, K.S.; Vlessidis, A.G.; Evmiridis, N.P., Dealuminated H-Y zeolites: Influence of the degree and the type of dealumination method on the structural and acidic characteristics of H-Y zeolites. *Industrial & Engineering Chemistry Research,* 39(2) 2000, 307-319.

[23] Marques, J.P.; Gener, I.; Ayrault, P.; Bordado, J.C.; Lopes, J.M.; Ribeiro, F.R.; Guisnet, M., Infrared spectroscopic study of the acid properties of dealuminated BEA zeolites. *Microporous and Mesoporous Materials,* 60(1-3) 2003, 251-262.

[24] Xia, J.C.; Mao, D.S.; Tao, W.C.; Chen, Q.L.; Zhang, Y.H.; Tang, Y., Dealumination of HMCM-22 by various methods and its application in one-step synthesis of dimethyl ether from syngas. *Microporous and Mesoporous Materials,* 91(1-3) 2006, 33-39.

[25] McDaniel, C.V.; Maher, P.K., Zeolite stability and ultrastable zeolites, Zeolite Chemistry and Catalysis (ACS Monograph 171), Rabo, J.A., Editor., American Chemical Society, Washington D. C., 1976.

[26] Chumbhale, V.R.; Chandwadkar, A.J.; Rao, B.S., Characterization of siliceous mordenite obtained by direct synthesis or by dealumination. *Zeolites,* 12(1) 1992, 63-69.

[27] Rachwalik, R.; Olejniczak, Z.; Jiao, J.; Huang, J.; Hunger, M.; Sulikowski, B., Isomerization of alpha-pinene over dealuminated ferrierite-type zeolites. *Journal of Catalysis,* 252(2) 2007, 161-170.

[28] Matias, P.; Lopes, J.M.; Ayrault, P.; Laforge, S.; Magnoux, P.; Guisnet, M.; Ribeiro, F.R., Effect of dealumination by acid treatment of a HMCM-22 zeolite on the acidity and activity of the pore systems. *Applied Catalysis A: General,* 365(2) 2009, 207-213.

[29] Garcia-Basabe, Y.; Rodriguez-Iznaga, I.; de Menorval, L.C.; Llewellyn, P.; Maurin, G.; Lewis, D.W.; Binions, R.; Autie, M.; Ruiz-Salvador, A.R., Step-wise dealumination of

natural clinoptilolite: Structural and physicochemical characterization. *Microporous and Mesoporous Materials,* 135(1-3) 2010, 187-196.

[30] Inagaki, S.; Shinoda, S.; Kaneko, Y.; Takechi, K.; Komatsu, R.; Tsuboi, Y.; Yamazaki, H.; Kondo, J.N.; Kubota, Y., Facile fabrication of ZSM-5 zeolite catalyst with high durability to coke formation during catalytic cracking of paraffins. *ACS Catalysis,* 3(1) 2012, 74-78.

[31] Rivera, A.; Farías, T.; de Ménorval, L.C.; Autié-Pérez, M.; Lam, A., Natural and sodium clinoptilolites submitted to acid treatments: Experimental and theoretical studies. *The Journal of Physical Chemistry C,* 117(8) 2013, 4079-4088.

[32] Tian, F.; Wu, Y.; Shen, Q.; Li, X.; Chen, Y.; Meng, C., Effect of Si/Al ratio on mesopore formation for zeolite Beta via NaOH treatment and the catalytic performance in α-pinene isomerization and benzoylation of naphthalene. *Microporous and Mesoporous Materials,* 173 2013, 129-138.

[33] Kerr, G.T., Chemistry of crystalline aluminosilicates.V. Preparation of aluminum-deficient faujasites. *Journal of Physical Chemistry,* 72(7) 1968, 2594-2596.

[34] Barthome.D; Beaumont, R., X, Y, Aluminum-deficient, and ultrastable faujasite-type zeolites. 3. Catalytic activity. *Journal of Catalysis,* 30(2) 1973, 288-297.

[35] Kubelkova, L.; Seidl, V.; Borbely, G.; Beyer, H.K., Correlations between wavenumbers of skeletal vibrations, unit-cell size and molar fraction of aluminum of Y-zeolites - Removal of non-skeletal Al species with H_2Na_2EDTA. *Journal of the Chemical Society-Faraday Transactions I,* 84 1988, 1447-1454.

[36] Rhodes, N.P.; Rudham, R., Catalytic studies with dealuminated Y-Zeolite. 1. Catalyst characterization and the disproportionation of ethylbenzene. Journal of the Chemical Society-Faraday Transactions, 89(14) 1993, 2551-2557.

[37] Fejes, P.; Hannus, I.; Kiricsi, I., Dealumination of zeolites with phosgene. *Zeolites,* 4(1) 1984, 73-76.

[38] Fejes, P.; Hannus, I.; JKiricsi, I.; Schöbel, G., Dealumination of zeolites by volatile reagents mechanism of and structural changes caused by dealumination, Studies in Surface Science and Catalysis: Zeolites; Synthesis, Structure, Technology and Application, Drzaj, B.;Hocevar, S.;Pejovnik, S., Editores., *Elsevier,* Amsterdam, 1985.

[39] Lok, B.M.; Izod, T.P.J., Modification of molecular-Sieves. 1. Direct fluorination. *Zeolites,* 2(2) 1982, 66-67.

[40] Lok, B.M.; Gortsema, F.P.; Messina, C.A.; Rastelli, H.; Izod, T.P.J., Zeolite modification - direct fluorination. *ACS Symposium Series,* 218 1983, 41-58.

[41] Klinowski, J.; Thomas, J.M.; Anderson, M.W.; Fyfe, C.A.; Gobbi, G.C., Dealumination of mordenite using silicon tetrachloride vapor. *Zeolites,* 3(1) 1983, 5-7.

[42] Hidalgo, C.V.; Kato, M.; Hattori, T.; Niwa, M.; Murakami, Y., Modification of mordenite by chemical vapor-deposition of metal chloride. *Zeolites,* 4(2) 1984, 175-180.

[43] Andera, V.; Kubelkova, L.; Novakova, J.; Wichterlova, B.; Bednarova, S., Aluminum-to-silicon ratio on the surface of zeolites dealuminated using silicon tetrachloride vapor and acid leaching. *Zeolites,* 5(2) 1985, 67-69.

[44] Grobet, P.J.; Jacobs, P.A.; Beyer, H.K., Study of the silicon tetrachloride dealumination of NaY by a combination of NMR and IR methods. *Zeolites,* 6(1) 1986, 47-50.

[45] Muller, M.; Harvey, G.; Prins, R., Comparison of the dealumination of zeolites Beta, mordenite, ZSM-5 and ferrierite by thermal treatment, leaching with oxalic acid and

treatment with SiCl₄ by H-1, Si-29 and Al-27 MAS NMR. *Microporous and Mesoporous Materials,* 34(2) 2000, 135-147.

[46] Beyer, H.K.; Belenykaja, I., A new method for the dealumination of faujasite-type zeolites, Studies in Surface Science and Catalysis: Catalysis by Zeolites, Imelik, B.;Naccache, C.;Taarit, Y.B.;Vedrine, J.C.;Coudurier, G.;Praliaud, H., Editores., *Elsevier,* Amsterdam, 1980.

[47] Ponthieu, E.; Grange, P.; Joly, J.F.; Raatz, F., Modification of offretite and omega-zeolites. 1. Structural characterization. *Zeolites,* 12(4) 1992, 395-401.

[48] Carrott, M.M.L.R.; Russo, P.A.; Carvalhal, C.; Carrott, P.J.M.; Marques, J.P.; Lopes, J.M.; Gener, I.; Guisnet, M.; Ribeiro, F.R., Adsorption of n-pentane and iso-octane for the evaluation of the porosity of dealuminated BEA zeolites. *Microporous and Mesoporous Materials,* 81(1-3) 2005, 259-267.

[49] Kim, J.; Choi, M.; Ryoo, R., Effect of mesoporosity against the deactivation of MFI zeolite catalyst during the methanol-to-hydrocarbon conversion process. *Journal of Catalysis,* 269(1) 2010, 219-228.

[50] Skeels, G.W.; Breck, D.W. Proc. 6th International Zeolite Conference. 1984: Butterworths

[51] Beyera, H.K.; Borbély-Pálnéa, G.; Wub, J., Solid-state dealumination of zeolites, Studies in Surface Science and Catalysis, Zeolites and Related Microporous Materials: State of the Art, Weitkamp, J.;Karge, H.G.;Pfeifer, H.;Hölderich, W., Editores., *Elsevier,* Amsterdam, 1994.

[52] Derouane, E.G.; Baltusis, L.; Dessau, R.M.; Schmitt, K.D., Quantitation and modification of catalytic sites In ZSM-5, Studies in Surface Science and Catalysis: Catalysis by Acids and Bases, Imelik, B.;Naccache, C.;Coudurier, G.;Ben Taarit, Y.B.;Vedrine, J.C., Editores., *Elsevier,* Amesterdam, 1985.

[53] Scherzer, J., De-aluminated faujasite-type structures with SiO_2-Al_2O_3 ratios over 100. *Journal of Catalysis,* 54(2) 1978, 285-288.

[54] Verboekend, D.; Mitchell, S.; Milina, M.; Groen, J.C.; Pérez-Ramírez, J., Full compositional flexibility in the preparation of mesoporous MFI zeolites by desilication. *Journal of Physical Chemistry C,* 115(29) 2011, 14193-14203.

[55] Machado, V.; Rocha, J.; Carvalho, A.P.; Martins, A., Modification of MCM-22 zeolite through sequential post-synthesis treatments. Implications on the acidic and catalytic behaviour. *Applied Catalysis A: General,* 445 2012, 329-338.

[56] Park, S.H.; Rhee, H.K., Shape selective properties of MCM-22 catalysts for the disproportionation of ethylbenzene. *Applied Catalysis A: General,* 219(1-2) 2001, 99-105.

[57] Triantafyllidis, K.S.; Vlessidis, A.G.; Nalbandian, L.; Evmiridis, N.P., Effect of the degree and type of the dealumination method on the structural, compositional and acidic characteristics of H-ZSM-5 zeolites. *Microporous and Mesoporous Materials,* 47(2-3) 2001, 369-388.

[58] Coutanceau, C.; DaSilva, J.M.; Alvarez, M.F.; Ribeiro, F.R.; Guisnet, M., Dealumination of zeolites. 7. Influence of the acid treatment of a HBEA zeolite on the framework composition and on the porosity. *Journal de Chimie Physique et de Physico-Chimie Biologique,* 94(4) 1997, 765-781.

[59] Triantafillidis, C.S.; Vlessidis, A.G.; Nalbandian, L.; Evmiridis, N.P., Effect of the degree and type of the dealumination method on the structural, compositional and

acidic characteristics of H-ZSM-5 zeolites. *Microporous and Mesoporous Materials,* 47(2–3) 2001, 369-388.
[60] De Baerdemaeker, T.; Yilmaz, B.; Muller, U.; Feyen, M.; Xiao, F.S.; Zhang, W.P.; Tatsumi, T.; Gies, H.; Bao, X.H.; De Vos, D., Catalytic applications of OSDA-free Beta zeolite. *Journal of Catalysis,* 308 2013, 73-81.
[61] Ivanov, D.P.; Pirutko, L.V.; Panov, G.I., Effect of steaming on the catalytic performance of ZSM-5 zeolite in the selective oxidation of phenol by nitrous oxide. *Journal of Catalysis,* 311 2014, 424-432.
[62] Carvalho, A.P.; Wang, Q.L.; Giannetto, G.; Cardoso, D.; Carvalho, M.B.; Ribeiro, F.R.; Nagy, J.B.; Asswad, J.E.H.A.; Derouane, E.G.; Guisnet, M., Effect of the conditions of hydrothermal treatments on the physicochemical characteristics of Y-zeolites. *Journal de Chimie Physique et de Physico-Chimie Biologique,* 87(2) 1990, 271-288.
[63] Janssen, A.H.; Koster, A.J.; de Jong, K.P., On the shape of the mesopores in zeolite Y: A three-dimensional transmission electron microscopy study combined with texture analysis. *Journal of Physical Chemistry B,* 106(46) 2002, 11905-11909.
[64] Xue, H.F.; Huang, X.M.; Zhan, E.S.; Ma, M.; Shen, W.J., Selective dealumination of mordenite for enhancing its stability in dimethyl ether carbonylation. *Catalysis Communications,* 37 2013, 75-79.
[65] Batonneau-gener, I.; Yonli, A.; Hazael-pascal, S.; Marques, J.P.; Lodes, J.M.; Guisnet, M.; Ribeiro, F.R.; Mignard, S., Influence of steaming and acid-leaching treatments on the hydrophobicity of HBEA zeolite determined under static conditions. *Microporous and Mesoporous Materials,* 110(2-3) 2008, 480-487.
[66] Carvalho, M.B.; Carvalho, A.P.; Ribeiro, F.R.; Florentino, A.; Gnep, N.S.; Guisnet, M., Dealumination of zeolites. 5. Influence of the hydrothermal treatment of offretite on its pore structure and acid properties. *Zeolites,* 14(3) 1994, 217-224.
[67] Meriaudeau, P.; Tuan, V.A.; Nghiem, V.T.; Lefevbre, F.; Ha, V.T., Characterization and catalytic properties of hydrothermally dealuminated MCM-22. *Journal of Catalysis,* 185(2) 1999, 378-385.
[68] Ma, D.; Deng, F.; Fu, R.Q.; Dan, X.W.; Bao, X.H., MAS MMR studies on the dealumination of Zeolite MCM-22. *Journal of Physical Chemistry B,* 105(9) 2001, 1770-1779.
[69] Kubo, K.; Iida, H.; Namba, S.; Igarashi, A., Effect of steaming on acidity and catalytic performance of H-ZSM-5 and P/H-ZSM-5 as naphtha to olefin catalysts. *Microporous and Mesoporous Materials,* 188 2014, 23-29.
[70] Kerr, G.T., Intracrystalline rearrangement of constitutive water in hydrogen zeolite Y. *Journal of Physical Chemistry,* 71(12) 1967, 4155-4156.
[71] Maher, P.K.; Hunter, F.D.; Scherzer, J. Proc. 2nd International Conference on Molecular Sieve Zeolites 1970. Worcester: American Chemical Society.
[72] Wang, Q.L.; Giannetto, G.; Torrealba, M.; Perot, G.; Kappenstein, C.; Guisnet, M., Dealumination of zeolites. 2. Kinetic-study of the dealumination by hydrothermal treatment of a NH₄NaY Zeolite. *Journal of Catalysis,* 130(2) 1991, 459-470.
[73] Breck, D.W.; Flaningen, E. Proc. Molecular Sieves. London: Society of Chemical Industry, 1968.
[74] Sohn, J.R.; Decanio, S.J.; Lunsford, J.H.; Odonnell, D.J., Determination of framework aluminum content in dealuminated Y-type zeolites - a comparison based on unit-cell size and wave-number of IR bands. *Zeolites,* 6(3) 1986, 225-227.

[75] Zi, G.; Yi, T., Influence of Si/Al ratio on the properties of faujasites enriched in silicon. *Zeolites*, 8(3) 1988, 232-237.

[76] Kerr, G.T., Determination of framework aluminum content in zeolite-X, zeolite-Y, and dealuminated-Y using unit-cell size. *Zeolites*, 9(4) 1989, 350-351.

[77] Cooper, D.A.; Hastings, T.W.; Hertzenberg, E.P., Process for preparing zeolite Y with increased mesopore volume, US 5601798 A, 1997.

[78] Wang, Q.L.; Giannetto, G.; Guisnet, M., Dealumination of Zeolites. 3. Effect of extra-Framework aluminum species on the activity, selectivity, and stability of Y-zeolites in normal-heptane cracking. *Journal of Catalysis*, 130(2) 1991, 471-482.

[79] Mirodatos, C.; Barthomeuf, D., Superacid sites in zeolites. *Journal of the Chemical Society-Chemical Communications*, (2) 1981, 39-40.

[80] Zhou, L.P.; Shi, M.T.; Cai, Q.Y.; Wu, L.; Hu, X.P.; Yang, X.M.; Chen, C.; Xu, J., Hydrolysis of hemicellulose catalyzed by hierarchical H-USY zeolites - The role of acidity and pore structure. *Microporous and Mesoporous Materials*, 169 2013, 54-59.

[81] Meyers, B.L.; Fleisch, T.H.; Ray, G.J.; Miller, J.T.; Hall, J.B., A multitechnique characterization of dealuminated mordenites. *Journal of Catalysis*, 110(1) 1988, 82-95.

[82] Stach, H.; Janchen, J.; Jerschkewitz, H.G.; Lohse, U.; Parlitz, B.; Zibrowius, B.; Hunger, M., Mordenite acidity - dependence on the Si/Al ratio and the framework aluminum topology. 1. Sample preparation and physicochemical characterization. *Journal of Physical Chemistry*, 96(21) 1992, 8473-8479.

[83] Tromp, M.; van Bokhoven, J.A.; Oostenbrink, M.T.G.; Bitter, J.H.; de Jong, K.P.; Koningsberger, D.C., Influence of the generation of mesopores on the hydroisomerization activity and selectivity of n-hexane over Pt/mordenite. *Journal of Catalysis*, 190(2) 2000, 209-214.

[84] Viswanadham, N.; Kumar, M., Effect of dealumination severity on the pore size distribution of mordenite. *Microporous and Mesoporous Materials*, 92(1-3) 2006, 31-37.

[85] Chung, K.H., Dealumination of mordenites with acetic acid and their catalytic activity in the alkylation of cumene. *Microporous and Mesoporous Materials*, 111(1-3) 2008, 544-550.

[86] Giudici, R.; Kouwenhoven, H.W.; Prins, R., Comparison of nitric and oxalic acid in the dealumination of mordenite. *Applied Catalysis A: General*, 203(1) 2000, 101-110.

[87] Leng, K.Y.; Wang, Y.; Hou, C.M.; Lancelot, C.; Lamonier, C.; Rives, A.; Sun, Y.Y., Enhancement of catalytic performance in the benzylation of benzene with benzyl alcohol over hierarchical mordenite. *Journal of Catalysis*, 306 2013, 100-108.

[88] Meriaudeau, P.; Tuel, A.; Vu, T.T.H., On the localization of tetrahedral aluminum in MCM-22 zeolite. *Catalysis Letters*, 61(1-2) 1999, 89-92.

[89] Wu, P.; Komatsu, T.; Yashima, T., Selective formation of p-xylene with disproportionation of toluene over MCM-22 catalysts. *Microporous and Mesoporous Materials*, 22(1-3) 1998, 343-356.

[90] Christidis, G.E.; Moraetis, D.; Keheyan, E.; Akhalbedashvili, L.; Kekelidze, N.; Gevorkyan, R.; Yeritsyan, H.; Sargsyan, H., Chemical and thermal modification of natural HEU-type zeolitic materials from Armenia, Georgia and Greece. *Applied Clay Science*, 24(1-2) 2003, 79-91.

[91] Cakicioglu-Ozkan, F.; Ulku, S., The effect of HCl treatment on water vapor adsorption characteristics of clinoptilolite rich natural zeolite. *Microporous and Mesoporous Materials*, 77(1) 2005, 47-53.

[92] Vasylechko, L.O.; Gryshchouk, G.V.; Kuz'ma, Y.B.; Zakordonskiy, V.P.; Vasylechko, L.O.; Lebedynets, L.O.; Kalytovs'ka, M.B., Adsorption of cadmium on acid-modified - Transcarpathian clinoptilolite. *Microporous and Mesoporous Materials*, 60(1-3) 2003, 183-196.

[93] Kouvelos, E.; Kesore, K.; Steriotis, T.; Grigoropoulou, H.; Bouloubasi, D.; Theophilou, N.; Tzintzos, S.; Kanelopous, N., High pressure N_2/CH_4 adsorption measurements in clinoptilolites. *Microporous and Mesoporous Materials*, 99(1-2) 2007, 106-111.

[94] Rivera, A.; Farias, T.; de Menorval, L.C.; Autie-Perez, M.; Lam, A., Natural and sodium clinoptilolites submitted to acid treatments: Experimental and theoretical studies. *Journal of Physical Chemistry C*, 117(8) 2013, 4079-4088.

[95] You, S.J.; Park, E.D., Effects of dealumination and desilication of H-ZSM-5 on xylose dehydration. *Microporous and Mesoporous Materials*, 186 2014, 121-129.

[96] Shetti, V.N.; Kim, J.; Srivastava, R.; Choi, M.; Ryoo, R., Assessment of the mesopore wall catalytic activities of MFI zeolite with mesoporous/microporous hierarchical structures. *Journal of Catalysis*, 254(2) 2008, 296-303.

[97] Xin, H.; Li, X.; Fang, Y.; Yi, X.; Hu, W.; Chu, Y.; Zhang, F.; Zheng, A.; Zhang, H.; Li, X., Catalytic dehydration of ethanol over post-treated ZSM-5 zeolites. *Journal of Catalysis*, 312 2014, 204-215.

[98] González, M.D.; Cesteros, Y.; Salagre, P., Comparison of dealumination of zeolites Beta, mordenite and ZSM-5 by treatment with acid under microwave irradiation. *Microporous and Mesoporous Materials*, 144(1–3) 2011, 162-170.

[99] Shekara, B.M.C.; Prakash, B.S.J.; Bhat, Y.S., Dealumination of zeolite BEA under microwave irradiation. *ACS Catalysis*, 1(3) 2011, 193-199.

[100] Kumar, S.; Sinha, A.K.; Hegde, S.G.; Sivasanker, S., Influence of mild dealumination on physicochemical, acidic and catalytic properties of H-ZSM-5. *Journal of Molecular Catalysis A: Chemical*, 154(1-2) 2000, 115-120.

[101] Gonzalez, M.D.; Cesteros, Y.; Salagre, P.; Medina, F.; Sueiras, J.E., Effect of microwaves in the dealumination of mordenite on its surface and acidic properties. *Microporous and Mesoporous Materials*, 118(1-3) 2009, 341-347.

[102] Kubota, Y.; Inagaki, S.; Takechi, K., Hexane cracking catalyzed by MSE-type zeolite as a solid acid catalyst. *Catalysis Today*, 226 2014, 109-116.

[103] Senderov, E.; Halasz, I.; Olson, D.H., On existence of hydroxyl nests in acid dealuminated zeolite Y. *Microporous and Mesoporous Materials*, 186 2014, 94-100.

[104] Inagaki, S.; Shinoda, S.; Kaneko, Y.; Takechi, K.; Komatsu, R.; Tsuboi, Y.; Yamazaki, H.; Kondo, J.N.; Kubota, Y., Facile fabrication of ZSM-5 zeolite catalyst with high durability to coke formation during catalytic cracking of paraffins. *ACS Catalysis*, 3(1) 2013, 74-78.

[105] Kubů, M.; Žilková, N.; Čejka, J., Post-synthesis modification of TUN zeolite: Textural, acidic and catalytic properties. *Catalysis Today*, 168(1) 2011, 63-70.

[106] Baran, R.; Millot, Y.; Onfroy, T.; Krafft, J.-M.; Dzwigaj, S., Influence of the nitric acid treatment on Al removal, framework composition and acidity of BEA zeolite investigated by XRD, FTIR and NMR. *Microporous and Mesoporous Materials*, 163 2012, 122-130.

[107] Yue, M.B.; Xue, T.; Jiao, W.Q.; Wang, Y.M.; He, M.-Y., Dealumination, silicon insertion and H-proton exchange of NaY in one step with acid ethanol solution. *Microporous and Mesoporous Materials,* 159 2012, 50-56.

[108] Gonzalez, M.D.; Cesteros, Y.; Llorca, J.; Salagre, P., Boosted selectivity toward high glycerol tertiary butyl ethers by microwave-assisted sulfonic acid-functionalization of SBA-15 and beta zeolite. *Journal of Catalysis,* 290 2012, 202-209.

[109] Gonzalez, M.D.; Salagre, P.; Taboada, E.; Llorca, J.; Cesteros, Y., Microwave-assisted synthesis of sulfonic acid-functionalized microporous materials for the catalytic etherification of glycerol with isobutene. *Green Chemistry,* 15(8) 2013, 2230-2239.

[110] Jin, Y.S.; Auroux, A.; Vedrine, J.C., Acidic ferrierite type zeolite. 2. Steamed and acid leached treatment. *Applied Catalysis,* 37(1-2) 1988, 21-33.

[111] Ghosh, A.K.; Kydd, R.A., An Infrared Study of the Effect of HF Treatment on the Acidity of ZSM-5. *Zeolites,* 10(8) 1990, 766-771.

[112] Garralon, G.; Fornes, V.; Corma, A., Faujasites dealuminated with ammonium hexafluorosilicate - Variables affecting the method of preparation. *Zeolites,* 8(4) 1988, 268-272.

[113] Vermeiren, W.; Gilson, J.P., Impact of zeolites on the petroleum and petrochemical industry. *Topics in Catalysis*, 52(9) 2009, 1131-1161.

[114] Dessau, R.M.; Valyocsik, E.W.; Goeke, N.H., Aluminum zoning in ZSM-5 as revealed by selective silica removal. *Zeolites,* 12(7) 1992, 776-779.

[115] Lietz, G.; Schnabel, K.H.; Peuker, C.; Gross, T.; Storek, W.; Volter, J., Modifications of H-ZSM-5 catalysts by NaOH treatment. *Journal of Catalysis,* 148(2) 1994, 562-568.

[116] Ogura, M.; Shinomiya, S.Y.; Tateno, J.; Nara, Y.; Kikuchi, E.; Matsukata, H., Formation of uniform mesopores in ZSM-5 zeolite through treatment in alkaline solution. *Chemistry Letters*, (8) 2000, 882-883.

[117] Groen, J.C.; Peffer, L.A.A.; Moulijn, J.A.; Pérez-Ramírez, J., Mesoporosity development in ZSM-5 zeolite upon optimized desilication conditions in alkaline medium. *Colloids and Surfaces a-Physicochemical and Engineering Aspects,* 241(1-3) 2004, 53-58.

[118] Groen, J.C.; Zhu, W.D.; Brouwer, S.; Huynink, S.J.; Kapteijn, F.; Moulijn, J.A.; Pérez-Ramírez, J., Direct demonstration of enhanced diffusion in mesoporous ZSM-5 zeolite obtained via controlled desilication. *Journal of the American Chemical Society,* 129(2) 2007, 355-360.

[119] Groen, J.C.; Abelló, S.; Villaescusa, L.A.; Pérez-Ramírez, J., Mesoporous Beta zeolite obtained by desilication. *Microporous and Mesoporous Materials,* 114(1-3) 2008, 93-102.

[120] Mokrzycki, L.; Sulikowski, B.; Olejniczak, Z., Properties of desilicated ZSM-5, ZSM-12, MCM-22 and ZSM-12/MCM-41 derivatives in isomerization of alpha-pinene. *Catalysis Letters,* 127(3-4) 2009, 296-303.

[121] Bonilla, A.; Baudouin, D.; Pérez-Ramírez, J., Desilication of ferrierite zeolite for porosity generation and improved effectiveness in polyethylene pyrolysis. *Journal of Catalysis,* 265(2) 2009, 170-180.

[122] Paixão, V.; Carvalho, A.P.; Rocha, J.; Fernandes, A.; Martins, A., Modification of MOR by desilication treatments: Structural, textural and acidic characterization. *Microporous and Mesoporous Materials,* 131(1-3) 2010, 350-357.

[123] Zhao, L.; Shen, B.J.; Gao, F.S.; Xu, C.M., Investigation on the mechanism of diffusion in mesopore structured ZSM-5 and improved heavy oil conversion. *Journal of Catalysis,* 258(1) 2008, 228-234.

[124] Matias, P.; Couto, C.S.; Graca, I.; Lopes, J.M.; Carvalho, A.P.; Ribeiro, F.R.; Guisnet, M., Desilication of a TON zeolite with NaOH: Influence on porosity, acidity and catalytic properties. *Applied Catalysis A: General,* 399(1-2) 2011, 100-109.

[125] Verboekend, D.; Vile, G.; Pérez-Ramírez, J., Hierarchical Y and USY zeolites designed by post-synthetic strategies. *Advanced Functional Materials,* 22(5) 2012, 916-928.

[126] Le Van Mao, R.; Ramsaran, A.; Xiao, S.Y.; Yao, J.H.; Semmer, V., PH of the sodium-carbonate solution used for the desilication of zeolite materials. Journal of Materials Chemistry, 5(3) 1995, 533-535.

[127] Groen, J.C.; Moulijn, J.A.; Pérez-Ramírez, J., Alkaline posttreatment of MFI zeolites. From accelerated screening to scale-up. *Industrial & Engineering Chemistry Research,* 46(12) 2007, 4193-4201.

[128] Pérez-Ramírez, J.; Verboekend, D.; Bonilla, A.; Abelló, S., Zeolite catalysts with tunable hierarchy factor by pore-growth moderators. *Advanced Functional Materials,* 19(24) 2009, 3972-3979.

[129] Holm, M.S.; Hansen, M.K.; Christensen, C.H., "One-pot" ion-exchange and mesopore formation during desilication. *European Journal of Inorganic Chemistry,* (9) 2009, 1194-1198.

[130] Groen, J.C.; Moulijn, J.A.; Pérez-Ramírez, J., Decoupling mesoporosity formation and acidity modification in ZSM-5 zeolites by sequential desilication-dealumination. *Microporous and Mesoporous Materials,* 87(2) 2005, 153-161.

[131] Fernandez, C.; Stan, I.; Gilson, J.P.; Thomas, K.; Vicente, A.; Bonilla, A.; Pérez-Ramírez, J., Hierarchical ZSM-5 zeolites in shape-selective xylene isomerization: Role of mesoporosity and acid site speciation. *Chemistry-An European Journal,* 16(21) 2010, 6224-6233.

[132] Abello, S.; Pérez-Ramírez, J., Accelerated generation of intracrystalline mesoporosity in zeolites by microwave-mediated desilication. *Physical Chemistry Chemical Physics,* 11(16) 2009, 2959-2963.

[133] Paixão, V.; Monteiro, R.; Andrade, M.; Fernandes, A.; Rocha, J.; Carvalho, A.P.; Martins, A., Desilication of MOR zeolite: Conventional versus microwave assisted heating. *Applied Catalysis A: General,* 402(1-2) 2011, 59-68.

[134] Groen, J.C.; Moulijn, J.A.; Pérez-Ramírez, J., Desilication: on the controlled generation of mesoporosity in MFI zeolites. *Journal of Materials Chemistry,* 16(22) 2006, 2121-2131.

[135] Groen, J.C.; Hamminga, G.M.; Moulijn, J.A.; Pérez-Ramírez, J., In situ monitoring of desilication of MFI-type zeolites in alkaline medium. *Physical Chemistry Chemical Physics,* 9(34) 2007, 4822-4830.

[136] Groen, J.C.; Jansen, J.C.; Moulijn, J.A.; Pérez-Ramírez, J., Optimal aluminum-assisted mesoporosity development in MFI zeolites by desilication. *Journal of Physical Chemistry B,* 108(35) 2004, 13062-13065.

[137] Groen, J.C.; Sano, T.; Moulijn, J.A.; Pérez-Ramírez, J., Alkaline-mediated mesoporous mordenite zeolites for acid-catalyzed conversions. Journal of Catalysis, 251(1) 2007, 21-27.

[138] Verboekend, D.; Chabaneix, A.M.; Thomas, K.; Gilson, J.P.; Pérez-Ramírez, J., Mesoporous ZSM-22 zeolite obtained by desilication: peculiarities associated with crystal morphology and aluminium distribution. *Crystengcomm,* 13(10) 2011, 3408-3416.

[139] van Miltenburg, A.; Pawlesa, J.; Bouzga, A.M.; Zilkova, N.; Cejka, J.; Stocker, M., Alkaline modification of MCM-22 to a 3D interconnected pore system and its application in toluene disproportionation and alkylation. Topics in Catalysis, 52(9) 2009, 1190-1202.

[140] Groen, J.C.; Peffer, L.A.A.; Moulijn, J.A.; Pérez-Ramírez, J., On the introduction of intracrystalline mesoporosity in zeolites upon desilication in alkaline medium. *Microporous and Mesoporous Materials,* 69(1-2) 2004, 29-34.

[141] Paixão Carvalho, V., Modificação de estruturas zeolíticas por desilicação, Master, Instituto Superior de Engenharia de Lisboa, Lisboa, 2008.

[142] Cizmek, A.; Subotic, B.; Aiello, R.; Crea, F.; Nastro, A.; Tuoto, C., Dissolution of high-silica zeolites in alkaline-solutions. 1. Dissolution of silicalite-1 and ZSM-5 with different aluminum content. *Microporous Materials,* 4(2-3) 1995, 159-168.

[143] Gregg, S.J.; Sing, K.S.W., Adsorption, surface area, and porosity. 2nd Ed,Academic Press, London ; New York, 1982.

[144] Wei, X.T.; Smirniotis, P.G., Development and characterization of mesoporosity in ZSM-12 by desilication. *Microporous and Mesoporous Materials,* 97(1-3) 2006, 97-106.

[145] van Miltenburg, A.; de Menorval, L.C.; Stocker, M., Characterization of the pore architecture created by alkaline treatment of HMCM-22 using Xe-129 NMR spectroscopy. *Catalysis Today,* 168(1) 2011, 57-62.

[146] Motuzas, J.; Julbe, A.; Noble, R.D.; Guizard, C.; Beresnevicius, Z.J.; Cot, D., Rapid synthesis of silicalite-1 seeds by microwave assisted hydrothermal treatment. *Microporous and Mesoporous Materials,* 80(1-3) 2005, 73-83.

[147] Li, Y.N.; Liu, S.L.; Xie, S.J.; Xu, L.Y., Promoted metal utilization capacity of alkali-treated zeolite: Preparation of Zn/ZSM-5 and its application in 1-hexene aromatization. *Applied Catalysis A: General,* 360(1) 2009, 8-16.

[148] Gopalakrishnan, S.; Lopez, S.; Zampieri, A.; Schwieger, W., Selective oxidation of benzene to phenol over H-ZSM-5 catalyst: role of mesoporosity on the catalyst deactivation. Zeolites and Related Materials: Trends, Targets and Challenges, *Proceedings of the 4th International FEZA Conference,* 174 2008, 1203-1206.

[149] Pérez-Ramírez, J.; Abelló, S.; Bonilla, A.; Groen, J.C., Tailored mesoporosity development in zeolite crystals by partial detemplation and desilication. *Advanced Functional Materials,* 19(1) 2009, 164-172.

[150] Modhera, B.K.; Chakraborty, M.; Bajaj, H.C.; Parikh, P.A., Influences of mesoporosity generation in ZSM-5 and zeolite Beta on catalytic performance during n-hexane isomerization. *Catalysis Letters,* 141(8) 2011, 1182-1190.

[151] Verboekend, D.; Groen, J.C.; Pérez-Ramírez, J., Interplay of properties and functions upon introduction of mesoporosity in ITQ-4 zeolite. *Advanced Functional Materials,* 20(9) 2010, 1441-1450.

[152] Monteiro, R.; Ania, C.O.; Rocha, J.; Carvalho, A.P.; Martins, A., Catalytic behavior of alkali-treated Pt/HMOR in n-hexane hydroisomerization Applied Catalysis A: General, 476 2014, 148-157.

[153] Martinez, C.; Doskocil, E.J.; Corma, A., Improved THETA-1 for light olefins oligomerization to diesel: Influence of textural and acidic properties. *Topics in Catalysis,* 57(6-9) 2014, 668-682.

[154] Mitchell, S.; Pérez-Ramírez, J., Mesoporous zeolites as enzyme carriers: Synthesis, characterization, and application in biocatalysis. *Catalysis Today,* 168(1) 2011, 28-37.

[155] Martins, L.M.D.R.S.; Martins, A.; Alegria, E.C.B.A.; Carvalho, A.P.; Pombeiro, A.J.L., Efficient cyclohexane oxidation with hydrogen peroxide catalysed by a C-scorpionate iron(II) complex immobilized on desilicated MOR zeolite. *Applied Catalysis A: General,* 464 2013, 43-50.

[156] Silva, A.R.; Guimaraes, V.; Carneiro, L.; Nunes, N.; Borges, S.; Pires, J.; Martins, A.; Carvalho, A.P., Copper(II) aza-bis(oxazoline) complex immobilized onto ITQ-2 and MCM-22 based materials as heterogeneous catalysts for the cyclopropanation of styrene. *Microporous and Mesoporous Materials,* 179 2013, 231-241.

[157] Madsen, C.; Jacobsen, C.J.H., Nanosized zeolite crystals - convenient control of crystal size distribution by confined space synthesis. *Chemical Communications,* (8) 1999, 673-674.

[158] Jacobsen, C.J.H.; Madsen, C.; Houzvicka, J.; Schmidt, I.; Carlsson, A., Mesoporous zeolite single crystals. *Journal of the American Chemical Society,* 122(29) 2000, 7116-7117.

[159] Kim, S.S.; Shah, J.; Pinnavaia, T.J., Colloid-imprinted carbons as templates for the nanocasting synthesis of mesoporous ZSM-5 zeolite. *Chemistry of Materials,* 15(8) 2003, 1664-1668.

[160] Egeblad, K.; Christensen, C.H.; Kustova, M.; Christensen, C.H., Templating mesoporous zeolites. *Chemistry of Materials,* 20(3) 2008, 946-960.

[161] Wei, X.T.; Smirniotis, P.G., Synthesis and characterization of mesoporous ZSM-12 by using carbon particles. *Microporous and Mesoporous Materials,* 89(1-3) 2006, 170-178.

[162] Kustova, M.Y.; Hasselriis, P.; Christensen, C.H., Mesoporous MEL-type zeolite single crystal catalysts. *Catalysis Letters*, 96(3-4) 2004, 205-211.

[163] Xin, H.C.; Zhao, J.; Xu, S.T.; Li, J.P.; Zhang, W.P.; Guo, X.W.; Hensen, E.J.M.; Yang, Q.H.; Li, C., Enhanced catalytic oxidation by hierarchically structured TS-1 zeolite. *Journal of Physical Chemistry C,* 114(14) 2010, 6553-6559.

[164] Chal, R.; Gerardin, C.; Bulut, M.; van Donk, S., Overview and industrial assessment of synthesis strategies towards zeolites with mesopores. *ChemCatChem,* 3(1) 2011, 67-81.

[165] Chen, L.H.; Li, X.Y.; Rooke, J.C.; Zhang, Y.H.; Yang, X.Y.; Tang, Y.; Xiao, F.S.; Su, B.L., Hierarchically structured zeolites: synthesis, mass transport properties and applications. *Journal of Materials Chemistry*, 22(34) 2012, 17381-17403.

[166] Garcia-Martinez, J.; Cazorla-Amoros, D.; Linares-Solano, A.; Lin, Y.S., Synthesis and characterisation of MFI-type zeolites supported on carbon materials. *Microporous and Mesoporous Materials,* 42(2-3) 2001, 255-268.

[167] Janssen, A.H.; Schmidt, I.; Jacobsen, C.J.H.; Koster, A.J.; de Jong, K.P., Exploratory study of mesopore templating with carbon during zeolite synthesis. *Microporous and Mesoporous Materials,* 65(1) 2003, 59-75.

[168] Sakthivel, A.; Huang, S.J.; Chen, W.H.; Lan, Z.H.; Chen, K.H.; Kim, T.W.; Ryoo, R.; Chiang, A.S.T.; Liu, S.B., Replication of mesoporous aluminosilicate molecular sieves

(RMMs) with zeolite framework from mesoporous carbons (CMKs). *Chemistry of Materials,* 16(16) 2004, 3168-3175.

[169] Sakithivel, A.; Huang, S.; Chen, W.; Kim, T.; Ryoo, R.; Chiang, A.S.T.; Chen, K.; Liu, S., Preparation of ordered mesoporous aluminosilicate using carbon mesoporous materials as template. *Recent Advances in the Science and Technology of Zeolites and Related Materials,* Pts A - C, 154 2004, 394-399.

[170] Yang, Z.X.; Xia, Y.D.; Mokaya, R., Zeolite ZSM-5 with unique supermicropores synthesized using mesoporous carbon as a template. *Advanced Materials,* 16(8) 2004, 727-732.

[171] Cho, H.S.; Ryoo, R., Synthesis of ordered mesoporous MFI zeolite using CMK carbon templates. *Microporous and Mesoporous Materials,* 151 2012, 107-112.

[172] Tao, Y.S.; Kanoh, H.; Kaneko, K., ZSM-5 monolith of uniform mesoporous channels. *Journal of the American Chemical Society,* 125(20) 2003, 6044-6045.

[173] Tao, Y.S.; Kanoh, H.; Kaneko, K., Uniform mesopore-donated zeolite Y using carbon aerogel templating. *Journal of Physical Chemistry B,* 107(40) 2003, 10974-10976.

[174] Xue, C.F.; Zhang, F.; Wu, L.M.; Zhao, D.Y., Vapor assisted "in situ" transformation of mesoporous carbon-silica composite for hierarchically porous zeolites. *Microporous and Mesoporous Materials,* 151 2012, 495-500.

[175] Kustova, M.; Egeblad, K.; Zhu, K.; Christensen, C.H., Versatile route to zeolite single crystals with controlled mesoporosity: in situ sugar decomposition for templating of hierarchical zeolites. *Chemistry of Materials,* 19(12) 2007, 2915-2917.

[176] Wang, H.; Pinnavaia, T.J., MFI zeolite with small and uniform intracrystal mesopores. *Angewandte Chemie-International Edition,* 45 2006, 7603-7606.

[177] Zhu, H.B.; Liu, Z.C.; Kong, D.J.; Wang, Y.D.; Xie, Z.K., Synthesis and catalytic performances of mesoporous zeolites templated by polyvinyl butyral gel as the mesopore directing agent. *Journal of Physical Chemistry C,* 112(44) 2008, 17257-17264.

[178] Tao, Y.S.; Kanoh, H.; Kaneko, K., Synthesis of mesoporous zeolite a by resorcinol-formaldehyde aerogel templating. *Langmuir,* 21(2) 2005, 504-507.

[179] Li, W.C.; Lu, A.H.; Palkovits, R.; Schmidt, W.; Spliethoff, B.; Schuth, F., Hierarchically structured monolithic silicalite-1 consisting of crystallized nanoparticles and its performance in the Beckmann rearrangement of cyclohexanone oxime. *Journal of the American Chemical Society,* 127(36) 2005, 12595-12600.

[180] Zhu, H.; Liu, Z.; Wang, Y.; Kong, D.; Yuan, X.; Xie, Z., Nanosized $CaCO_3$ as hard template for creation of intracrystal pores within silicalite-1 crystal. *Chemistry of Materials,* 20(3) 2008, 1134-1139.

[181] Zhang, B.J.; Davis, S.A.; Mendelson, N.H.; Mann, S., Bacterial templating of zeolite fibres with hierarchical structure. *Chemical Communications,* (9) 2000, 781-782.

[182] Dong, A.G.; Wang, Y.J.; Tang, Y.; Ren, N.; Zhang, Y.H.; Yue, J.H.; Gao, Z., Zeolitic tissue through wood cell templating. *Advanced Materials,* 14(12) 2002, 926-929.

[183] Zhang, B.J.; Davis, S.A.; Mann, S., Starch gel templating of spongelike macroporous silicalite monoliths and mesoporous films. *Chemistry of Materials,* 14(3) 2002, 1369-1375.

[184] Valtchev, V.; Smaihi, M.; Faust, A.C.; Vidal, L., Dual templating function of Equisetum arvense in the preparation of zeolite macrostructures. Recent Advances in

the Science and Technology of Zeolites and Related Materials, Pts A - C, 154 2004, 588-592.
[185] Valtchev, V.P.; Smaihi, M.; Faust, A.C.; Vidal, L., Equisetum arvense templating of zeolite beta macrostructures with hierarchical porosity. *Chemistry of Materials,* 16(7) 2004, 1350-1355.
[186] Song, C.E.; Lee, S.G., Supported chiral catalysts on inorganic materials. *Chemical Reviews,* 102(10) 2002, 3495-3524.
[187] Fan, Q.H.; Li, Y.M.; Chan, A.S.C., Recoverable catalysts for asymmetric organic synthesis. *Chemical Reviews,* 102(10) 2002, 3385-3465.
[188] Valtchev, V.; Mintova, S., Layer-by-layer preparation of zeolite coatings of nanosized crystals. *Microporous and Mesoporous Materials,* 43(1) 2001, 41-49.
[189] Valtchev, V., Silicalite-1 hollow spheres and bodies with a regular system of macrocavities. *Chemistry of Materials,* 14(10) 2002, 4371-4377.
[190] Valtchev, V., Core-shell polystyrene/zeolite A microbeads. *Chemistry of Materials,* 14(3) 2002, 956-958.
[191] Chu, N.B.; Wang, J.Q.; Zhang, Y.; Yang, J.H.; Lu, J.M.; Yin, D.H., Nestlike hollow hierarchical MCM-22 microspheres: Synthesis and exceptional catalytic properties. *Chemistry of Materials,* 22(9) 2010, 2757-2763.
[192] Holm, M.S.; Egeblad, K.; Vennestrom, P.N.R.; Hartmann, C.G.; Kustova, M.; Christensen, C.H., Enhancing the porosity of mesoporous carbon-templated ZSM-5 by desilication. *European Journal of Inorganic Chemistry,* (33) 2008, 5185-5189.
[193] Pérez-Ramírez, J.; Christensen, C.H.; Egeblad, K.; Christensen, C.H.; Groen, J.C., Hierarchical zeolites: enhanced utilisation of microporous crystals in catalysis by advances in materials design. *Chemical Society Reviews,* 37(11) 2008, 2530-2542.
[194] Zhang, S.Z.; Chen, S.L.; Dong, P., Synthesis, Characterization and hydroisomerization performance of SAPO-11 molecular sieves with caverns by polymer spheres. *Catalysis Letters,* 136(1-2) 2010, 126-133.
[195] Na, K.; Choi, M.; Ryoo, R., Recent advances in the synthesis of hierarchically nanoporous zeolites. *Microporous and Mesoporous Materials,* 166 2013, 3-19.
[196] Tao, Q.; Xu, Z.Y.; Wang, J.H.; Liu, F.L.; Wan, H.Q.; Zheng, S.R., Adsorption of humic acid to aminopropyl functionalized SBA-15. *Microporous and Mesoporous Materials,* 131(1-3) 2010, 177-185.
[197] Beck, J.S.; Vartuli, J.C.; Kennedy, G.J.; Kresge, C.T.; Roth, W.J.; Schramm, S.E., Molecular or supramolecular templating - Defining the role of surfactant chemistry in the formation of microporous and mesoporous molecular-sieves. *Chemistry of Materials,* 6(10) 1994, 1816-1821.
[198] Gu, F.N.; Wei, F.; Yang, J.Y.; Lin, N.; Lin, W.G.; Wang, Y.; Zhu, J.H., New strategy to synthesis of hierarchical mesoporous zeolites. *Chemistry of Materials,* 22(8) 2010, 2442-2450.
[199] Choi, M.; Cho, H.S.; Srivastava, R.; Venkatesan, C.; Choi, D.H.; Ryoo, R., Amphiphilic organosilane-directed synthesis of crystalline zeolite with tunable mesoporosity. *Nature Materials,* 5(9) 2006, 718-723.
[200] Choi, M.; Srivastava, R.; Ryoo, R., Organosilane surfactant-directed synthesis of mesoporous aluminophosphates constructed with crystalline microporous frameworks. *Chemical Communications*, (42) 2006, 4380-4382.

[201] Srivastava, R.; Choi, M.; Ryoo, R., Mesoporous materials with zeolite framework: remarkable effect of the hierarchical structure for retardation of catalyst deactivation. *Chemical Communications*, (43) 2006, 4489-4491.

[202] Garcia-Martinez, J.; Li, K.; Krishnaiah, G., A mesostructured Y zeolite as a superior FCC catalyst - from lab to refinery. *Chemical Communications*, 48(97) 2012, 11841-11843.

[203] Christensen, C.H.; Johannsen, K.; Schmidt, I.; Christensen, C.H., Catalytic benzene alkylation over mesoporous zeolite single crystals: Improving activity and selectivity with a new family of porous materials. *Journal of the American Chemical Society*, 125(44) 2003, 13370-13371.

[204] Sun, Y.Y.; Prins, R., Friedel-Crafts alkylations over hierarchical zeolite catalysts. *Applied Catalysis A: General*, 336(1-2) 2008, 11-16.

[205] Musilova, Z.; Zilkova, N.; Park, S.E.; Cejka, J., Aromatic transformations over mesoporous ZSM-5: advantages and disadvantages. *Topics in Catalysis*, 53(19-20) 2010, 1457-1469.

[206] Xu, L.; Wu, S.J.; Guan, J.Q.; Wang, H.S.; Ma, Y.Y.; Song, K.; Xu, H.Y.; Xing, H.J.; Xu, C.; Wang, Z.Q.; Kan, Q.B., Synthesis, characterization of hierarchical ZSM-5 zeolite catalyst and its catalytic performance for phenol tert-butylation reaction. *Catalysis Communications*, 9(6) 2008, 1272-1276.

[207] Park, H.J.; Heo, H.S.; Jeon, J.K.; Kim, J.; Ryoo, R.; Jeong, K.E.; Park, Y.K., Highly valuable chemicals production from catalytic upgrading of radiata pine sawdust-derived pyrolytic vapors over mesoporous MFI zeolites. *Applied Catalysis B-Environmental*, 95(3-4) 2010, 365-373.

[208] Suzuki, K.; Aoyagi, Y.; Katada, N.; Choi, M.; Ryoo, R.; Niwa, M., Acidity and catalytic activity of mesoporous ZSM-5 in comparison with zeolite ZSM-5, Al-MCM-41 and silica-alumina. *Catalysis Today*, 132(1-4) 2008, 38-45.

[209] Siddiqui, M.A.B.; Aitani, A.M.; Saeed, M.R.; Al-Khattaf, S., Enhancing the production of light olefins by catalytic cracking of FCC naphtha over mesoporous ZSM-5 catalyst. *Topics in Catalysis*, 53(19-20) 2010, 1387-1393.

[210] Lei, Q.; Zhao, T.B.; Li, F.Y.; Zhang, L.L.; Wang, Y., Catalytic cracking of large molecules over hierarchical zeolites. *Chemical Communications*, (16) 2006, 1769-1771.

[211] Christensen, C.H.; Schmidt, I.; Christensen, C.H., Improved performance of mesoporous zeolite single crystals in catalytic cracking and isomerization of n-hexadecane. *Catalysis Communications*, 5(9) 2004, 543-546.

[212] Mei, C.S.; Wen, P.Y.; Liu, Z.C.; Liu, H.X.; Wang, Y.D.; Yang, W.M.; Xie, Z.K.; Hua, W.M.; Gao, Z., Selective production of propylene from methanol: Mesoporosity development in high silica HZSM-5. *Journal of Catalysis*, 258(1) 2008, 243-249.

[213] Park, D.H.; Kim, S.S.; Wang, H.; Pinnavaia, T.J.; Papapetrou, M.C.; Lappas, A.A.; Triantafyllidis, K.S., Selective petroleum refining over a zeolite catalyst with small intracrystal mesopores. *Angewandte Chemie-International Edition*, 48(41) 2009, 7645-7648.

[214] Xiao, F.S.; Wang, L.F.; Yin, C.Y.; Lin, K.F.; Di, Y.; Li, J.X.; Xu, R.R.; Su, D.S.; Schlogl, R.; Yokoi, T.; Tatsumi, T., Catalytic properties of hierarchical mesoporous zeolites templated with a mixture of small organic ammonium salts and mesoscale cationic polymers. *Angewandte Chemie-International Edition*, 45(19) 2006, 3090-3093.

[215] Derouane, E.G.; Schmidt, I.; Lachas, H.; Christensen, C.J.H., Improved performance of nano-size H-BEA zeolite catalysts for the Friedel-Crafts acetylation of anisole by acetic anhydride. *Catalysis Letters*, 95(1-2) 2004, 13-17.

[216] Reichinger, M.; Schmidt, W.; van den Berg, M.W.E.; Aerts, A.; Martens, J.A.; Kirschhock, C.E.A.; Gies, H.; Grunert, W., Alkene epoxidation with mesoporous materials assembled from TS-1 seeds - Is there a hierarchical pore system? *Journal of Catalysis*, 269(2) 2010, 367-375.

[217] Moushey, D.L.; Smirniotis, P.G., n-Heptane hydroisomerization over mesoporous zeolites made by utilizing carbon particles as the template for mesoporosity. *Catalysis Letters*, 129(1-2) 2009, 20-25.

[218] Guo, L.; Fan, Y.; Bao, X.J.; Shi, G.; Liu, H.Y., Two-stage surfactant-assisted crystallization for enhancing SAPO-11 acidity to improve n-octane di-branched isomerization. *Journal of Catalysis*, 301 2013, 162-173.

[219] Garcia-Martinez, J.; Johnson, M.; Valla, J.; Li, K.H.; Ying, J.Y., Mesostructured zeolite Y-high hydrothermal stability and superior FCC catalytic performance. *Catalysis Science & Technology*, 2(5) 2012, 987-994.

[220] Parton, R.; Uytterhoeven, L.; Martens, J.A.; Jacobs, P.A.; Froment, G.F., Synergism of ZSM-22 and Y-zeolites in the bifunctional conversion of n-alkanes. *Applied Catalysis*, 76(1) 1991, 131-142.

[221] Mao, R.L.V., Hybrid catalyst containing a microporous zeolite and a mesoporous cocatalyst forming a pore continuum for a better desorption of reaction products. *Microporous and Mesoporous Materials*, 28(1) 1999, 9-17.

[222] Kinger, G.; Majda, D.; Vinek, H., n-Heptane hydroisomerization over Pt-containing mixtures of zeolites with inert materials. *Applied Catalysis A: General*, 225(1-2) 2002, 301-312.

[223] Le Van Mao, R.; Al-Yassir, N.; Nguyen, D.T.T., Experimental evidence for the pore continuum in hybrid catalysts used in the selective deep catalytic cracking of n-hexane and petroleum naphthas. *Microporous and Mesoporous Materials*, 85(1-2) 2005, 176-182.

[224] Tompsett, G.A.; Conner, W.C.; Yngvesson, K.S., Microwave synthesis of nanoporous materials. *ChemPhysChem*, 7(2) 2006, 296-319.

[225] Jacob, J.; Chia, L.H.L.; Boey, F.Y.C., Thermal and nonthermal interaction of microwave-radiation with materials. *Journal of Materials Science*, 30(21) 1995, 5321-5327.

[226] Chu, P.; Dwyer, F.G.; Vartuli, J.C., Crystallization method employing microwave radiation, US 4778666, 1988.

[227] Arafat, A.; Jansen, J.C.; Ebaid, A.R.; Vanbekkum, H., Microwave preparation of zeolite-Y and ZSM-5. *Zeolites*, 13(3) 1993, 162-165.

[228] Park, S.E.; Chang, J.S.; Hwang, Y.K.; Kim, D.S.; Jhung, S.H.; Hwang, J.S., Supramolecular interactions and morphology control in microwave synthesis of nanoporous materials. *Catalysis Surveys from Asia*, 8(2) 2004, 91-110.

[229] Newalkar, B.L.; Olanrewaju, J.; Komarneni, S., Direct synthesis of titanium-substituted mesoporous SBA-15 molecular sieve under microwave-hydrothermal conditions. *Chemistry of Materials*, 13(2) 2001, 552-557.

[230] Newalkar, B.L.; Olanrewaju, J.; Komarneni, S., Microwave-hydrothermal synthesis and characterization of zirconium substituted SBA-15 mesoporous silica. *Journal of Physical Chemistry B,* 105(35) 2001, 8356-8360.

[231] Girnus, I.; Jancke, K.; Vetter, R.; Richtermendau, J.; Caro, J., Large ALPO 4-5 crystals by microwave-heating. *Zeolites,* 15(1) 1995, 33-39.

[232] Jhung, S.H.; Chang, J.S.; Hwang, J.S.; Park, S.E., Selective formation of SAPO-5 and SAPO-34 molecular sieves with microwave irradiation and hydrothermal heating. *Microporous and Mesoporous Materials,* 64(1-3) 2003, 33-39.

[233] Katsuki, H.; Furuta, S.; Komarneni, S., Microwave versus conventional-hydrothermal synthesis of NaY zeolite. *Journal of Porous Materials,* 8(1) 2001, 5-12.

[234] Slangen, P.M.; Jansen, J.C.; vanBekkum, H., The effect of ageing on the microwave synthesis of zeolite NaA. *Microporous Materials,* 9(5-6) 1997, 259-265.

[235] Maher, P.K.; McDaniel, C.V., Process for increasing the thermal stability of synthetic faujasite, US 3374056, 1968.

[236] Bowes, E.; Pelrine, B.P., Treatment with oxalate, US4388177, 1983.

[237] Apelian, M.R.; Degnan, T.F.; Fung, A.S.; Kennedy, G.J., *Process for the dealumination of zeolite Beta,* US5200168, 1993.

[238] Apelian, M.R.; Degnan, T.F., Process for the dealumination of mordenite, US5238677, 1993.

[239] Apelian, M.R.; Bell, W.K.; Fung, S.L.A.; Haag, W.O.; Venkat, C.R., Selective surface dealumination of zeolites and use of the surface-dealuminated zeolites, CA2094947, 2003.

[240] Mingxing, S.; Jianweil, S.; Yunfei, L., Method for preparing modified beta zeolite catalyst used for alkylation reaction of isobutane/butylene, CN1124891, 2003.

[241] Timken, H.K.C.; Chester, A.W.; Ardito, S.; Hagemeister, M.P., Using zeolite selectively deluminated to remove nonframework aluminum; making alkyl naphthalenes for example, US6747182, 2004.

[242] Timken, H.K.C.; Bull, L.M.; Harris, T.V., Extremely low acidity ultrastable Y zeolite catalyst composition and process, US 6860986, 2005.

[243] Jacobsen, C.J.H.; Herbst, K.; Schmidt, I.; Pehrson, S.; Dahl, S., Method of preparing zeolite single crystals with straight mesopores, EP1284237, 2006.

[244] Wang, L., Hydrocracking catalyst containing beta and Y zeolites, and process for its use to make jet fuel or distillate, US7585405, 2009.

[245] Jan, D.Y.; Miller, R.M.; Koljack, M.P.; Bauer, J.E.; Bogdan, P.L.; Lewis, G.J.; Gajda, G.J.; Koster, S.C.; Gatter, M.G.; Moscoso, J.G., Hydrocarbon conversion processes using a catalyst comprising a UZM-8HS composition, US7638667, 2009.

[246] Sobolevskiy, A.; Rossin, J.A.; Knapke, M.J., High temperature catalyst and process for selective catalytic reduction of NOx in exhaust gases of fossil fuel combustion, US7695703, 2010.

[247] Wang, L., Hydrocracking catalyst containing Beta and Y zeolites, and process for its use to produce naphtha, CA2627337, 2012.

[248] Jan, D.Y., Catalysts for improved cumene production and method of making and using same, US8524966, 2013.

[249] Kaliaguine, S.; Do, T.O., Mesoporous zeolitic material with microporous crystalline mesopore walls, US6669924, 2003.

[250] Chaumonnot, A.; Coupe, A.; Sanchez, C.; Euzen, P.; Boissiere, C.; Grosso, D., Material with hierarchized porosity, comprising silicium, CN100522810, 2009.
[251] Christensen, C.H.; Zhu, K.; Kegnaes, M.; Egeblad, K., Fabrication of hierarchical zeolites, US7824657, 2010.
[252] Chaumonnot, A.; Coupe, A.; Sanchez, C.; Boissiere, C., Crystallised material with hierarchised porosity and containing silicon, EP2197794, 2011.
[253] Chaumonnot, A.; Coupe, A.; Sanchez, C.; Boissiere, C.; Martin, M., Mesostructured material having a high aluminium content and composed of spherical particles of specific size, EP2274236, 2012.
[254] Chaumonnot, A.; Pega, S.; Sanchez, C.; Boissiere, C., Amorphous silicon-containing material with hierarchical and organized porosity, US8236419, 2012.
[255] Wu, J., Extra mesoporous Y zeolite, US8361434, 2013.
[256] Chang, L.; Yonghui, D.; Dongyuan, Z., Three-dimensional interconnected hierarchical-structured zeolite molecular sieve material and preparation method thereof, CN102390843, 2013.
[257] Garcia-Martinez, J.; Johnson, M.M.; Valla, I., Introduction of mesoporosity in low Si/Al zeolites, US8486369, 2013.
[258] Chaumonnot, A.; Coupe, A.; Sanchez, C.; Boissiere, C., Crystallized material with hierarchical porosity containing silicon, US8685366, 2014.
[259] Li, K.H.; Valla, J.; Garcia-Martinez, J., Realizing the commercial potential of hierarchical zeolites: new opportunities in catalytic cracking. *Chemcatchem,* 6(1) 2014, 46-66.
[260] Garcia-Martinez, J.; Dight, L.B.; Speronello, B.K., Methods for enhancing the mesoporosity of zeolite-containing materials, US8685875, 2014.
[261] Serrano, D.P.; Escola, J.M.; Pizarro, P., Synthesis strategies in the search for hierarchical zeolites. *Chemical Society Reviews,* 42(9) 2013, 4004-4035.
[262] Garcia-Martinez, J.; Johnson, M.M., Methods of recovery of pore-forming agents for mesostructured materials, US8206498, 2012.
[263] Verboekend, D.; Pérez-Ramírez, J., Towards a sustainable manufacture of hierarchical zeolites *ChemSusChem,* 7(3) 2014, 651-651.

In: Comprehensive Guide for Mesoporous Materials. Volume 3   ISBN: 978-1-63463-318-5
Editor: Mahmood Aliofkhazraei   © 2015 Nova Science Publishers, Inc.

Chapter 8

NOVEL MESOPOROUS MATERIALS FOR ELECTROCHEMICAL ENERGY STORAGE

J. Santos-Peña[1], J. Ortiz-Bustos[2], S. G. Real[3], A. Benítez de la Torre[4], C. Medel[4], J. Morales[4], M. Cruz[4] and R. Trócoli[4]

[1]Laboratoire de Physicochimie des Matériaux et Electrolytes pour l'Energie, Campus de Grandmont, Université de Tours, France,
[2]Department of Chemical and Environmental Technology, ESCET, Universidad Rey Juan Carlos, Móstoles, Spain
[3]Instituto de Investigaciones Fisicoquímicas Teóricas y Aplicadas, Facultad de Ciencias Exactas, Universidad Nacional de La Plata, La Plata, Argentina
[4]Departamento de Química Inorgánica e Ingeniería Química, Instituto Universitario de Química Fina y Nanoquímica, Universidad de Córdoba, Córdoba, Spain

ABSTRACT

Since the discovery of the environmental issues associated to the use of fosil-fuel based energy, special attention has been paid to the development of electrochemical energy storage devices.In this context, lithium ion batteries and supercapacitors have been proposed for satisfying the future energy demands. However, the current state of the art for both devices indicates that advanced materials are required to a truly implantation of both technologies for stationary and mobile applications. Mesoporous materials can play an important role in these advances. They mainly show unique textural and morphological properties that can facilitate the electrochemical reactions involved in a lithium ion battery or can improve the necessary electrolyte/electrode surface interaction in a supercapacitor. After an introduction exposing the different mechanism of energy storage for lithium ion batteries and supercapacitors, we will focus on how the interesting properties of the mesoporous materials can influence the performance of themost common electrode materials. Finally, we will show our more recent research in the field, namely mesoporous $LiFePO_4$ and template mesoporous carbons for lithium ion battery and supercapacitor electrodes, respectively.

INTRODUCTION

Lithium ion batteries, supercapacitors and fuel cells are clean and rechargeable energy alternatives to the traditional primary fossil fuel-based technologies [1,2]. The development of the current energy market is focused on two essential characteristic of these devices: energy density (related to charge and potential) and power density (related to energy, reaction kinetics and system resistance). For instance, for an electric vehicle, whereas the first characteristic determines the degree of autonomy (running time) of the car, the second one is related to the power properties (acceleration, braking...). Environmentally speaking, fuel cells are the most attractive devices since the products of the electrochemical reactions involved, mainly water, are not toxic. However, according to the classical Ragone plot (Figure 1) these are devices with very low power densities. Lithium ion batteries show unique advantages in terms of energy density (100-150 $Wh \cdot kg^{-1}$) due to the high voltages that the current technology can provide (>3.8V by unit cell). However, these batteries fail in guaranteeing high power devices. Electrochemical capacitors fill the gap between devices providing high energy but low power (batteries) and those furnishing high power but low energy (traditional capacitors) [3]. Conventional capacitors can sometimes show a specific power higher than 1000 $kW \cdot dm^{-3}$ but very low energy density. Thus a combination of batteries and electrochemical capacitors represents a good compromise for a hybrid or an electric vehicle. Table I [4] resume these different technological properties of batteries, supercapacitors and capacitors.

The lithium ion battery concept starts with the works of Armand and Whittingham [5 and references therein]. The concept was a response to the various drawbacks that the use of lithium metal produced in the development of the first rechargeable lithium batteries. The growth of dendritic lithium and difficulty in handling the reactive and low melting-pointmetal accounted for safety and industrial scaling concerns. The lithium ion batteries (formerly named rocking-chair batteries) substitute lithium ions for lithium metal and save these technological limitations.

During the charge of a lithium ion battery, lithium ions are deinserted (or deintercalated) from a 3D (or 2D) inorganic material, LiM (positive plate), containing an easily reducible/oxidizable transition metal ion. Electrons cross the external circuit and at the interface of the electrolyte/negative electrode, lithium ions are inserted (or intercalated) and electrons are accepted by a host H (negative plate, typically graphite) simultaneously. During the discharge, if the process is reversible, the schema of reaction is reversed. A lithium salt dissolved in an organic carbonate is the most usual electrolyte. The schema of a lithium ion battery is shown in Figure 2. Therefore, a chemical energy is stored during the charge of the device and after solicitation (discharge) the chemical reaction between positive and negative plates provides electrical energy.

Different systems (as we will show later) have the ability of reversibly react with lithium. The number of mole of lithium involved by mole of cathode or anode will determine the capacity (charge stored) of the final system. Therefore, lithium-rich compounds (Li_xM, HLi_y) are extensively researched. However and since lithium ion devices deal with inorganic solids, the reversibility strongly depends on each particular architecture, on the transition metal implied, and on the intrinsic ionic and electronic properties of the network. Regarding the power, lithium ion diffusion remains a challenge for practically all the materials since lithium ions diffuses very slowly and can limit the fast delivery of energy. All these features and other specific for each electrode material/electrolyte couple, determine that the lithium ion battery lacks, at this moment, of a cycling efficiency higher than 10000 cycles [1].

Table I. Technological characteristics of batteries and capacitors

Characteristic	Battery	Supercapacitor	Capacitor
Storage mechanism	Chemical	Electrostatic/Chemical	Electrostatic
Energy (Wh·kg^{-1})	~20-150	1-10	<0.1
Power (Wh·kg^{-1})	<1000	5000-10000	>>10000
Discharge time	0.3-3h	Seconds to minutes	10^{-6}-10^{-3} s
Charging time	1-5 h	Seconds to minutes	10^{-6}-10^{-3} s
Efficiency	0.7-0.85	0.85-0.99	~1.0
Cycle life (cycles)	~1500	>10 years	>>10 years
Vmax (V)	High	<3V	Low

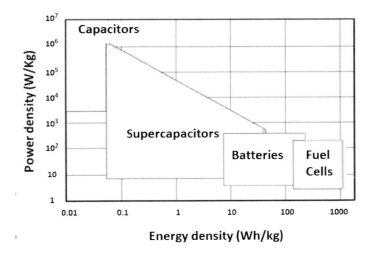

Figure 1. Ragone plot of various electrochemical energy storage devices.

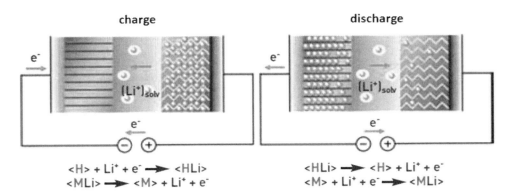

Figure 2. Scheme for a lithium ion battery based on a host (H) acting as negative electrode and a material (M) as positive electrode.

Because of the particular charge storage mechanism, electrochemical capacitors are known to show a long cycling life (one million of cycles or more) [2,3]. Other technological advantages are: fast charge/discharge time (a few miliseconds to a few seconds), high coulombic efficiency and environmental goodness if no corrosive electrolytes are present in the device. In an electrochemical capacitor, charge is stored in the electrode/electrolyte interface. Depending on the mechanism of charge storage, two different kinds of capacitors have been suggested [4]. When charge stored concerns the thin double layer formed by the electrolyte ions distributed parallel to a polarized electrode surface (Figure 3), they receive the name of electrochemical double layer capacitors (EDLC). The charge is in this case strongly dependent on the surface available to the electrolyte ions and on the electrolyte concentration. EDLC typical electrodes are carbonaceous materials: activated carbons, carbide-derived carbons, carbon fiber and aerogels…

The second mechanism for charge storage explains the use of transition metal oxides and polymers as electrochemical capacitors electrodes. Unlike for EDLC the charge is stocked by means of fast redox reactions (usually called pseudo-faradaic reactions) taking place at or near the electrode material surface. For this reason they are called electrochemical pseudo-capacitors. The major difference between this mechanism and that of EDLC is that a real charge transfer occurs in the electrode/electrolyte surface and that a redox reaction modifies the structure and properties of the electrode material, at least at a surface level. Consequently and unlike the EDLC, cycling life of these systems strongly depends on the influence of the redox reaction on the structural stability of the electrode material.

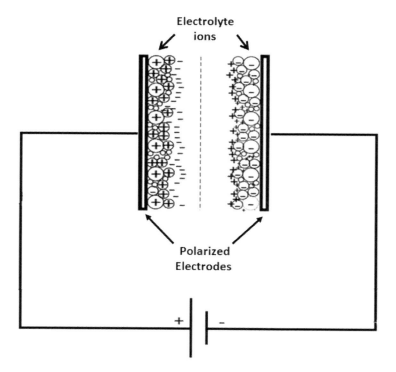

Figure 3. Scheme of an electrochemical double layer capacitor (EDLC).

When the values of capacitance surpass 100 F·g^{-1}, whatever the mechanism of charge storage, the devices are called supercapacitors or ultracapacitors. Symmetric or asymmetric devices contain, respectively, the same or different materials as positive and negative electrodes. Hybrid systems contain a capacitor material in the negative plate and a lithium ion electrode in the positive one. Electrolytes can consist of aqueous or organic solutions, with different electrochemical voltage windows and therefore, maximal values of energy and power.

For both, lithium ion batteries and supercapacitors, not only the electrode material or the electrolyte are important. The electrode/electrolyte interface governs several of the electrochemical parameters that define the final device. Thus, extension of the faradaic or seudo-faradic reaction, charge transfer from electrolyte to electrode, double layer formation, dissolution of material upon cycling and several other factors are strongly connected with such interface and will determine the energy, power and cycling efficiency of the global system. The use of mesoporous materials [6,7] accounts for a net improvement of the properties of the surface exposed to the electrolyte, in terms of increased surface area, monomodal distribution of pores, among other textural characteristics, as well the fact that they are constituted generally of nanoparticles, decreasing the lithium ion path and increasing particles connectivity.

According to the IUPAC definition, mesoporous materials contain pores in the size range of 2-50 nm. From the discovery of Mobil Oil company [8] that allowed the development of mesoporous silica, and later, metal oxides, the use of soft or hard templates for designing new mesoporous materials has been employed. Soft templates are based on organic surfactants, polymers or even biological viruses which are flexible in shape. Hard templates are solids with pores, channels or connected hollow space. Silica and anodized aluminium oxide are typical examples of such templates. However, the use of the templates does not guarantee obtaining a mesoporous material. The literature shows that the template approach is useful for at least obtaining nanoparticles with different morphologies and excellent electrochemical properties, though.

In the next sections we will show the contribution of mesoporous materials to these appealing electrochemical energy storage devices and we will illustrate with a few examples our current work on this research field.

MESOPOROUS MATERIALS FOR LITHIUM ION BATTERIES

As mentioned above the lithium ion positive electrode contains initially lithium ions in the network. This is the case of widely studied $LiCoO_2$, $LiMn_2O_4$, $LiFePO_4$, etc. However, these chemical formulae represent a challenge for the direct preparation of their mesoporous versions. In fact, the template-assisted synthesis involves solvents (in the soft template approach) or leaching agents (in the hard template technique) that can dissolve the lithium ions. Therefore, only a few reports exist on mesoporous lithiated transition metal oxides or phosphate. Table 2 collects these examples and reports the advantages shown by these systems compared with micro- or macroporous solids. For instance, Bruce et al. [9] needed to treat mesoporous Co_3O_4 (obtained from KIT-6 hard template) with LiOH for preparing mesoporous $LiCoO_2$. The latter exhibited remarkable less capacity fade than a normal

LiCoO$_2$. Furthermore, the reported mesoporous LiFePO$_4$ materials consist basically on phosphate occluded in the mesopores of templated carbons [10]. Therefore, these materials are mesoporous C/LiFePO$_4$ composites, with a high content in carbon.

The preparation of lithium-free mesoporous systems for negative electrode purposes or low potential positive electrodes should be easier (Table 2). Nevertheless, it should be noted that too often, the template approach provides nanotube or nanowire-shaped materials and not mesoporous systems. SBA-15 and KIT-6, both mesoporous silica, can be used as template for the preparation of mesocarbons (and hybrid derivatives) [11] or β-MnO$_2$ [12], respectively.

In the first case a carbon named CMK-3 with remarkable high stable capacity (three times that of graphite) has been prepared. The 3D ordered structure, with uniform pore size, high surface area and large pore volume accounts for these excellent properties. The interconnectivity of the pore walls is another prominent morphology feature. Hollow interior of porous carbons provide extra space for the electrolyte accommodations. These properties facilitate electrolyte contact and electric and ionic diffusion diffusion/transfer and thereby result in a large capacity and favorable rate performance. Similar characteristics are shown by mesoporous β-MnO$_2$ [12]. Mesoporous voids available in this structure facilitate the interfacial Li$^+$ transfer and intercalation. The ordered mesopores decrease the effective diffusion path and increase the surface area for insertion and extraction of Li$^+$, and meanwhile act as a buffer between the blocks to alleviate the volume expansion caused by repeated ionic intercalation. Additionally it hinders the undesirable conversion of β-MnO$_2$ to LiMn$_2$O$_4$ upon cycling.

Table 2. Selected reports on mesoporous materials used as lithium ion battery electrodes

Mesoporous material	Preparation	Observed advantages	Ref
LiCoO$_2$	Thermal treatment of mesoporous Co$_3$O$_4$ with LiOH	Limited capacity fading	[9]
C/LiFePO$_4$	Thermal treatment of mesoporous carbon with LiFePO$_4$ precursor	Excellent rate capability	[10]
TMC	Thermal treatment of sugar impregnated in SBA-15 and KIT-6 mesoporous silica in presence of sulphuric acid	High stable capacities	[11]
β-MnO$_2$	Thermal decomposition of Mn(NO$_3$)$_2$ impregnated in KIT-6	Increased reversibility	[12]
Anatase TiO$_2$	Hydrolysis of TiCl$_4$ in the presence of Pluronic P-123	Fast capacity and lithium- insertion charging	[13]
Sn	Electrodeposition of lyotropic liquid crystals based on SnSO$_4$	High reversibility	[14]
SnO$_2$	Calcination of SnCl$_2$/P123 solutions	Enhanced capacity retention	[15]
FePO$_4$	Precipitation of iron phosphate templated by CTAB	Increased reversibility	[17]

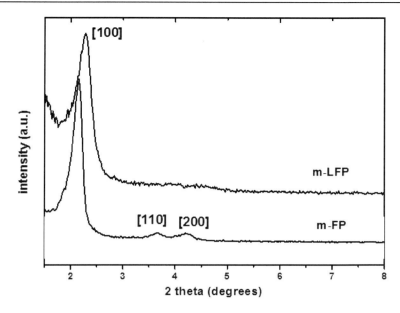

Figure 4. Small–angle XRD patterns for iron phosphate based mesoporous materials.

The porous nanostructure also accounts for the fast kinetics of solid-state reaction in mesoporous titanium oxide [13], and the suppression of nanotin aggregation in mesoporous tin [14] or tin oxides [15]. The three systems can readily be prepared by using a block copolymer as template and using chemical or electrochemical approaches.

One of the most interesting current lithium ion electrodes is LiFePO$_4$, a phospholivine structure showing stable potential plateau at 3.5V and relatively high theoretical capacity, 170 mAh·g^{-1} [16]. Our synthesis approach consisted of preparing mesoporous iron phosphate, which was then chemically lithiated and crystallized *in vacuo* to obtain an electroactive material. Details on the preparation of mesoporous iron phosphate (m-FP) can be found elsewhere [17]. An amount of 0.7 g of iron phosphate (m-FP) was immersed in a 1 M LiI solution in acetonitrile, the system immediately changes to red color. After stirring at r.t. for 18h a solidwas recovered by filtration and washed with acetonitrile first and then water. The sample thus obtained was named m-LFP. The phosphate was crystallized by heating 0.4 g of m-LFP at 550°C for 3 h under dynamic vacuum. The resulting sample was named m-LFP550.

Figure 4 shows the XRD patterns at a small angle for the different mesoporous materials prepared in this work. The mesoporous phosphate, m−FP, exhibited an ordered pore distribution, with peaks corresponding to the [100], [110] and [200] planes at 2.1°, 3.6° and 4.1° (2Θ). Cell parameter a_0 was calculated from Figure 4, using the expression $a_0 = 2d_{100}/\sqrt{3}$. a_0 was 4.1 nm for m−FP. Lithiation of this sample resulted in the cell contracting from 4.1 to 3.88 nm. This contraction was accompanied by an increased pores disorder, suggested by the broadening in the peak for the [100] plane (Figure 4).

The two mesoporous phosphates (m−FP and m−LFP) were subjected to nitrogen adsorption/desorption measurements. The isotherms for both samples (Figure 5) were type IV in the IUPAC classification and exhibited a pronounced hysteresis loop which was assigned to a well-defined mesoporous system.

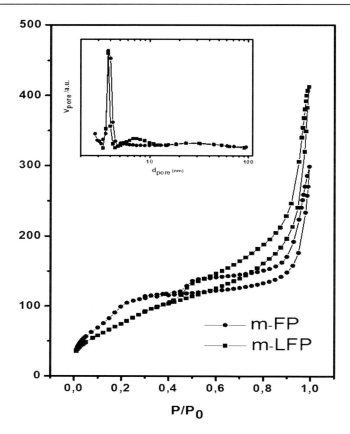

Figure 5. Nitrogen sorption isotherms for mesoporous samples synthesized, at 77K. The inset shows the corresponding BJH pore size distribution curve as calculated from the adsorption branch of the mesoporous samples.

Table 3. Textural properties of the phosphate materials prepared in this work

Sample	S_{BET} (m^2·g^{-1})	V_{pore} (cm^3·g^{-1})	$V_{micropore}$ (cm^3·g^{-1})	$<d_{pore}>$ (nm)
m−FP	395	0.373	0.090	4.7
m−LFP	136	0.261	0.036	8.7
m−LFP550	13	0.142	0.005	47

Table 3 collects the specific surface areas, derived from Figure 5. Under the assumption that the solid consisted of spherical particles, a diameter of 30 nm was calculated from S_{BET} and the phosphate density. Samples m−FP and m−LFP exhibited a significant increase in S_{BET} values, which were 3.5−4.5 times higher to that obtained for amorphous iron phosphate and its lithiated product. Also, reaction with LiI halved the surface area. The pore size distribution of both samples is shown as an inset in Figure 2b. Sample m−FP gave a rather symmetric peak centred around 3.7 nm. A similar peak in the same range was also observed for m−LFP; the profile, however, was somewhat more complex owing to the presence of a weak, broad peak centered at 8.0 nm.

The mesopore diameter was estimated from [18]

$$d_{pore} = cd_{100}[\rho V_p/(1+\rho V_p)]^{1/2} \qquad [1]$$

where V_p is the mesopore volume, ρ the density of pore walls, c a geometric parameter equal to 1.213 for cylindrical pores, and d_{100} the spacing corresponding to the [100] plane. If we assume pore to be cylindrical and pore walls to be built from $FePO_4 \cdot 2H_2O$ (average density $\approx$ 2.79 g·cm^{-3}), then d_{pore} is 3.55 nm for m−FP. This value is close to that derived from the isotherms. However, the corresponding value for m−LFP is 3.27 nm or 3.05 nm depending on whether pore walls consist of $LiFePO_4$ or $FePO_4 \cdot 2H_2O$ −could not obtain the value for $LiFePO_4 \cdot 2H_2O$−, respectively. These values are somewhat lower than those obtained from adsorption measurements. In any case, the pore system observed by XRD is more ill−defined than that of sample m-FP, which is consistent with a more heterogenous pore size distribution as suggested by the curve profile (see inset of Figure 5).

TEM images of the m−LFP sample confirmed its porous structure. Figure 6a reveals that the sample is made up of rounded-shape nanoparticles of heterogeneous size that keep an excellent connection. A closer examination of these particles (Figure 6b and inset) reveals a distribution of mesopores. In a first approach, we investigated the electroactivity of non-crystalline lithium iron phosphates (m−LFP). The electrolyte used was 1 M $LiPF_6$ in ethylene carbonate (EC) and dimethyl carbonate (DMC) in a 1:1 w/w ratio. To our knowledge, this is the first time amorphous "$LiFePO_4$" and "$LiFePO_4 \cdot 2H_2O$" have been tested as electrodes in lithium half−cells with an organic electrolyte.

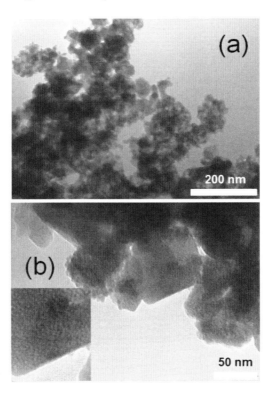

Figure 6. Selected TEM images of the m−LFP sample at different magnifications: (a) x 92000, (b) x 340000. Inset corresponds to a magnification of the area (b).

Figure 7 (top) shows the potentiostatic curves obtained between 3 and 4 V versus the Li/Li$^+$ redox couple. During charge, the simplified reaction expected to take place in a LiFePO$_4$ electrode is a two–phase process described by [16]

$$LiFe^{II}PO_4 \rightarrow Fe^{III}PO_4 + Li^+ + e^- \qquad (1)$$

No similar scheme has been reported for LiFePO$_4$·2H$_2$O. However, the reverse reaction has been found to take place at 3 V and confirmed to follow the scheme [17]

$$Fe^{III}PO_4 \cdot 2H_2O + Li^+ + e^- \rightarrow LiFe^{II}PO_4 \cdot 2H_2O \qquad (2)$$

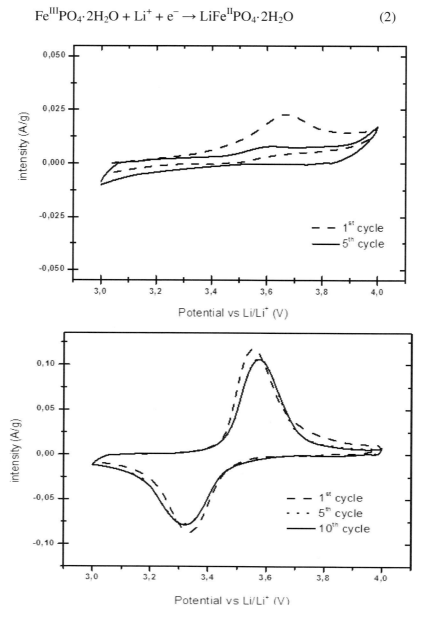

Figure 7. Potentiostatic tests on Li/LiPF$_6$(EC,DMC) half–cells containing (top) m-LFP and (bottom) m-LFP550 as positive electrodes. Voltage steps: 10 mV·10 min^{-1}.

However, based on the curves in Figure 7 (top), such material is not electrochemically active versus lithium; in fact, no oxidation or reduction bands were observed during charge and discharge processes, respectively. These results unambiguously indicate that lithium ions require a well–defined pathway for deinserting/inserting in the LiFePO$_4$ structure (note that the iron phosphate units in a–LFP and m–LFP are disordered in the long range); therefore, the LiFe^{II}PO$_4$·2H$_2$O phase must be created in situ in the electrode in order to be electrochemically active.

The m–LFP systems were heated under dynamic vacuum at 550°C for 3h in order to crystallize a compound named m–LFP550. Unfortunately, the XRD patterns contained peaks of Li$_3$PO$_4$ along with that expected for LiFe^{II}PO$_4$. A possible explanation for the occurrence of this salt is that the necessary heating for crystallization induces the partial dissociation of the lithium iron phosphate. Finally, the occurrence of a strong peak located close to 45° for m–LFP550 and easily ascribed to metallic iron, is noteworthy. The thermograms for the m–LFP system (not shown therein) revealed the presence of a carbothermal process below 600°C involving residual carbon resulting from incomplete pyrolysis of CTAB and acetates. This residual carbon can reduce the potential iron oxides formed during the heating (similarly to a–LFP550) and results in the conspicuous presence of metallic iron in sample m–LFP550. An estimation of the weight ratio of the phases present in m–LFP550 was 4% Fe, 26.6% Li$_3$PO$_4$ and 69.4% LiFePO$_4$.

The nitrogen adsorption/desorption isotherms for m–LFP550 indicated that the thermal treatment provokes collapsing of the pores and, as a result, the m–LFP system does no retain its mesoporosity. However, some textural properties like pores volume or specific area surface, which are both beneficial properties for electrochemical purposes, are significantly high for m–LFP550 (Table 3). Finally, the examination of both materials by means of TEM microscopy (Figure 8) showed that m–LFP550 is constituted by round shaped nanoparticles (average radii smaller than 100 nm) leaving pores between them. These observations agree with the textural parameters collected in Table 3.

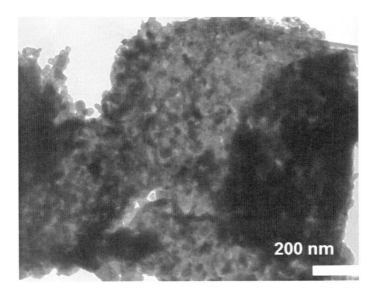

Figure 8. TEM image of the m–LFP550 sample.

Unlike the amorphous material, the crystalline m−LFP550 exhibited electroactivity (Figure 7 (bottom)) in the lithium half-cells. Thus, the potentiostatic tests on m−LFP550 based cell revealed a strong oxidation peak at ca. 3.55 V and mainly associated to reaction (**1**), although the oxidation of iron present in this sample is also possible in this voltage range. During the reduction process, a cathodic wave centered at 3.35 V was observed. The difference between the centers of the two peaks (ΔE_p) was 200 mV, which is consistent with a LiFePO$_4$–based lithium ion electrode [16]. Peak profiles were strongly retained upon cycling.

Selected galvanostatic charge/discharge curves for the LiFePO$_4$–based electrode are shown in Figure 9. In agreement with the potentiostatic results, the first charge curves exhibited a voltage pseudoplateau located at 3.5 V for m−LFP550. The charge capacities obtained were 107 mAh·g^{-1}, being lower compared to that reported for carbon-coated lithium iron phosphates [19-26]. However, these capacities reach 153 mAh·g^{-1} when referred to the mass of LiFePO$_4$ (70% of the material). In addition, the first discharge curve indicated a limited reversibility of reaction (**1**). The profile contains a wide pseudoplateau, close to 3.35 V. The discharge capacity, 58 mAh·g^{-1} for m−LFP550 is limited, although referred to the mass of LiFePO$_4$, the capacities shift to 82 mAh·g^{-1}. The successive charge profiles of the m−LFP550 electrode remain similar. However, after 25 cycles, the capacities provided are in 51 mAh·g^{-1} (73 mAh·g^{-1} if referred to LiFePO$_4$) which is below those found in carbon-coated systems [17].

Electrochemical impedance spectroscopy was applied on m−LFP550 electrode at OCV. The corresponding Nyquist plot, shown in Figure 10, was found to fit well a Randles classic model. The electrolyte resistance is only 1 Ω which proves a good electrolyte impregnation in the porous and high surface area structure. In line with this property, the values found for C_{dl} were very high for m−LFP550 (190 µF). Furthermore, the value of R_{ct} is 29 Ω for m-LFP550 due to the presence of a big amount of insulating Li$_3$PO$_4$. Therefore, the observed ΔE_p in the potentiostatic and the galvanostatic tests reflects the presence of a kinetic factor resulting from lithium insertion. The estimated value of the lithium ion diffusion coefficient is $8.8 \cdot 10^{-15}$ cm^2·s^{-1} for m−LFP550. This value is somewhat lower than those found in the phosphate based materials [17] and can explain to some extent their low capacities.

In spite of some advantageous textural properties and particle size, other features account for the limited performances shown by the phosphates under study. Regarding the technological advances in the development of lithium iron phosphates, it seems that the most important feature is the lack of an adequate, conductive carbon coating onto the phosphate particles [17]. In fact, the performance of mesoporous lithium iron phosphates reported in the literature, none of which showing pores order, depends greatly of the carbon content [10]. For instance, at least a 3.8% weight ratio is required for a suitable electrochemical response. Thus, further work on our samples should be focused on the production, after or during the formation of the mesoporous lithium iron phosphate, of a carbon film by filling the pores with a carbon precursor and by heating the whole system at mild temperatures.

MESOPOROUS MATERIALS FOR ELECTROCHEMICAL CAPACITORS

Regarding mesoporous pseudocapacitor electrodes, the presence of mesopores endows RuO$_2$ and MnO$_2$ with a cycling profile similar to an ideal capacitor and guarantees cycling stability of the capacitance [19]. However the wide literature on the application of mesoporous materials for electrochemical capacitors deals with the use of EDLC, mainly based on carbonaceous materials [7,20-24]. Formula **(2)** correlates the typical double layer parameters (thickness (t), surface (A) and dielectric constant (ε)) to the capacitance, C, provided for an electrode.

$$C = \varepsilon \varepsilon_0 A/t \qquad [2]$$

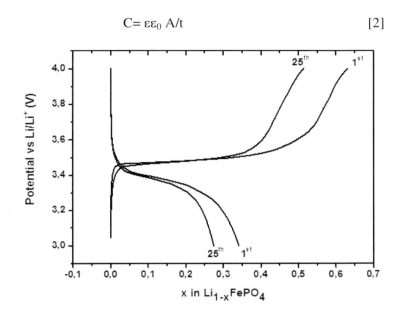

Figure 9. Selected galvanostatic charge/discharge cycles of Li/LiPF$_6$(EC,DMC) half−cells containing m−LFP550 as positive electrodes, under a C/7 regime.

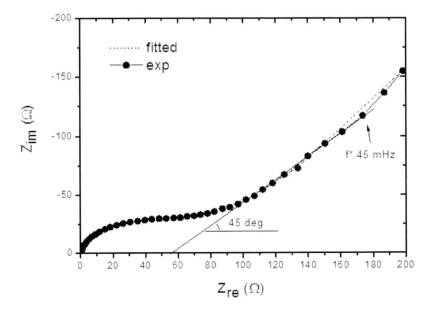

Figure 10. Nyquist plots recorded for m−LFP550 as positive electrode at OCV.

Whereas the electrochemical double layer thickness, for concentrated electrolytes, is 5-10 Å thick, an electrode provides an average capacitance of 10-20 µF·cm^{-2}. Thus, a material with a specific surface area of 1000 m^2·g^{-1} and a capacitance of 10 µF·cm^{-2} associates a capacitance of 100 F·g^{-1}. Apparently, according to (2) the higher the surface the bigger capacitance values. However, recent studies on activated carbons have shown that there is no linear relationship between the surface specific area and the capacitance [7,20-22,24-27]. In fact some studies suggested that when aqueous electrolytes are involved in the devices, pores smaller than 0.5 nm were not accessible to hydrated ions. A pore size distribution in the range 2-5 nm, which is larger than the size of two solvated ions, was identified as a way to improve the energy density and power capability [24,28-30]. Later, Chmiola et al. [31] showed that ions can desolvate and accommodate in the subnanometer (0.5 nm <d$_{pore}$<1 nm) porosity leading to dramatic increase of the capacitance. Formula (3) has been proposed to determine the value of the capacitance in the mesopores (d$_{pore}$> 2 nm):

$$C = \varepsilon\varepsilon_0 A/b \ln(b/b-d)) \quad [3]$$

where b and d are the pore radius and the distance of approach of the ion to the carbon surface respectively. As a matter of fact, the claimed capacitance originated from nanopores [31-36] has decreased the interest in investigating mesoporous materials. However, Frackowiack et al. [20] pointed that even if the micropores play an essential role for ions adsorption, the mesopores are necessary for their quick transportation to the bulk of the materials. They stated that under polarization, the solvated ions migrate easily in the mesoporous (playing the role of corridors) until reaching the entrance of micropores, being then desolvated to become adapted to the pore size. In a typical activated carbon, the situation differs. The pathway of solvated ions to reach the active surface is very tortuous and long, with several bottle-necks. Table 4 resumes these approaches to the pore effect on the capacitive properties of the electrodes.

According to this, mesoporous carbons have been prepared with the hard template technique (by using MCM-48, SBA-15, MSU-1 among others), introducing preferentially sucrose into the voids as carbon precursor [20,23,27,39-42]. The resulting Templated Mesoporous Carbons (TMC) have single modal distribution of pores, centered in 2.5-4 nm range. One of these TMC (CMK-3, described in the section devoted to lithium ion electrodes) excels in capacitive properties (170 F·g^{-1} [20] or 200 F·g^{-1} [23] in organic or aqueous electrolyte respectively).

Table 4. Pore effect on the capacitor behavior of carbonaceous materials

Pore diameter, φ (nm)	Fonction	Ref.
> 50 (macropores)	Electrolyte ions reservoirs	[7,20-22,24-27]
2-50 (mesopores)	Decrease of the charge transfer resistance and electrode tortuosity	[20,23,27,39-42]
< 2 (micropores)	Maximal capacitance due to fitting between electrolyte ions and pore size	[24,28-30]
<1 (nanopores/ ultramicropores)	Increased capacitance due to fitting between desolvated electrolyte ions and pore size	[31-36]

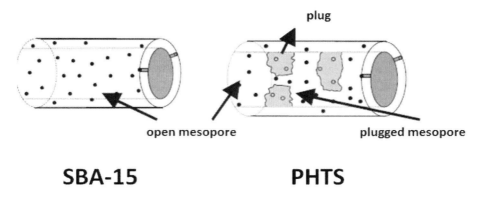

Figure 11.Schematic representation of the pore structure in SBA-15 and PHTS.

Plugged Hexagonal Templated Silica (PHTS) (Figure 11) consists of hexagonally ordered mesopores with diameters that are similar to those of SBA-15 [43]. It has thick pore walls (3-6 nm) perforated with micropores making PHTS a combined micro- and mesoporous material. PHTS also possesses microporous amorphous nanoparticles (plugs) in the uniform mesoporous channels resulting in higher micropore volumes. The pillaring effect of the nanoparticles gives PHTS a higher mechanical stability compared to pure SBA-15.

For the synthesis of PHTS, 15 g of TEOS are added to a solution containing 4 g of P123 ($EO_{20}PO_{70}EO_{20}$ copolymer), 130 ml of distilled water and 20 ml of concentrated HCl (1TEOS: 2.77HCl:192 H_2O:0.008 P123 molar ratios). After heating at 60°C under stirring for 7.5h, the mixture is transferred to an oven at 80°C and get old for 15.5 h. Finally a white solid is recovered, filtered, washed with distilled water and dried at r.t. In order to remove the template, the product is heated under air at 550°C for 6 h with a gradient of 1°C·min^{-1}.

Mesostructured Cellular Foam (MCF) is obtained by addition of a swelling agent (oftenmesitylene (1,3,5-trimethylbencene) to the synthesis of SBA-15 [43]. The swelling agent causes an enlargement of the micelle resulting in a sponge-like foam with three-dimensional structure with large uniform spherical cells (15-50 nm), accessible via large windows (5-20 nm). MCF is a very open structure with large uniform pore diameters and large pore volumes. It has thick pore walls resulting in a high hydrothermal stability. For the synthesis of MCF, 4 g of P123 and 20 mL of concentrated HCl are dissolved in 130 ml of distilled water. 0.0467 g of NH_4F and 4.6 mL of mesitylene are added to the solution and the whole stirred at 35-40°C. After one hour, 9.14 ml of TEOS (1 TEOS: 5.87 HCl: 194 H_2O: 0.017 P123: 0.031 NH_4F: 0.815 mesitylene molar ratios) are added and the mixture is stirred for 20 h. After transferring to an oven at 100°C under vacuum for 24 h, the mixture is allowed to cool. The product is filtered and three times washed with distilled water. Template is finally removed by the same treatment used for PHTS.

For the preparation of templated mesoporous carbons, 1.25 g of sucrose and 0.145 g of sulfuric acid are dissolved in 5 ml of distilled water. 1 g of the mesoporous silica is added to this solution for one hour, assisted by ultrasounds. The mixture is dried in an oven at 100°C for 12 h and then at 160°C for 24h. The formed carbon/silica composite is graphitized by a thermal treatment at 900°C under a nitrogen atmosphere for 20h. After cooling, the silica contained in the powder is removed by treating with HF, centrifuging and thoroughly washing with distilled water. Figure 12 shows the TEM images of the different mesoporous systems prepared in this work: PHTS, CPHTS, MCF, CMCF. Although both mesoporous silica show large pores their morphologies differ. Whereas MCF shows an expected foam shape the PHTS shows a surprising flaky form, different to that reported [43]. Furthermore, a morphological resemblance between the pristine silica and its replica is only found for PHTS and CPHTS. CMCF contains more compact and thicker particles tan MCF does. Finally, graphitic zones, defined by graphite fringes, are easily found in CMCF (inset in Figure 12).

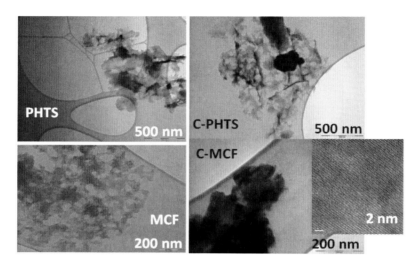

Figure 12. TEM images of the mesophases prepared in this work.

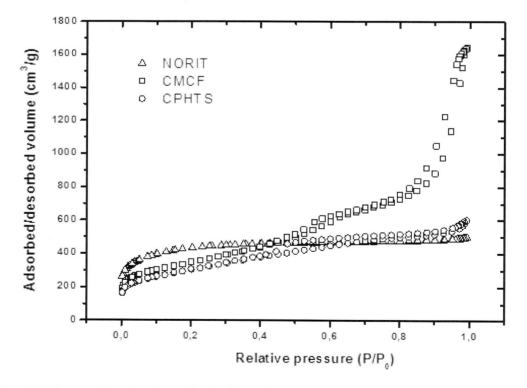

Figure 13. Nitrogen adsorption/desorption isotherms for the template mesoporous carbons prepared in this work and NORIT carbon.

Nitrogen adsorption/desorption isotherms for TMC (Figure 13) confirm the mesoporous character of these solids, (type IV in the IUPAC definition). However, the chemical and thermal process involved in their fabrication has largely contributed to significant variations from the pristine silica profiles [43]. For instance, C-PHTS losses the second desorption step according to the vanishing of the plugs. Besides, in the preparation of C-MCF pores of large size are formed and isotherms contain a new step in both branches. For comparison purposes the Figure 13 also shows the isotherm for NORIT, a microporous carbon which is a standard EDLC material. Its isotherm belongs to the IUPAC type I, corresponding to the adsorption/desorption of one monolayer of nitrogen in the system nanopores. As a consequence, no hysteresis is observed for this material.

C-PHTS shows a very narrow pore distribution centered in 2.5 nm (Figure 14), a value close to the average pore size collected in Table 5. Although C-MCF mainly developed two very narrow pore sizes in the 2.25-3.05 nm range, pores as large as 14.6 nm are also formed. Therefore, C-MCF shows the larger average pore size. Similar to C-PHTS, NORIT carbon shows a narrow distribution centered in 2.4 nm. Other interesting properties, the specific surface area and the pores volume, are collected in Table 5. There is a relevant increase of the first parameter from the pristine silica to the TMC (850 and 640 $m^{2}\cdot g^{-1}$ for PHTS and MCF respectively) but the surface area remains smaller than that of NORIT. The observed pore volumes also confirm the mesoporous character of C-PHTS and C-MCF.

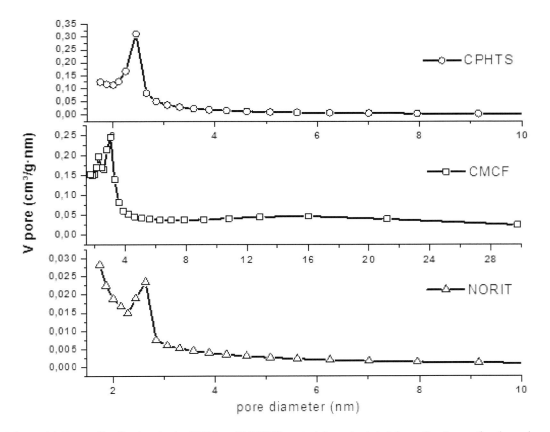

Figure 14. Pores distribution in the TCM and NORIT materials, calculated from the desorption branch by using the BJH method.

Table 5. Physicochemical parameters of the carbons studied in this work

Sample	S_{BET} (m$^2 \cdot$g^{-1})	V_{pore} (cm$^3 \cdot$g^{-1})	$V_{micropore}$ (cm$^3 \cdot$g^{-1})	$<d_{pore}>$ (nm)	D/G
NORIT	1400	0.77	0.39	2.2	3.00
C-MCF	1215	2.54	0.10	8.1	3.06
C-PHTS	1050	0.93	0.07	3.2	2.24

Raman spectroscopy is a powerful tool for the characterization of carbonaceous materials. Selected Raman spectra for the TMC and NORIT are collected in Figure 15. The spectra corresponding to the TMC can be fitted to four components, the maximum of which appear at 1194, 1340, 1510 and 1580 cm^{-1}, respectively [23]. NORIT spectra contain an additional weak component at 1442 cm^{-1}. The two bands at 1580 cm^{-1} and 1340 cm^{-1} are assigned to the E_{2g} and A_{1g} graphite modes and called G and D band, respectively. The D band is related to the breakage of symmetry occurring at the edges of graphite sheets. Therefore, the relative intensity of D and G bands is associated with the structural disorder: the smaller the D/G ratio the higher the conductivity. Furthermore, the bands at 1194, 1442 and 1597 cm^{-1} are assigned to tetrahedral and sp^3-type carbon with smaller conductivity. The values of D/G ratio for the various carbons are collected in Table 5 and reveal good conductivity properties for the C-PHTS material.

Electrode mixture was prepared by mixing the carbon, a binder (teflon) and carbon black (conductive additive) in 80:10:10 weight ratios. Preliminary electrochemical tests have been carried out on aqueous K_2SO_4 solutions with the aims of determining the stability voltage window for the three carbon materials. Choice of a neutral sulphate is based on the suitable properties of such salts to replace corrosive electrolytes (KOH, H_2SO_4) of common use in EDLC. Counter electrode consists of a platinum gauze and reference electrode was Ag/AgCl ($E^0_{(25°C)}$= +0.199V vs NHE). Figure 16 shows the adaptation of a typical first voltammetric cycle (CV, I/V profile) in 0.5M K_2SO_4 solution to a capacitance/V profile. The values of capacitance are determined by using the formula C=I/(dV/dt), where I is the observed intensity developed between the working and the counter electrode, and dV/dt corresponds to the constant voltage steps. The NORIT-based cell shows the typical box shape of an EDLC and its electrochemical stability limits are -0.9 and +0.8V vs Ag/AgCl. The TMCs prepared in this work are less stable and the anodic limit is set at +0.55V. Furthermore, every TMC cycles contain broad bands at -0.2V which are unambiguously associated to the presence of functional groups. Furthermore the average capacitance of the carbons is 80, 75 and 65 F·g^{-1} for NORIT, C-MCF and C-PHTS respectively. These values agree with a lesser value of surface area and smaller content in micropores of the TMCs compared to NORIT.

Information about the groups in TMC surfaces can be obtained by X-rays photoelectron spectroscopy (XPS). C 2p XPS spectra (Figure 17) clearly indicate a contribution of oxidized carbon species in the TMC profile, C-MCF material showing the largest one. Therefore, its I/V profile is a mirror of the presence of carbon functionalities. It has been stated that the electrochemical stability of such groups can be limited and thus, electrochemical cycling can result in a loss of capacity. This is the reason why the activated carbons, after activation, suffer processes to remove such functionalities.

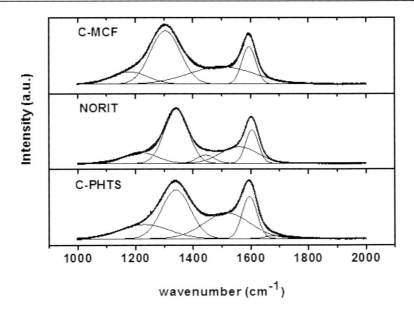

Figure 15. RAMAN spectra (included fitting with four or five components) of the carbons studied in this work.

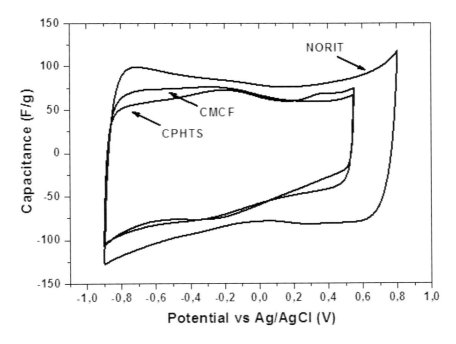

Figure 16. First voltammetric cycle of the carbons studied in this work in 0.5M $K_2SO_{4(aq)}$. Voltage steps = 20 mV·s^{-1}. Counter electrode: Pt. Reference electrode: Ag/AgCl.

In spite of the lower values found for TMC capacitances, the porous active layer seems to be more accessible for such surfaces than in NORIT. This assumption has been checked by carrying out electrochemical impedance spectroscopy (EIS) experiment. Figure 18 shows a Bode diagram (therein as log ($1/Z_{im}·\omega$) vs log (ω), where Z_{im} and ω correspond to imaginary part of the system impedance and the angular frequency, respectively). The low frequencies values of the log ($1/Z_{im}\cdot\omega$) ratio correspond to the highest capacitance that a system can develop and correspond to the double layer capacitance times the total pore surface area. The higher the frequency where the capacitance is maximal the thinner the active layer for the development of the electrochemical double layer.

As the path that the electrolyte ions must cross until reach the micropores is tortuous, the activated carbon NORIT provides the smaller capacitance at higher frequencies (Figure 18). Unlike, C-MCF and C-PHTS establishes more rapidly the electrochemical double layer in agreement with the presence of mesopores where electrolyte ions diffusion is facilitated. However, if the electrolyte ions are allowed to penetrate deeply (in the Bode plot, at the lowest frequencies) the active layer offered by the activated carbon provides a superior capacitance.

As EIS experiments approach the equilibrium, thecapacitance values calculated by EIS are always larger than that obtained by CV. Despite their different specific surface area and conductivities, NORIT and C-PHTS develop a similar value of "equilibrium" capacitance, consistent with some similar pore properties. However, referred to the total pore surface area the reported double layer capacitance are similar for both mesoporous carbons, 6.2 $\mu F\cdot cm^{-2}$, whereas NORIT provides 5.7 $\mu F\cdot cm^{-2}$ [23].

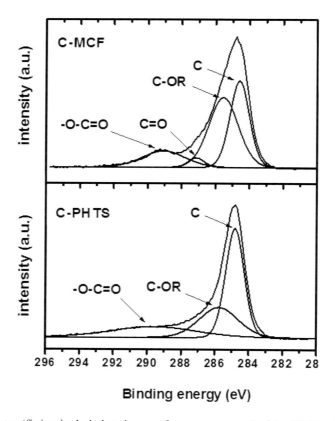

Figure 17. XPS spectra (fitting included to three or four components) of the TMCs studied in this work.

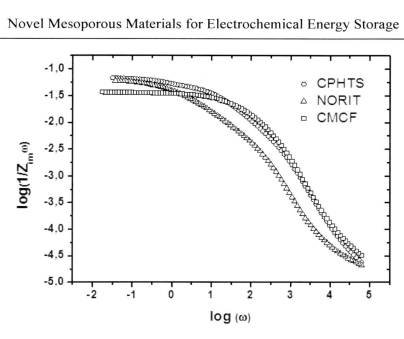

Figure 18. Bode diagramm for the TMCs and NORIT studied in this work.

Prompted by these interesting results we built symmetric devices containing as positive and negative electrode similar amounts of the same carbon. The voltage limits were fixed by those presented by the mesoporous carbons, i.e., -0.9 V (when acting as negative electrode) and +0.55 V (when acting as positive electrodes). Cycling in the 0-1.45V was initially done under galvanostatic regime of 0.5 A/g. Figure 19 shows the first (after 5 activation cycles) and 500th galvanostaticcycle for the NORIT and TMCs. For an ideal capacitor, with constant capacitance value, the profile of a V/t curve should be a straight line with slope equal to I/C. Therefore, the NORIT carbon behaves closer to ideally. Up to 1.3V, both TMC also behave ideally but for higher voltages a curvature is noticed, consistent with their CV profiles at voltages close to +0.55V and -0.9V (Figure 16). There are two other important features in the Figure 19. First, the charge/discharge time is longer for NORIT, which means that more charge is stocked in the activated carbon than in the TMC-based electrodes, in agreement with the higher value of capacitance. Secondly, the ohmic drop after the charge is stronger for NORIT. Consequently the TMC-based devices should provide higher powers.

After 500 cycles NORIT shows the same features (I/t profile and charge/discharge time). However, both TMC electrodes showed decreasing crossing charge time, indicating a loss of capacitance upon cycling. Furthermore, C-MCF still deviates from the ideal behavior beyond 1.3V and, unexpectedly, C-PHTS approximates it with the cycling.

The three systems were voltammetrically cycled after stopping at selected galvanostatic cycles (Figure 20). The CV curves are consistent with an initial activation of the NORIT electrode, according to a thicker active layer constituted of micropores, that requires more time to be accessed by the electrolyte ions in both, positive and negative plates. After such activation, the CV profile is close to the box shape until 2000 cycles, keeping a constant area (and therefore, a constant crossed charge) under the anodic or the cathodic waves. Conversely, TMCs showed a different behavior, with decreasing area under the curves and deviating from the ideal behavior upon cycling. This deviation is especially remarkable for the C-MCF based electrode in agreement with the curves in Figures 16 by virtue of a higher content in surface functionalities.

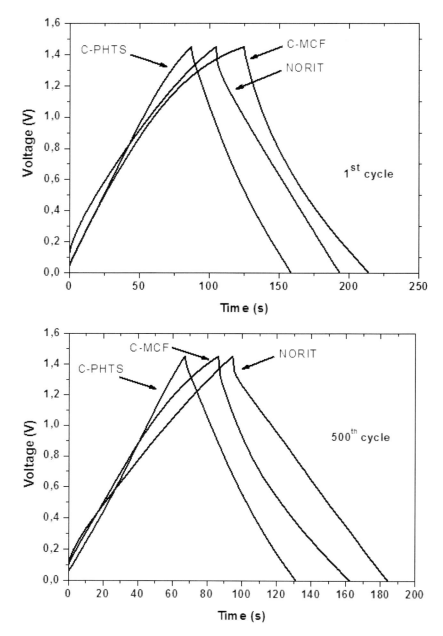

Figure 19. First and 500th galvanostatic cycle (i=0.5 A/g) of the symmetric capacitor based on NORIT and TMC-based electrodes.

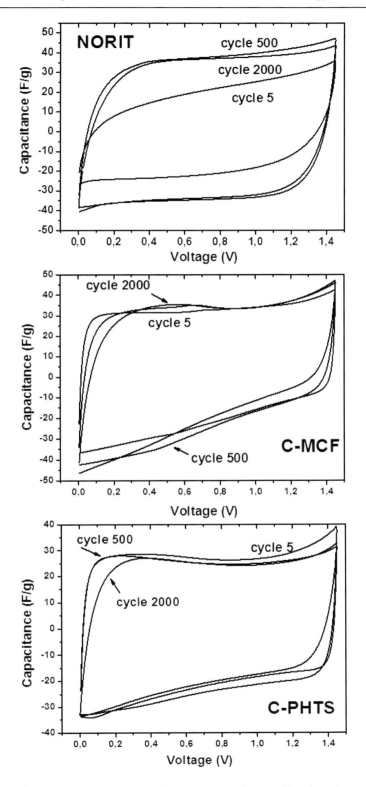

Figure 20. Cyclic voltammogramms corresponding to symmetric capacitors based on carbon electrodes after 5, 500 and 2000 cycles under 0.5 A/g.

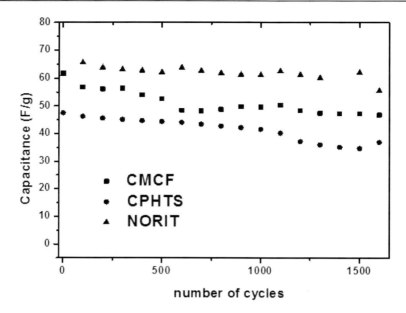

Figure 21. Capacitance versus number of galvanostatic cycle (i=0.5 A/g) of the symmetric capacitors with NORIT and TCM-based electrodes.

Evolution of the capacitance versus the number of cycles is shown in Figure 21. Seemingly the carbon conductivities are not affecting the electrochemical behavior of our electrodes. The rank C_{NORIT}> C_{C-MCF}> C_{C-PHTS} is maintained up to 2000 cycles and follows the surface area and the micropore volume trend. The presence of functional surface groups can explain the deterioration of electrochemical properties of the TMC, i.e., these groups are unstable upon cycling and the seudocapacitance associated should decrease in a parallel manner.

Although the observed capacitance trend is against the use of our TMC as electrodes, it is noteworthy that the power values should be higher for these materials. As smaller charge transfer resistance and faster electrolyte ions diffusion is expected for the mesoporous systems, the power provided by mesophases-based devices should be higher. Table 6 collects the capacitance, equivalent series resistance (ESR), real energy and maximal power values for the three devices at the first and after 500 cycles. As expected, the TMC-based capacitors show bigger values of power than the activated carbon, initially. However, upon cycling their capacitances decrease and as a consequence, energies lower (-11% at the 500th cycle).

Table 6. Energetic parameters of the supercapacitor symmetric devices studied in this work

Electrode	1st cycle				500th cycle	
	Capacitance (F/g)	ESR (Ω)	E_{real} (Wh/kg)	P_{max} (kW/kg)	Capacitance (F/g)	E_{real} (Wh/kg)
NORIT	61	17.4	8.9	26	62	8.9
C-MCF	62	7.5	9.0	64	52	7.7
C-PHTS	47	9.0	7.3	53	44	6.5

These results prompted us to continue our study of mesoporous materials building devices with organic electrolytes, where large electrolyte ions prefer to confine in mesostructures and its diffusion is hindered in the micropores. Preliminary results, in the form of cyclic voltammetries in 1.5M NEt$_4$BF$_4$ (CH$_3$CN) electrolytic solutions are shown in Figure 22. Symmetric devices based in the three carbons showed similar electrochemical behaviors. These preliminary results let imagine that TMCs would be advantageous in this kind of relatively high voltage capacitors and further efforts are now being carried out in order to evaluate and to optimize their electrochemical performances by choosing a suitable electrolyte salt.

Finally Table 7 collects previous results for comparison with those shown in this work.

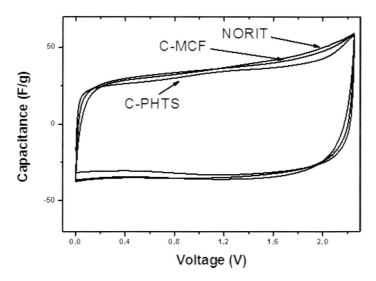

Figure 22. Fifth voltammetric cycle of the symmetric capacitors based on carbons (NORIT, C-MCF, C-PHTS) in 1.5M NEt$_4$BF$_4$ (CH$_3$CN). Voltage steps = 20 mV·s^{-1}.

Table 7. Electrochemical parameters of TMC used as electrodes in supercapacitors

Ref.	Synthesis (carbon precursor, template)	Surface area (m^2/g)	Pore volume (cm^3/g)	<d$_{pore}$> (nm)	Higher capacitance and measurement characteristics	Pseudo-capacitance
20	Propylene, sucrose or pitch inserted in MCM-48, SBA-15 or MSU-1	630-2000	0.20-0.58	2.4-3.7	202 F/g (1M H$_2$SO$_4$) 160 F/g (6M KOH) 115 F/g (1.4M NEt$_4$BF$_4$)	NO
23	Sugar inserted in SBA-15	623-1525	0.63-0.80	1.9-3.6	132 F/g (1M H$_2$SO$_4$)	YES
27	PTSA inserted in HMS	1330-2352	1.01-1.39	1.8-3.5	0.14 F/m^2 (1M H$_2$SO$_4$)	NO
41	Poly(VDC/MA) carbonisation	1033-2923				YES

CONCLUSION

The use of template-based techniques allows the preparation of various mesoporous systems with control of pore size and specific surface area. Additionally, such materials consist of small size particles, show large surface-to-volume ratio and favorable structural stability and sometimes 2D or 3D pores connectivity. These properties of the mesoporous materials are advantageous for electrochemical purposes. Surface area is especially important as faradaic or pseudofaradaic reactions take place at or near to the electrode surface. For EDLC, the surface plays a unique role in the development of the electrochemical double layer. Nanosize of the particle is effective to guarantee a decrease in the lithium ion path and to ensure a high reactivity against lithium, endowing the material with new reaction mechanism in a few occasions. Mesopores are also essential to allow electrolyte ions to diffuse from the electrolyte container to the micropores where the double layer responsible of highest capacitances is formed.

Our own work on mesoporous materials have proven that they can also be used as precursors for lithium ion electrodes with adequate textural properties or as electrochemical capacitor electrodes with suitable power performances. Unambiguously, improved electrode performances such as higher overall capacity/capacitance, better high-rate capability and longer cycling life can be obtainedby using mesoporous materials. Therefore, advanced mesoporous materials account for the development of future electrochemical energy storage devices.

ACKNOWLEDGMENTS

This work was supported by MICINN (MAT 2010-16440). The authors are indebted to Dr. Manuel Mora (Department of Organic Chemistry, University of Córdoba) for recording the Raman spectra and F. Gracia (Transmission Electronic Microscopy and Scanning Electronic Microscopy Services, S.C.A.I.) of the University of Córdoba for SEM and TEM measurements.

REFERENCES

[1] Scrosati, B; Garche, J. *J. Power Sourc.*, 2010, 195, 2419-2430.
[2] Conway, BE. *Electrochemical supercapacitors*, Plenum Publishing, New York, 1999, Kötz, R; Carlen, M. *Electrochim. Acta*, 2000, 45, 2483-2498;
[3] *Interface*, 2008, 17 (1), 33-53.
[4] Pandolfo, T; Ruiz, V; Sivakkumar, S; Nerkar J. in: (F. Beguin and E. Frackowiak Eds) Supercapacitors. Materials, Systems and Applications, Wiley V-CH (Weinheim, Germany), 2013, 69-109 (ISBN 978-3-527-32883-3).
[5] Tarascon, JM; Armand, M. *Nature.*, 2001, 414, 359-367.
[6] Cheng, F; Tao, Z; Liang, J; Chen. J. *Chem. Mater.*, 2008, 20, 667-681.
[7] Simon, P; Gogotsi, Y. *Nature Mat.*, 2008, 7, 845-854.

[8] Hoffmann, F; Cornelius, M; Correll, J; Fröba, M. *Angew. Chem. Int. Ed.*, 2006, 45, 3216-3251.
[9] Jaio, F; Shaju, KM; Bruce, PG. *Angew. Chem. Int. Ed.*, 2005, 44, 6550-6553.
[10] Yu, F; Zhang, J; Yang, Y; Song,G. *J. Power Sourc.*, 2009, 189, 794-797.
[11] Wang, T; Liu, X; Zhao, D; Jiang, Z. *Chem. Phys. Lett.*, 2004, 389, 327-331.
[12] Luo, JY; Zhang, JJ; Xia, YY. *Chem. Mater.*, 2006, 18, 5618-5623.
[13] Kavan, L; Rathousky, J; Grätzel, M; Shklover, V; Zukal, A. *J. Phys. Chem. B.*, 2000, 104, 12012-12020.
[14] Whitehead, AH; Elliott, JM; Owen, JR. *J. Power Sourc.* 1999, 81-82, 33-38.
[15] Wang, T; Ma, Z; Xu, F; Jiang, Z. *Electrochem. Commun.*, 2003, 5, 599-602.
[16] Padhi, AK; Nanjundaswany, KS; Masquelier, C; Okada, S; Goodenough, JB. *J. Electrochem. Soc.* 1997, 144, 1609-1613; Franger, S; Le Cras, F; Bourbon, C; Benoit, C; Soudan, P; Santos-Peña, J. in: S.G. Pandalai (Ed.), Recent Research Development in Electrochemistry, Transworld Research Network Ed. (Kerala, India)., 2005, 225-256 (ISBN 81-7895-183-5).
[17] Santos–Peña, J; Soudan, P; Otero–Areán, C; Turnes–Palomino, G; Franger, S. *J. Solid State Electrochem.*, 2006, 10, 1-9.
[18] Dabadie, T; Ayral, A; Guizard, C; Cot, L; Lacan, P. *J. Mater. Chem.*, 1996, 6, 1789-1794.
[19] Lokhande, CD; Dubal, DP; Joo, O-S. *Current Appl. Phys.*, 2011, 11, 255-270.
[20] Vix-Gurtel, C; Frackowiak, E; Jurewicz, K; Friebe, M; Parmentier, J; Béguin, F. *Carbon.*, 2005, 43, 1293-1302.
[21] Shi, H. *Electrochim. Acta.*, 1996, 41, 1633.
[22] Qu, D; Shi, H. *J. Power Sources.*, 1998, 74, 99.
[23] Lufrano, F; Staiti, P. *Int. J. Electrochem. Sci.*, 2010, 5, 903-916.
[24] Barbieri, O; Hahn, M; Herzog, A; Kotz, R. *Carbon.*, 2005, 43, 1303-1310.
[25] Endo, M; Maeda, T; Takeda, T; Kim, Y.J; Koshiba, K; Hara, H; Dresselhaus, MS. *J. Electrochem. Soc.*, 2001, 148, A910.
[26] Shiraishi, S; Kurihara, H; Shi, L; Nakayama, T; Oya, A. *J. electrochem. Soc.*, 2002, 149, A855.
[27] Centeno, TA; Sevilla, M; Fuertes, AB; Stoekli, F. *Carbon.*, 2005, 43, 3012-3015.
[28] Kim, CH; Pyun, SI; Shin, HC. *J. Electrochem. Soc.*, 2002, 149, A93.
[29] Frackowiak, E; Beguin, F. *Carbon.*, 2001, 39, 937.
[30] Salitra, G; Soffer, A. Eliad, L; Cohen, Y; Aurbach, D. *J. Electrochem. Soc.*, 2000, 147, 2486.
[31] Chmiola, J; Yushin, G; Gogotsi, Y; Portet, C; Simon, P; Taberna, P.L. *Science.*, 2006, 313, 1760-1763.
[32] Dash, R; Chmiola, J; Yushin, G; Gogotsi, Y; Laudisio, G; Singer, J; Fischer, J; Kuycheyev, S. *Carbon.*, 2006, 44, 2489.
[33] Laudisio, G; Dash, R; Singer, J.P; Yushin, G; Gogotsi, Y; Fischer, JE. *Langmuir.*, 2006, 8945.
[34] Mysyk, R; Raymundo-Piñeiro, E; Beguin, F. *Electrochem. Comm.*, 2009, 11, 554.
[35] Mysyk, R; Raymundo-Piñeiro, E; Pernak, J; Beguin, F. *J. Phys. Chem. C.*, 2009, 113, 13443.
[36] Mysyk, R; Gao, Q; Raymundo-Piñeiro, E; Beguin, F. *Carbon*, 2012, 50, 3367.

[37] Fuertes, AB; Lota, G; Centeno, TA; Frackowiack, E. *Electrochim. Acta.*, 2005, 50, 2799.
[38] Fuertes, AB. *J. Mater. Chem.*, 2003, 13, 3085.
[39] Tamai, H; Kouzu, M; Morita, M; Yasuda, H. *Electrochem. Solid State Lett.*, 2003, 6, A214.
[40] Yamada, Y; Tanaike, O; Liang, TT; Hatori, H; Shiraishi, S; Oya, A. *Electrochem. Solid-State Lett.*, 2002, 5, A283.
[41] Yoon, S; Lee, J; Hyeon, T; Oh, SM. *J. Electrochem. Soc.*, 2000, 147, 2507.
[42] Jun, S; Joo, S.H; Ryoo, R; Kruk, M; Jaroniec, M; Liu, Z; Ohsuna, T; Terasaki, O. *J. Am. Chem. Soc.*, 2000, 122, 10712.
[43] Meynen, V; Cool, P; Vasant, EF. *Microp. Mesop. Mater.*, 2009, 125, 170-223.

Chapter 9

OPTICAL PROPERTIES OF MESOPOROUS SILICON/SILICON OXIDE AND THE LIGHT PROPAGATION IN THESE MATERIALS

Joël Charrier and Parastesh Pirasteh
Université Européenne de Bretagne, CNRS FOTON, UMR,
Lannion, France

Mesoporous silicon owns a large range of refractive indices and is a material which offers many advantages for integrated photonic circuits. The light propagation will be presented for waveguides manufactured from porous silicon or oxidized porous silicon layers. Optical properties of these materials will also be discussed as a function of the oxidation degree, different functionalization steps of these porous layers. Some different types of waveguides will be described. Surface and volume scattering losses of these mesoporous materials will be modeled and discussed in order to determine the principal contributions to optical losses.

I. INTRODUCTION

Porous Silicon (PSi) has a large range of refractive indices [1] and is a material which offers many advantages for optical waveguides. Theses waveguides are composed of a porous silicon guiding layer of low porosity (high refractive index) surrounded by porous silicon cladding layer(s) with higher porosity (low refractive index). Once obtained, they are then oxidized to extend the operating wavelength range into the visible spectrum. Such an approach has been used to form multilayer optical planar waveguides [2, 3], strip loaded rib waveguides [2] or buried waveguides [4, 5, 6, 7]. These applications are very promising for gaz detectors [8, 9] and all-silicon optoelectronic integrated devices [10]. Previous works are summarized in Table 1.

Table 1. Different types of optical waveguides based on porous silicon

Type of optical waveguides	References
Multilayer waveguides	[2, 3]
Strip loaded rib	[2]
Buried waveguides	[4-7]

Moreover, sensors based on porous silicon nanostructures, having large internal surface areas and widely tuneable refractive indices, have been developed. Different types of optical devices have been reported such as single layer interferometers, waveguides, Bragg mirrors, microcavities, rugate filters [1-7]... An inherent feature of porous silicon material is its open pore network. Normally, this can be used to modify the properties of optical devices in-situ, by filling the pores with other materials. Moreover, the easy oxidation of the internal surface of Porous Silicon (PSi) allows organic molecules to be covalently attached. Previously cited PSi photonic structures have shown a very good sensitivity in the label-free sensing by the modification of the effective refractive index of a porous silicon layer [8-13]. In fact, many molecular probes were used for these sensor systems. We can tell proteics probes apart like antibodies [14], biotin [15], protein A [16], KAA peptide [14] and TWTCP [17] and nucleic probes like DNA and RNA fragments [18].

In this chapter, we report qualitative and quantitative studies of the propagation losses obtained in planar PSi waveguides and oxidized PSi waveguides. We describe the physical and optical properties of porous silicon (PSi) and oxidized porous silicon (OPSi) layers, and then the fabrication of planar and buried waveguides will be presented. The light propagation in these structures and their functionalization will be described.

II. OPTICAL AND PHYSICAL PROPERTIES

Porous silicon layers were obtained by electrochemical anodisation of heavily doped (5mΩ.cm) p-type (100)-oriented silicon wafers. The electrolyte solution was HF (50%): H_2O:ethanol (2:1:2) at room temperature. To achieve multilayer planar optical waveguides, layers of different porosities (and thus different refractive indices) were formed by abruptly changing the anodisation current. The experimental data are summarized in tables 2 and 3.

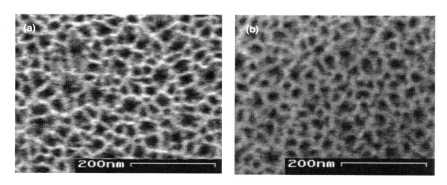

Figure 1. SEM micrographs of a surface of (a) porous silicon layer and (b) oxidised porous silicon layer.

Figure 1 shows the SEM micrographs of the surface of PSi layer (Figure 1-a) and OPSi (Figure 1-b). These micrographs allowed the size and density of pores to be estimated. In fact, these values contained an estimate error because of the micrograph contrast.

The pore size was measured directly from these micrographs and its distribution was deduced (Figure 2). It could be seen that the porous layer was composed of a pore network (in black) separated by silicon crystallites (in white). For PSi, pore size varied from 7 to 30nm and was centered around a mean value equal to 18nm. This distribution was almost symmetrical with a low root mean square which revealed the homogeneity in pore size values. The distribution was always almost symmetrical for OPSi but the mean value was different and equal to 13nm. The value corresponding to the maximum was equal to 11.5nm. It also appeared that oxidation induced a small sharpening of the distribution. By comparing figures 1-a and 1-b, we can see that after oxidation the pore shape is conserved in spite of size reduction.

Moreover, the wall size which separate pores can also be evaluated from these micrographs. It was about 6nm before oxidation (PSi sample) and about 9nm after that (OPSi sample). This increase is due to the volume expansion of silicon transformed into oxidized silicon. This expansion was assumed to be isotropic through the volumeand was described by a factor of 2.27 [19]. However the pores did not collapse because of the initial porosity which was more important than volume expansion and the structure was always open on the surface. So the density was nearly the same before and after oxidation and it was about 10^{11} pores/cm^2.The poredensity is the same and the pore size decreased so the porosity after oxidation of course was lower than the porosity before oxidation. After oxidation, the porosity was estimated to 33%.

Furthermore, volume expansion due to oxidation was isotropic and one could observe an increase in the thickness of layer. For a single layer whose the thickness was equal to 5µm before oxidation, it was nearly 5.8µm thick after oxidation.

On the other hand, values deduced from SEM micrographs were compared with those measured from the TEM cross section of the samples. Due to the complex network of the porous silicon texture, it is difficult to determine a precise pore distribution from TEM observation. However the values measured by TEM showed a good correlation with the mean pore sizes determined by SEM.

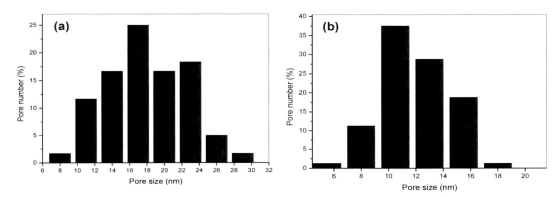

Figure 2. Distribution of pore size of (a) por (b) oxidised porous silicon layer.

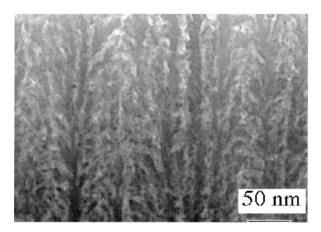

Figure 3. TEM micrograph of a typical cross section of oxidised porous silicon layer.

Both PSi and OPSisamples were studied by TEM. The nanostructures seemed very similar before and after oxidation. Figure 3 represents the cross section TEM micrograph of sample B. The microstructure is anisotrop: the pores are perpendicular to the surface of the PSi layer. The pores were considered as been cylindrical with numerous side branches which have almost the same size as the pore mean diameter. The pore size can be simulated by taking into account or not the side branches.

Figure 3 shows a schematic drawing of the pores. The cylinder with D_1 diameter corresponds to the principal part of the pore and D_2 also includes the side branches. We called D_1 and D_2 the intern and extern diameters respectively. They were of 18 and 34nm before oxidation (Table 2). These values changed, of course, after oxidation and decreased to 13 and 30nm.

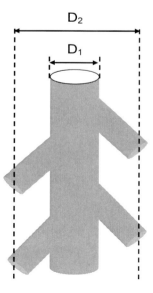

Figure 4. Schematic representation of principal pore with its branches.

Table 2. Pore diameters estimated from TEM microgaphs of Porous Silicon and Oxidised Porous Silicon layers

	D_1 (nm)	D_2 (nm)
Porous Silicon	18	34
Oxidised Porous Silicon	13	30

Table 3. Refractive index measured by m-lines and calculated by Bruggeman model of Porous Silicon and Oxidised Porous Silicon layers

	Porosity (%)	Calculated refractive index at 1550nm	Measured refractive index at 1550nm
Porous Silicon	60	1.84	1.82
Oxidised Porous Silicon	33	1.31	1.32

The refractive indexes of PSi and OPSi sample were measured at 1550nm using the m-lines method. The results are shown in Table 3. The knowledge of refractive index permits to study the effective medium. The index decreased after oxidation and its value was between that of air (n=1) and that of silica (n=1.46). The refractive index was simulated by the Bruggeman model before and after oxidation [20]. This model took into account the two mediums: silicon and vacuum for PSi layer and silica and vacuum for OPSi layer (sample B), considering respectively their porosity (60 and 33%). The calculated results were approximately equal to those measured (Table 2). So we can consider that the index of OPSi sample can be simulated by an effective medium which is porous silica.

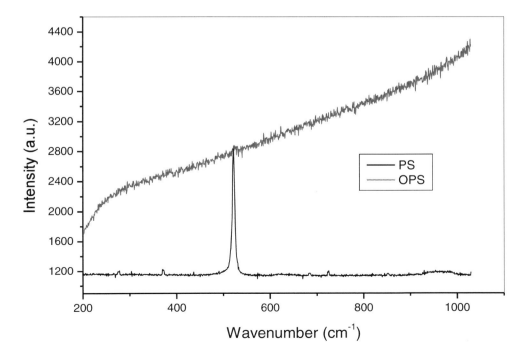

Figure 5. Micro-Raman spectra of Porous Silicon and Oxidised Porous Silicon.

Moreover, the samples were studied by micro-Raman spectroscopy in order to evaluate the oxidation rate. We studied different points along the cross-section at different depths of the porous layers. The characteristic peak of silicon is centered at 520.2cm^{-1} which we can see in the Fig 5 for the PS layer. It totally disappeared for the OPSi layer. So we could consider a nearly complete transformation of PS into OPSi by using this thermal oxidation process. Therefore the absorption coefficient of OPSisample is negligible compared to that of PSisample, particularly in the visible wavelength. The base line of the Raman spectrum rises for OPSisample due to photoluminescence of PS layers after oxidation.

III. STUDY OF THE LIGHT PROPAGATION IN THE WAVEGUIDE BASED ON POROUS SILICON LAYERS

III.1. Fabrication of Planar Waveguides

To achieve multilayer planar optical waveguides, layers of different porosities (and thus different refractive indices) were formed by abruptly changing the anodisation current. The 5μm thickness of each of the guiding and cladding layers was controlled by the anodisation time. The guiding (upper) layer and the cladding (lower) layer were formed by successively applying current densities of 50 and 80 mA/cm^2 in order to obtain porosities about 60% and 65% respectively. This process was followed by a brief plasma etching in order to remove the superficial parasitic film [21]. Once obtained, they could be oxidized to extend the operating wavelength range into the visible wavelengths. The prepared structures were oxidized at 900°C for 60 minutes in wet O_2 after a preoxidisation step at 300°C for one hour [22]. The principal optical waveguides based on porous silicon are summarized in Table 1.

The characteristics of PSiW and OPSiW are shown in table 4. The refractive indices were measured by the m-line method [23] and the porosity was deduced using the Bruggeman model [20]. In the case of OPSW it has been assumed that there was complete oxidation of the porous layer [24].

III.2. Fabrication of Buried Waveguides

The buried waveguides are obtained by selective anodisation where we use a masking layer. The mask is composed of a thin SiO_2 layer recovered by a very thin a-Si layer. The pattern consists of a periodic group of small straight open windows (4-10μm) and great ones (300μm). Two kinds of buried waveguides have been carried out. The first one (sample A) is a structure with a gradient index in both guiding and cladding layers. In this case, the applied current intensity is constant for each layer during the anodisation (50 and 80mA/cm²). For the second one (sample B), we tried to avoid the formation of this gradient by controlling the applied current intensity according to surface evolution [25, 26, 27]. Some of the waveguides are thermally oxidized. Figure 6 presents a typical cross section view of our buried waveguides obtained by optical microscopy. The characteristics of waveguides are represented in Table 3.

Table 4. Characteristics of waveguides

Current density (mA/cm^2)	PSi Porosity (%)	PSi Refractive Index at 1550nm	OPSi Porosity (%)	OPSi Refractive index at 1550nm
50	60	1.84	30	1.30
80	70	1.70	45	1.25

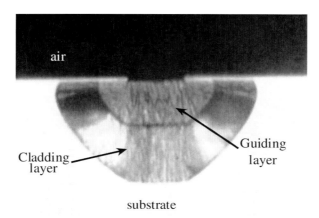

Figure 6. A typical cross section view of buried waveguides obtained by optical microscopy.

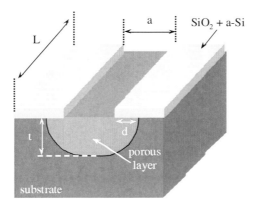

Figure 7. Schematic representation of localized anodisation through a masking layer: a is the width of the open window (a + 2d is the maximal width of porous area), L the length of waveguide and t the thickness of the layer.

To explain the surface evolution during anodisation, and the fabrication of the two kinds of buried waveguides, Figure 7 presents schematically, the localized formation of porous silicon through a masking layer where a is the initial opening window. The formation of porous silicon under the masking layer varies with the thickness of layer (t) by the relation: d = 0.8×t [26]. Therefore, during the anodisation, the interfacial reaction area S increases following the eqn (1) and it is equal to the sum of all the opening area.

$$S = L\left(\sum_{i=1}^{m} a_i + m \times t(\mu m) \times 0.8\pi\right) \quad (1)$$

where a_i is the width of the open window (μm), L the length of waveguide (μm), m the number of the open windows and t the thickness of the layer.

At a constant current intensity I, the current density J depends on the evolution of the surface (interfacial reaction area S)

$$J_k = \frac{I_k}{S} \quad (2)$$

where k = 0 or 1 corresponding to the case respectively of guiding layer and cladding layer. As far as during anodisation the surface increases, the current density decreases. This reduction of current density induces a gradual rise of refractive index in the layer.

The presence of the great opening windows could reduce this effect. In fact, if $\sum_{i=0}^{m} a_i \gg m \times t(\mu m) \times 0.8\pi$, the variation of the surface is less important and the index gradient is attenuated.

III.3. Study of the Light Propagation in Planar Waveguides

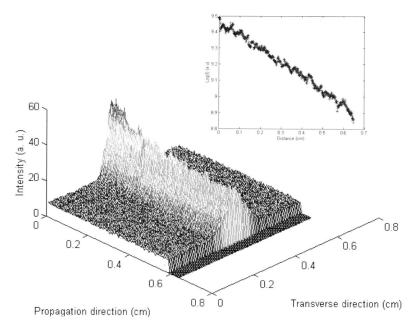

Figure 8. Two-dimension light intensity for OPSiW; inset: logarithmic of light intensity (I) propagating in the OPSiW as a function of the propagation distance at 633nm.

Optical losses were measured by studying the scattered light from the surface of the waveguide [28, 29]. Laser light was single-mode fiber-coupled into the waveguide. The intensity of the scattered light was recorded with a digital camera placed above the sample. Transverse scanning along the light propagation direction enabled us to obtain the 2-D light intensity distribution of the waveguide modes. Figure 8 and its inset show respectively, a typical 2-D intensity distribution of the visualized light and its longitudinal variation. This last curve was obtained by integrating the data along each sampling transverse line. The light intensity decreased exponentially with the z-propagation distance.

In this study, the attenuation values were the average of several measurements performed on several samples. The polarization of the coupled light into the waveguide was not controlled.

In addition, light propagation was observed at the output of the waveguides by near field profiles of guided modes at 1550nm.

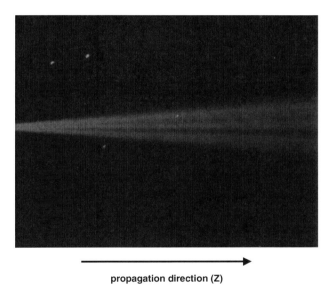

Figure 9. Picture of scattered light from OPSiW at 633nm: broadening of the beam.

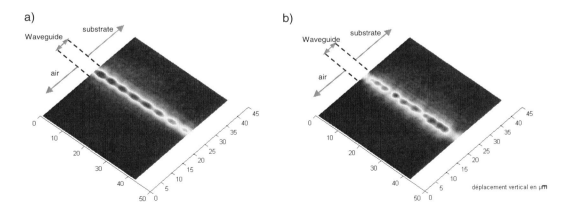

Figure 10. Near fields of propagated modes for OPSiW at 1550nm.

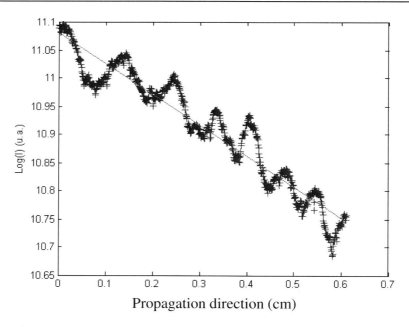

Figure 11. Logarithmic of light intensity (I) propagating in the OPSiW as a function of the propagation distance at 633nm: effect of waviness on the propagated light intensity.

One advantage of the loss measurement experimental set-up was that it permitted light propagation within the waveguide to be observed. In fact, r fields of propagated, the scattered light from the surface of the sample, recorded by the camera, represented an image of light propagation within the waveguide. Figure 10 represents a typical picture of light propagation in an OPSiW at 633nm. As we can see, the beam diverges and splits into several rays. This light behavior was observed for both of PSiW (Porous Si waveguide) and OPSiW (Oxidized Porous Silicon Waveguide). Such a behavior was also reported for liquid crystal waveguides [30]. In order to check this feature, the guided modes were observed at the end of samples using the near field method. The typical observation results at 1550nm are presented in figure 11. The light distribution shows a succession of high and low intensity areas. In the case of PSiW, the high intensity areas are less numerous and more separated. Therefore, both observations, from the surface and at the end of samples, are compatible. In fact, the beam broadens due to diffraction [30].

Moreover, the pictures taken from the surface of samples reveal another effect of light propagation in our waveguides. In some cases, a pseudo-periodic variation of propagated light intensity has been observed according to the propagation direction. This behavior was observed in 1-D light intensity distribution as a function of distance by light intensity maxima and minima as illustrated in Figure 11. The spatial interval between two intensity maxima or two minima was about 500μm. We think that this intensity variation is due to spatial periodicity of substrate doping fluctuation. In fact, during the formation of silicon substrates, concentric areas with higher or lower doping concentration were formed (Figure 12). These striations remained after the formation of the porous layers and they induced slow variations in porosity particularly in thickness, called waviness [31]. Their spatial periodicity was larger than visible light wavelength and was about 250μm for p-type doped (5mΩ.cm) silicon substrate [31].

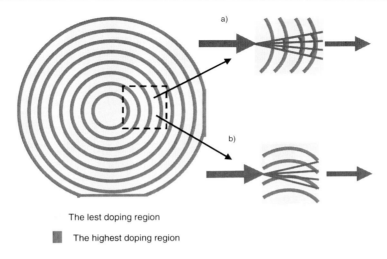

Figure 12. Schematic representation of silicon substrate with doping fluctuations.

Table 5. Measured values of optical losses as a function of the wavelength for PSiW and OPSiW

Wavelength (nm)	Losses of PSi waveguides (dB/cm)	Losses of OPSi Waveguides (dB/cm)
546	-	2.4 ± 0.5
633	-	1.8 ± 0.5
980	-	0.5 ± 0.3
1550	5 ± 0.5	0.6 ± 0.3

Depending on cleavage directions, light propagation can be perpendicular or parallel to the striations (Figures 11a and 11b). This possibility explains why this variation in light intensity variation was not always observed. When light propagation is quasi-parallel to striations, the intensity decreased monotonously as illustrated in the inset of Figure 1. However, in both these two cases, the slope of the curves had the same value.

The slope of the curve represented in inset of Figure 8 corresponds to the attenuation coefficient α (cm^{-1}). This attenuation coefficient mainly depends on material absorption (α_{abs}) and also on surface scattering (α_{surf}, due to interface roughness) and volume scattering (α_{vol}, due to the porous aspect of the material). Thus, the overall attenuation is equal to the sum of all these optical loss contributions. In our case, because of the cladding layer thickness (about 5µm), the losses due to substrate leakymodes have been estimated to 10^{-4} dB/cm^{-1} which is considered to be negligible [32].

The losses were measured as a function of the wavelength for PSiW and OPSiW. The measured values are shown in table 5. Before oxidation, light propagation was not observed in the visible range at the output of the PSiW. Light propagation was observed for a very small distance from the input of the waveguide (a few mm) but the measurement of losses was not possible. We suppose that the coupled light was totally attenuated due to strong material absorption at these wavelengths. However, the losses were measured at 1550nm and were about 5dB/cm.

After oxidation, thre was an important decrease in losses about 0.6dB/cm at 1550nm. The measured loss value was more important for the visible domain than for the NIR. The losses decreased as the wavelength increased. However, we can note a slight increase in losses between 980 and 1550nm.

III.3.1. Description of Optical Loss Models

Three reasons permit to explain the origin of optical losses: (i) absorption, (ii) surface scattering and (iii) volume scattering. The value of losses due to absorption was calculated using the absorption coefficient obtained from publications [33, 34]. In order to discuss the experimental results, scattering losses due to surface and volume were modelled. In the following paragraph, we will call E the polarization case when the electric field was parallel to the pores, and H the case when the electric field was orthogonal to the pores. From these models, the principal trends are described.

(i) Surface Scattering Losses

Surface scattering losses are related to the root mean square (rms) deviation of the planarity at the guiding layer interfaces (σ). In our case, the correlation length of surface variation was important compared to wavelength [31]. The power lost by surface scattering at the two interfaces of the guiding layer is given by a model developed by P. K. Tien [35]. In this model, we assume that the roughness σ is the same value for both the air/guiding (due to plasma etching) layer and guiding layer /cladding layer interfaces. Surface losses calculated as a function of roughness (σ) for the first two modes (m=0 and m=1) are shown in figure 13. In addition, Figure 14 represents surface losses versus wavelength for three values of σ. We consider that the curves obtained in figure 13 and 14, in the case of OPSiW, are approximately the same as those obtained for PSiW. For a fixed wavelength, the surface losses in figure 13 increase importantly with both the value of σ and with the mode order. However, for a fixed value of σ, in Figure 14, losses increase slightly with wavelength.

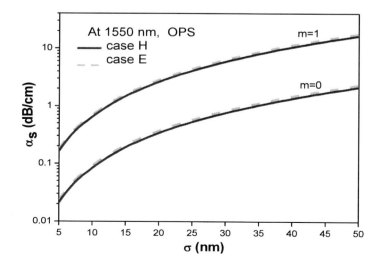

Figure 13. Calculated surface scattering losses as a function of σ (rms) for the two first modes and for OPSiW.

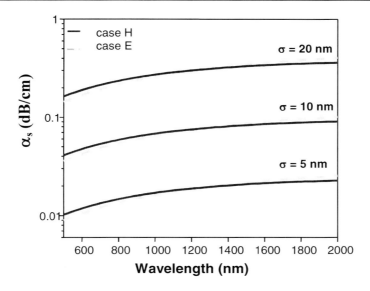

Figure 14. Calculated surface scattering losses as a function of the wavelength for different values of σ (rms) and for OPSiW.

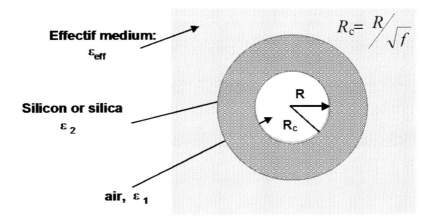

Figure 15. Schematic representation of the medium to calculate the volume scattering losses.

(ii) Volume Scattering Losses

In order to study the volume scattering losses, the morphological anisotropy of the porous silicon skeleton must be considered. The pores are perpendicular to the surface of the PSi layer [24, 37]. The pores can be considered as cylinders with numerous side branches. To take into account the structural anisotropy in volume scattering, A. Kirchner et al. [36] considered a composite medium of two materials with dielectric constants ε_1 and ε_2. The composite medium is assumed to consist of infinitely extended, parallel oriented and randomly placed cylinders with radius R and dielectric constant ε_1 embedded within a host material with dielectric constant ε_2. The random medium is also characterized by f, the volume fraction occupied by the cylinders. In our case, the cylinders are the pores, f the porosity and the host material is either silicon in the case of PSiW or silica in the case of OPSiW.

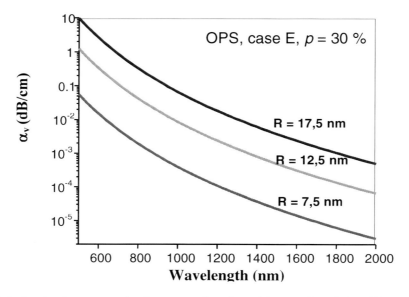

Figure 16. Calculated volume scattering losses as a function of the wavelength for different values of pore size and for OPSiW.

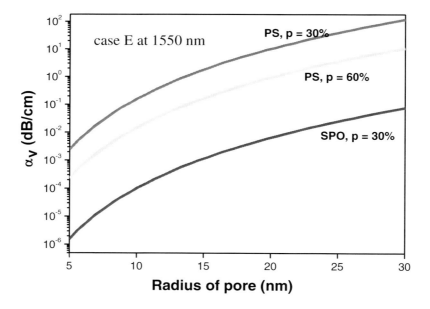

Figure 17. Calculated volume scattering losses as a function of the pore size for different porosities and for PSiW and OPSiW

The underlying idea of any effective medium theory of disordered systems is to focus on one scattering particular and to replace the surrounding random medium by an effective homogeneous medium. So the multiple scattering effects at different centers were simulated using a coated cylinder as the basic scattering unit. The radius of the coated cylinder was $R_c = R/f^{1/2}$ as illustrated in Figure 15.

Figure 16 shows the calculated variation of volume scattering losses as a function of wavelength for different values of R in the case of OPSiW. We can notice that the losses decrease when the wavelength increases and they increase with the value of R for a fixed wavelength. Figure 17 represents the variation of losses as a function of radius R for different porosities and different materials (Porous Silicon or Oxidized Porous Silicon (OPSi)) at 1550nm. For a 30%porosity, the OPSiW losses are lower than those of PSiW. This difference is due to the fact that the refractive index contrast between silica and air is lower that between silicon and air at fixed wavelength. The effect of scattering is less important in the cases of low index contrasts. Note also that in our cases for a given material, volume scattering losses decrease as the porosity increases. This loss variation is due to a decrease in the effective medium index as the porosity increases.

III.3.2. Discussion

For PSiW, optical losses are due principally to absorption. In fact, waveguides are manufactured from heavily doped (5mΩ.cm) p-type silicon wafers. Thus free carrier absorption can account for the majority of the losses observed in the P^+ as-anodised porous silicon waveguides. For comparable porosities, the absorption coefficient was measured equal to $1cm^{-1}$ (greater than 4 dB/cm) by PDS [33]. The surface and volume scattering losses exist but their contribution was lower (respectively 0.5 (see figure 6 for σ = 30nm [19] and m = 0) and 1 dB/cm approximately (see figure 10 for R = 25 nm [24]).

After oxidation, the absorption coefficient is negligible [34]. We have illustrated in figure 18 the measured values of losses (solid circles) as a function of wavelength for OPSiW. The dashed lines correspond to simulated losses using the described method for volume scattering and short-dashed lines represent those for surface scattering in the case of OPSiW. The solid line is equal to the sum of the two dashed curves. The best fit of the experimental results is obtained for σ = 24 nm and R = 16 nm. These curves permit volume losses and surface losses to be separated. Obviously, the trends that we are discussing are very crude. However, they establish a guideline for the complex problem of light propagation trends in nanostructured silicon. We can notice that in the visible spectrum, the volume losses are more important than surface losses. On the contrary, in NIR, almost all of the losses are due to surface scattering.

The chosen values of σ and R are realistic. Indeed, micrographs obtained by TEM have permitted us to determine the pore sizes of the oxidized porous layer. The pores can be considered cylindrical with numerous sides. The mean value of the effective radius including side branches is approximately equal to 15nm [24].

III.4. Study of the Light Propagation in Buried Waveguides

Because of their less complex structure, planar waveguides were firstly studied (Figure 19). For the PSi waveguides, the measured losses were about 5dB/cm at 1550nm. Here, absorption is the principal source of losses because of the heavily doped silicon substrate [33]. After oxidation, in the same wavelength, losses decrease strongly and reach a value equal to 0.6dB/cm. In fact, the transformation of PSi to OPS after oxidation strongly decreases the coefficient of absorption of the material. The combination of an experimentally and theoretically study of this problem [6] allowed us to distinguish the principal origin of the losses in this case: for OPS waveguides almost all of the losses in the NIR are due to surface scattering.

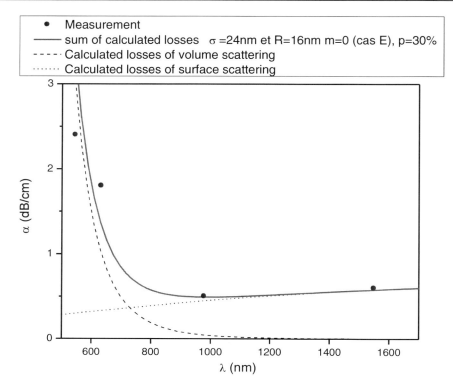

Figure 18. Measured and calculated losses as a function of the wavelength for OPSiW. The solid circles represent the measurements, the dashed lines correspond to simulated losses using the described method for volume scattering and short-dashed lines represent those for surface scattering in the case of OPSiW.

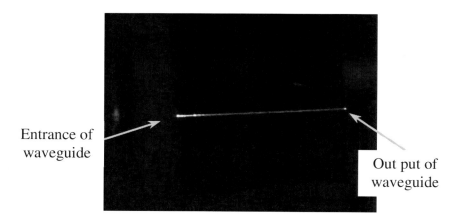

Figure 19. View of scattered light in an OPS waveguide at 1550nm obtained by a NIR camera placed above the sample.

Then, the losses at 1550nm were measured on the two types of buried waveguides (sample A and sample B). The results of losses before and after oxidation are presented in table 5.

Table 5. Optical losses of planar and buried waveguides at 1550nm (± 0.3dB/cm)

	Planar waveguide	5	0.6
Optical losses (dB/cm)	sample A	6	2
	sample B	5	1

It is important to note that the first type of these waveguides (sample A) contains aindex gradient in both guiding and cladding layers even if this gradient is not so important in our case due to the presence of the great opening areas (300μm). This gradient is due to surface evolution during anodisation. Nevertheless the surface variation could be minimized by the presence of the large opening area (see the appendix below).The measured losses of waveguides formed on the small opening area (4-10μm) are about 6dB/cm and they decrease after oxidation to 2dB/cm. These values are greater than those of planar waveguides at the same wavelength. In addition, the same trend of losses decrease after oxidation is conserved.

Here-above, we saw that for the PSiplanar waveguides, the principal contribution of losses is the intrinsic absorption due to heavily doped silicon substrate. In the case of buried waveguides, we should consider not only the absorption effect, but also the presence of few defects due to photolithography process and overall the existence of the edges. Indeed, there are more interfaces between the guiding and cladding layers because of the edges, therefore the participation of surface scattering becomes more important. This could explain the difference of losses between planar and buried waveguides. On the other hand, the surface scattering is the main losses source for the oxidized waveguides in NIR wavelengths. Therefore, it should be accentuated because of the presence of the edges.

Moreover, the near-field observation of guided modes at 1550nm was carried out. A typical result of this observation, before and after oxidation, is shown in Figure 20. It reveals that this waveguide is nearly single mode for both non oxidized and oxidized waveguides. We can also observe a small difference of shape of the propagation zone between these two cases. The modal shape is more elliptical for no oxidized waveguides while it is more circular for oxidized ones; but in the both cases, there is a slight radiative field towards air/guiding layer interface. Moreover, the propagation zone is located near the guiding/cladding layer interface which confirms the gradient of refractive index in each layer and it also explains the increase of losses due to surface scattering in these waveguides.

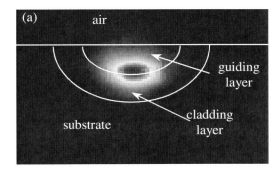 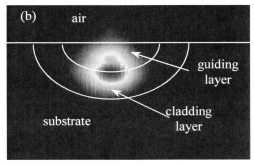

Figure 20. Near field observations at 1550nm of guided mode for a waveguide of 6μm initial open window in the case of (a) PSi buried waveguide (b) OPSi buried waveguide – Sample A.

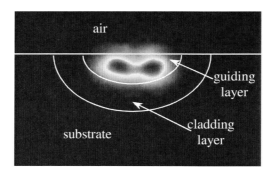

Figure 21. Near field pattern of guided mode at 1550nm in the Oxidized PS buried waveguide – Sample B.

As we have already seen, we tried to avoid the formation of the gradient in each layer by controlling the applied current intensity during anodisation. Therefore, to compensate for the increase of the surface in contact with electrolyte, we increased step by step the current intensity to obtain a constant refractive index (see appendix). The measured losses of these waveguides (sample B) are about 5dB/cm and they decrease after oxidation to 1dB/cm. These values are nearly the same as those of planar waveguides at the same wavelength. The near-field observation of guided modes on this sample (Figure 21) permits a better understanding of this effect. In fact, the propagation zone in this waveguide is rather placed at the center of guiding layer. Therefore the effect of surface scattering becomes less important, so the losses are lower here than those of sample A. It reveals also this waveguide is no longer single mode since the guiding area seems to be much larger.

IV. FUNCTIONALIZATION OF PSi WAVEGUIDES

IV.1. Planar Waveguides

An inherent feature of porous silicon material is its open pore network. This can be used to modify the waveguide properties, by filling the pores with other materials such as liquids, dyes or polymers [39]. We study the optical loss of Oxidised Porous Silicon Waveguide (OPSiW) impregnated by dyes. The used dyes are the Congo Red (CR), which is a pH indicator, and the Disperse Red 1 (DR1) well known for its non-linear optical properties. Theses materials can be employed respectively for sensor or optical applications [40, 41]. The optical loss is modified by introducing of these dyes in the pores of OPSiW.

Before impregnation, the dye molecules are dissolved in N,N-Dymethylformamide (DMF) for CR and in Tetrahydrofurane (THF) for DR1 dyes. Different CR concentrations in solution and 20g/l concentration in solution for DR1 were used to achieve the composite materials. For the two dye-impregnation, the method was quite similar. The solution was deposed on the surface of sample by using a micro-syringe. After wetting, the sample was dried under vacuum. The penetration of dyes was studied by micro-Raman spectroscopy and the results showed that the dried dyes cover uniformly the pore walls [42, 43].

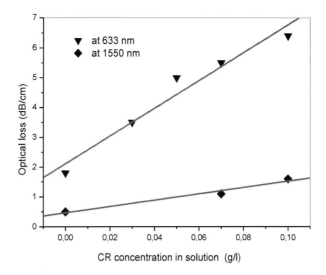

Figure 22. Optical loss as a function of the CR concentration in solution measured on impregnated OPSiW at 633 and at 1550nm.

Before impregnation, the optical loss of OPSiW was equal to 1.8 ± 0.5 dB/cm at 633 and 0.6 ± 0.3dB/cm at 1550nm. In the visible, the loss depends principally on the volume scattering whereas, in the NIR, it depends particularly on the surface scattering due to interface roughness.

OPSiW impregnated with different concentrations of CR in solution (0.03, 0.05, 0,07 and 0.1 g/l) are studied at 633 and at 1550nm. The results of the measured loss are represented in Figure 22. The loss values of impregnated OPSW are always higher than those of no-impregnated one. We can also observe, notably at 633nm, that the attenuation increases nearly linearly with the CR concentration in solution. This rise is associated with the introduced quantity of dye. Furthermore, the optical loss is more important at visible than that at NIR wavelengths. Moreover, for high concentrations (for example few g/l), no light propagation is observed at 633nm whereas it exists in infrared.

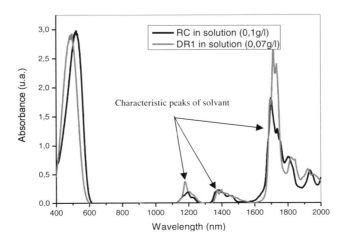

Figure 23. Absorbance spectra of DR1 in solution with THF and CR in solution with DMF.

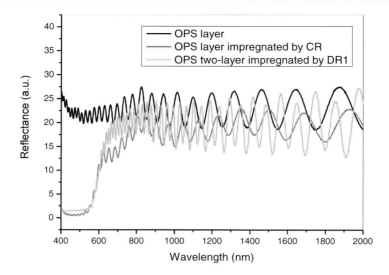

Figure 24. Reflectance spectra of OPS layer, OPS layer impregnated by CR and OPS two-layer impregnated by DR1.

At first, the absorption contribution was studied. In Figure 23, the absorption spectrum of CR in DMF is presented. At NIR, the absorption peaks of solution belong to DMF. However, the absorbance of CR in solution increases greatly for the wavelengths below 600nm.

On the other hand, the reflectance spectra of an OPSi layer, before and after impregnation by CR, are presented in figure 24. We can see that the reflectance of composite falls for the wavelengths in the visible. This drop should be due firstly to the absorption coefficient and secondly to the interface roughness of composite material. In fact, the absorption coefficient of CR could rise that of the composite. We should notice that the CR in composite material is in dry state which explains the wavelength range difference of absorption contribution between composite spectrum and that of CR in solution [42]. These results suggest that the attenuation due to absorption is the main contribution in the visible and explain the loss difference between impregnated and not impregnated OPSiW at 633nm. Knowing that in the visible, the surface scattering contribution to loss exists but it is less important than that of absorption. This could explain that the loss increases nearly linearly with the CR concentration in Figure 22.

Furthermore, measured loss values, in Figure 22, increase also at 1550nm with the CR concentrations but the composite material is weakly absorbent at this wavelength. Therefore the absorption is not really the dominant cause of loss here and the principal loss contribution, is the surface scattering [35]. In fact, the presence of non penetrated dried dye at the surface of the waveguide confirmed by microscopic observations could increase slightly the loss due to surface scattering.

The other studied dye is the DR1 chosen for its non-linear optical properties. In order to obtain an important non linear effect in future applications, the DR1 concentration in solution is important and is equal to 20 g/l. The absorbance spectrum of DR1 in solution with THF and also reflectance that of an oxidised porous silicon layer impregnated by these dyes, are represented respectively in Figures 23 and 24. The same trends as the composite material with CR are observed. After evaporation of the solvent (THF), the composite material is absorbent in visible.

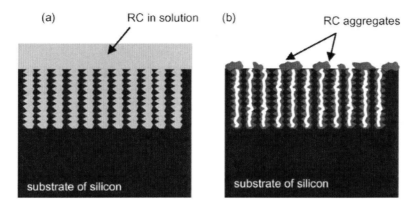

Figure 25. Schematic representations of the OPS layer impregnated with CR (a) before evaporation of the solvent (b) after evaporation of the solvent: presence of CR aggregates on the surface of the layer.

At 633nm, the light propagation has been observed just for a sample of about 2mm of length (along the propagation direction). Beyond this length of sample, there is no light propagation. Considering the important concentration of DR1 in solution, it could be due to absorption of DR1dried dye at this wavelength. However, the light propagation at 1550nm, is observed for the samples of length more important (about 15mm).

The measured loss value of OPSiW impregnated by DR1 is equal to 2.4 dB/cm at 1550nm which is the used wavelength for telecommunication applications. This value is more important than that of non impregnatedOPSiW. We think also that dried DR1 dyes on the surface of the samples act as an additional source of surface scattering (Figure 25).

IV.1. Burried Waveguides

The buried porous silicon waveguideswre impregnated by the DR1dye molecules by using TetraHydroFurane (THF) as solvent [18]. A solution of 10g/l was used to manufacture the composite materials. The solution was deposited on the surface of the sample using a micro-syringe. After wetting, the solvent was evaporated under vacuum.

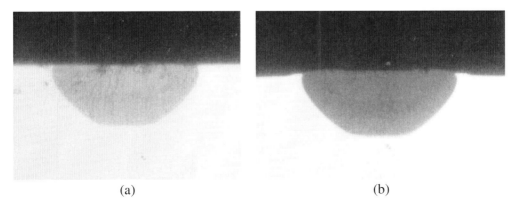

Figure 26. Typical cross section views obtained by optical microscopy: (a) poroussilicon buried waveguides and (b) oxidized poroussilicon buried waveguides.

The 2D-profile of oxidation rates and the pore filling with DR1 dyes were studied by micro-Raman mapping using a T64000 Horiba-Jobin-Yvon spectrometer. The measurements were performed (using as excitation the 514nm wavelength radiation of an Argon/Krypton laser) to probe the Raman lines characteristic of silicon before and after oxidation and those of the DR1 dyes. Raman acquisitions were done by scanning the cross section of the waveguides: after each acquisition step, the sample was shifted using a X-Y micro-translation stage with 0.8µm step.

Figure 26(a) represents a typical cross section view of a buried waveguide obtained by optical microscopy before oxidation. It shows the PS morphology formed locally through a masking layer. It can be seen that the pores have a radial orientation. In a previous study [44], we have shown that the variation of PS/substrate interface area during the localized anodisation induces a porosity gradient according to the vertical cross section of the structure. Moreover, the porosity in the same layer is probably not homogeneous. Indeed, the etching rate is lower in the <111> crystallographic direction than in the <100> direction [37]. This means that the porosity is higher in the <111> direction. Which is clearly confirmed on the micrograph of figure 1a when the cladding layer is thinner on both sides than at the bottom, and the pore ramifications on both sides are more numerous.

After oxidation at 900°C, the cross-section of the buried waveguide shown Figure 26(b) reveals a volume expansion of porous layers. It is well known that the oxide layer are subjected to volume expansion resulting from the transformation of Si into SiO_2 [24]. In this specific case, the transformation of Si into SiO_2 is not expected to take place because the volume expansion (56%) of SiO_2 is lower than the initial porosity of the PS layers (60% for the guiding layer and 65% for the cladding layer). This is confirmed by the measurements of layer refractive indices (Table 3). They indicate that oxidized silicon at this temperature does not occupy the whole available volume and strongly suggest that the layers have remained porous.

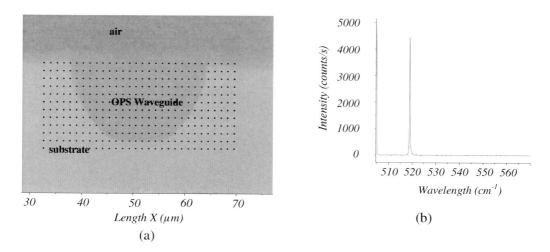

Figure 27. (a) Measurement principle of Raman spectra obtained at the cross section of the buried waveguide: the dots represent the measurement places, (b) micro-Raman spectrum of silicon substrate.

Optical Properties of Mesoporous Silicon/Silicon Oxide ...

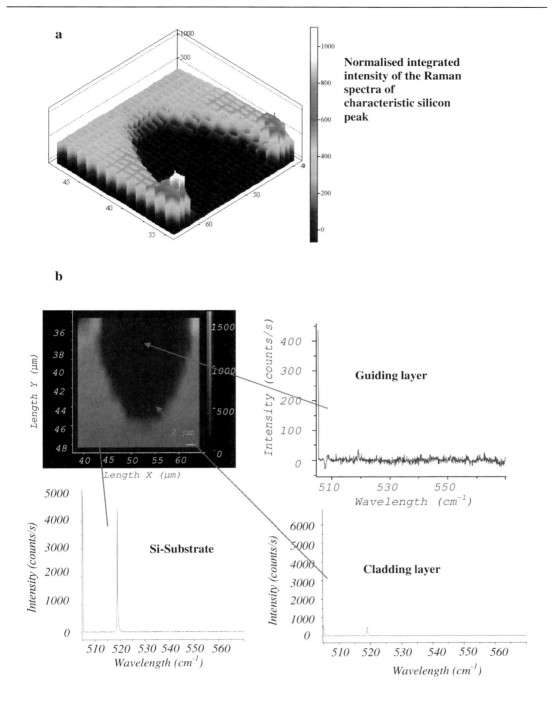

Figure 28. 2D-profile of normalized integrated intensity of the Raman spectra of a characteristic silicon peak for a buried waveguide after oxidation.

Moreover, the oxidized porous samples were studied by micro-Raman spectroscopy in order to evaluate the oxidation rate. We analyzed different locations of the porous layers according to X-Y cross-section of the waveguides as shown in Figure 27(a). The characteristic Raman peak of silicon centered at 520.2cm^{-1} is shown in the inset of Figure

27(b). The relative oxidation rate was deduced from the integrated intensities of the characteristic peak of silicon. If this peak totally disappeared then one could assume that porous silicon was almost completely transformed into Oxidized Porous Silicon (OPSi) using the thermal oxidation process.

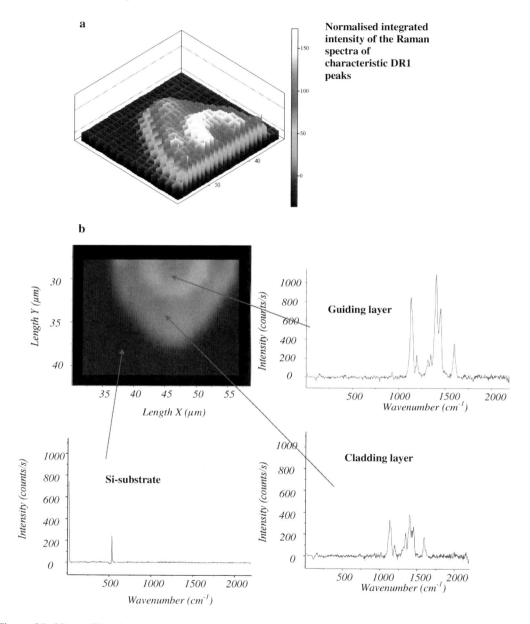

Figure 29. 2D-profile of normalised integrated intensity of the Raman spectra of characteristic DR1 peaks for an oxidized porous silicon buried waveguide after impregnation.

The 2D-profile of the oxidation rate deduced from the Raman integrated intensities for OPS waveguides is shown in Figure 28. This profile has the same shape as the cross section of the waveguide. For the guiding layer, the characteristic peak of silicon totally disappeared so that the PS layer was totally transformed into porous silica. On the other hand, for the cladding layer, principally at the edges of the layer (at the cladding layer / substrate interface), the Raman intensity peak due to silicon, although weak, was still present. The presence of this peak revealed a partial oxidation of the cladding layer. Its intensity increased from guiding layer / cladding layer interface to cladding layer / substrate interface.

The buried waveguides filled by DR1 dyes were also analyzed by micro-Raman spectrometry to determine the filling 2D-profile. The DR1 dyes gave rise to an intense characteristic Raman lines in the 1100-1700 cm^{-1} range(Figure 29). The relative DR1 filling rate was deduced from their integrated intensities in this wavenumber range. Figure 29 shows the 2D-profile of the DR1 filling rate. This 2D-profile resembles the shape of the waveguide, the guiding and cladding layers on this figure being clearly identified. Moreover, the profile is homogeneous in each layer. Nevertheless, the integrated intensity of DR1 dyes is higher in the guiding layer than in the cladding layer. As a result, both layers can be assumed to be homogenous both after oxidation and after filling of the OPS layers using DR1 dyes.

Moreover, propagation loss investigations at 1550nm were performed by measuring scattered light intensity at the surface of the waveguides. Figure 30 shows a top view of the scattered light from a 4-μm-wide buried waveguide. We can see a good lateral confinement of the modes that propagate in the waveguide. The results of losses for PSW and OPSW are shown in figure 31 as a function of the width of the pattern opening. We can notice that the losses decrease strongly after oxidation and attain a value equal to 0.6 ± 0.3 dB/cm in the case of a 300μm-width buried waveguide. In fact, the transformation of PSi to OPSi after oxidation strongly decreases the absorption coefficient of the material. The combination of an experimental and theoretical study allowed us to distinguish the main origin of the losses. We have previously demonstrated that for OPSW, almost all of the losses in the NIR are due to surface scattering [45].

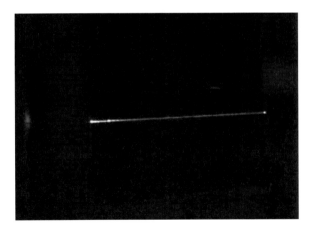

Figure 30. Top view of the light scattered along the oxidized porous silicon 4-μm-wide wave guide at 1550nm.

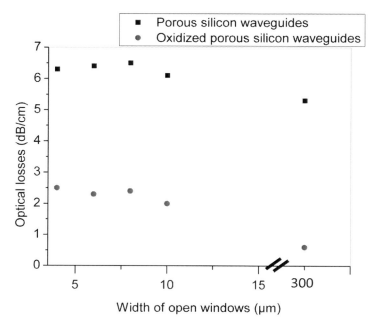

Figure 31. Variation in propagation losses as a function of the width of the opening for non-oxidised and oxidized porous silicon buried waveguides.

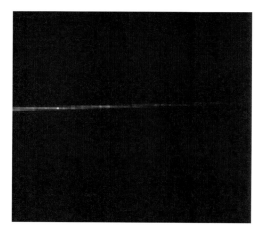

Figure 32. Top view of the light scattered along the oxidized porous silicon waveguide after DR1 impregnation.

Moreover, for the lowest widths (4 to 10μm), the loss values were nearly constant and were around 2.5 ± 0.5 dB/cm whereas there was an important reduction in losses for the 300μm-width. This can be mainly explained by considering that for small widths, the overlap between guided modes and interface was more important as the modal confinement becomes poorer. This explains the difference in losses between large and small waveguides. This loss difference is estimated to be around 1 and 1.5 respectively for unoxidized and oxidized waveguides.

After the filling of oxidized waveguides by DR1 dyes, the loss measurements were also performed at 1550 nm. Figure 32 shows a top view of the scattered light from buried waveguides filled by DR1 dyes. We can see that, in several places, the light intensity scattered by the surface increases and the light intensity decreases exponentially with z-propagation distance and confirms this observation by a sporadic rising in the decay exponential curve. By taking this noise into account, the losses increase after filling and are around 6 ± 0.8 dB/cm. Knowing that the composite material is weakly absorbing at this wavelength; the absorption is not really the major cause of losses here. We believe that after evaporation of the solvent, the presence of no-penetrated dried dyes at the surface of the waveguide, confirmed by microscopic observations, increases losses due to surface scattering and explains the increase in the light which is scattered by the surface.

CONCLUSION

We have studied oxidation and pore filling by DR1 or CR dyes and loss contributions in planar or buried waveguides manufactured from mesoporous layers.

It has been shown that in NIR, absorption is the principal source of losses for the PSi planar waveguides because of the heavily doped silicon substrate, whereas for OPSiplanar waveguides, the participation of surface scattering has the most important role. For the buried waveguides, we should also consider the presence of the edges and the defaults due to photolithography process which induces supplementary losses. Moreover, the losses are always lower in the case of oxidized waveguides.

The micro-Raman experiments were performed to control the oxidation rate. The guiding layer was considered to be almost completely oxidized whereas the area of the cladding layer near the Si substrate was only partially oxidized.

The same method was used to study the 2D-profile of buried waveguide filled by DR1 dyes. For each layer, the profile was homogeneous but the amount of dyes was more important in the layers of low porosity. But the optical losses increased after filling of oxidized waveguides due to a rise in surface scattering due to the presence of dyes at the surface of waveguide.

Lastly, for buried waveguides, by modifying the elaboration conditions, the propagation zone could be shifted to center of guiding layer. So, the propagated field becomes less sensitive to the roughness and the losses could be decreased.

These results with low values of losses ($\approx$ 1dB/cm) are very promising for all-silicon opto-electronic integrated devices and sensor applications by light transmission.

REFERENCES

[1] Pellegrini, V; Tredicucci, A; Mazzoleni, C; Pavesi, L. *Phys. Rev. B*, 52, R14328 (1995)
[2] Pavesi, L; Panzarini, G; andreani, LC. *Phys. Rev. B*, 58, 15794 (1998)
[3] Mulloni, V; Pavesi, L. *Appl. Phys. Lett.*, 76, 2523 (2000)
[4] Bruyant, A; Lerondel, G; Reece, PJ; Gal, M. *Appl. Phys. Lett.*, 82, 3227 (2003)
[5] Ghulinyan, M; Oton, CJ; Bonetti, G; Gaburro, Z; Pavesi, L. *J. Appl. Phys.*, 93, 9724 (2003)
[6] Zheng, WH; Reece, P; Sun, BQ; Gal, M. *Appl. Phys. Lett.*, 84, 3519 (2004)

[7] Ishikura, N; Fujii, M; Nishida, K; Hayashi, S; Diener, J; Mizuhata, M; Deki, S. *Optical Materials*, 31, 102 (2008)
[8] Wei, X; Kang, C; Liscidini, M; Rong, G; Retterer, ST; Patrini, M; Sipe, JE; Weiss, SM. *J. Appl. Phys.*, 104, 123113 (2008)
[9] Guillermain, E; Lysenko, V; Orobtchouk, R; Benyattou, T; Roux, S; Pillonnet, A; Perriat, P. *Appl. Phys. Lett.*, 90, 241116 (2007)
[10] HuiminOuyang, M Christophersen; Viard, R; Miller, BL; Philippe, M Fauchet. *Adv. Funct. Mat.*, 15, 1851 (2005)
[11] Saarinen, J; Weiss, S; Philippe, M; Fauchetand JE. *Sipe, Optics Express*, 13, 3754 (2005)
[12] Starodub, VM; Fedorenko, LL; Sisetskiy, AP; Starodub, NF. *Sensors, and Actuators B*, 58, 409 (1999).
[13] Perelman1, LA; Schwartz, MP; Wohlrab, AM; VanNieuwenhze, MS; Sailor, MJ. *phys. stat. sol.* (a), 204, 1394 (2007).
[14] Tinsley-Bown, A; Smith, RG; Hayward, S; anderson, MH; Koker, L; Green1, A; Torrens, R; Wilkinson, A-S; Perkins, EA; Squirrell, DJ; Nicklin, S; Hutchinson, A; Simons, AJ; Cox, TI. *phys. stat. sol.* (a), 202, 1347 (2005).
[15] Janshoff, A; Dancil, KS; Steinem, C; Greiner, DP; Lin, VS-Y; Gurtner, C; Motesharei, K; Sailor, MJ; Ghadiri, MR. *J. Am. Chem. Soc.*, 120, 12108 (1998)
[16] Dancil, K-P. S., Greiner, D. P., and Sailor, M. J. *J. Am. Chem. Soc.*, 121, 7925(1999).
[17] Chan, SHorner, ScR; Fauchet, PM; Millar, BL. *J. Am. Chem. Soc.*, 123, 11797 (2001).
[18] Chan, S; Fauchet, PM; Li, Y; Rothberg, LJ. Proceedings of SPIE (Micro-, and Nanotechnology for Biomedical, and Environmental Applications, Raymond P. Mariella, Jr., Editor), 3912, 23 (2000)
[19] Barla, K; Herino, R; Bomchil, G. « Stress in oxidised porous silicon layers », *J. Appl. Phys.*, 59 (2), 439 (1986).
[20] Aspnes, DE. *Thin Solid Film*, 89, 249 (1982)
[21] Chamard, V; Dolino, G; Muller, F. *J. Appl. Phys,* 84 (12), 6659 (1998).
[22] Gupta, P; Colvin, VL; George, SM. *Phys. Rev.B*, 37, 8234 (1988)
[23] Tien, PK; Ulrich, R. *J. Opt. Soc. America*, 60, 1325 (1970).
[24] Pirasteh, P; Charrier, J; Soltani, A; Haesaert, S; Haji, L; Godon, C; Errie, N. n, *Applied Surface Science*, 253, 1999 (2006,)
[25] Charrier, J; Lupi, C; Haji, L; Boisrobert, C. *Mat. Sci. In Semiconductor Processing*, 3, 357-361 (2000)
[26] Charrier, J; Guendouz, M; Haji, L; Joubert, P. *Phys. Stat. Sol.* (a), 182, 431-436 (2000)
[27] Pirasteh, P. Phd thesis, university of Rennes1 – France (2005)
[28] Okamura, Y; Yoshinaka, S; Yamamoto, S. *Applied Optics*, 22, N°23, 3892 (1983)
[29] Okamura, Y; Sato, S; Yamamoto, S. *Applied Optics*, 24, N°24, 57 (1985)
[30] Beeckman, J; Neyst, K; Hutsebaut, X; Cambournac, C; Haelterman, M. *Optics Express*, 12, N°6, 1011 (2004)
[31] Lerondel, G; Romestain, R; Barret, S. *Journal Applied Physics*, 81, N° 9, 6171 (1997)
[32] Ferrand, P. Phd thesis, university of Grenoble 1 – France (2001)
[33] Xie, YH; Hybersten, MS; Wilson, WL; Ipri, SA; Larven, GE; Dons, E; Brown, WL; Dons, E; Wein, BE; Kortan, AR; Watson, GP; Liddle, AJ. *Physical Review B*, 49, N°8, 5386 (1994)

[34] Gaillet, M; Guendouz, M; Ben Salah, M; Le Jeune, B; Le Brun, G. *Thin Solid Film*, 410, 455 (2004)
[35] Tien, PK. *Applied Optics*, 10, N°11, 2395 (1971)
[36] Kirchner, A; Busch, K; Soukoulis, CM. *Physical Review B*, 57, N°1, 277 (1998)
[37] Lehmann, V. "Electrochemistry of Silicon", Ed Wiley-VCH (2002)
[38] Setzu, S; Lerondeland, G; Romestain, R. *Journal Applied Physics*, 84, N° 6, 3129 (1998)
[39] Charrier, J; Pirasteh, P; Pedrono, N; Joubert, P; Haji, L; Bosc, D. *Phys. Stat. Sol. (a)*, 197, N°288 (2003)
[40] Scott, BJ; Wirnsberger, G; Stucky, GD. *Chem. Mater*, 13, 3140 (2001)
[41] Ziss, J. (Ed.), *Molecular Nonlinear optics: Physics, and Devices*, Academic Press, New York (1994)
[42] Rivolo, P; Pirasteh, P; Chaillou, A; Joubert, P; Kloul, M; Bardeau, JF; Geobaldo, F. *Sensors & Actuators B*, 100, 99-102 (2004).
[43] Guendouz, M; Kloul, M; Haesaert, S; Joubert, P; Bardeau, JF; Bulou, A. *Phys. Stat. Sol. (a)*, 2, 3453 (2005).
[44] Charrier, J; Guendouz, M; Haji, L; Joubert, P. *Phys. Sta. Sol. (a)*, 182 (2000) 431
[45] Pirasteh, P; Charrier, J; Dumeige, Y; Haesaert, S; Joubert, P. *Journal of Applied Physics*, 101 (7) (2007) 083110-1 083110-6

Chapter 10

BEHAVIOR OF UNMODIFIED MCM-41 TOWARDS CONTROLLED RELEASE OF PRO-DRUG MOLECULE, CYSTEINE

*Anjali Patel** and *Varsha Brahmkhatri*
Department of Chemistry, Faculty of Science,
M. S. University of Baroda, Vadodara, India

ABSTRACT

Mesoporous silica material, MCM-41 was synthesized and characterized by various physico-chemical techniques such as FT-IR, X-ray diffraction, N_2 adsorption- desorption, SEM and TEM. Cysteine was selected as a pro-drug molecule and the potential of MCM-41 as a drug delivery system was explored. The study on release profile was also carried out for cysteine loaded sample in stimulated body fluid (SBF) at room temperature. It was found that MCM-41 was able to release cysteine in a controlled manner and exhibit better effect than N-acetylcysteine in solution. 100% Cysteine release was observed in 12h.

Keywords: drug delivery, mesoporous material, MCM-41, cysteine

1. INTRODUCTION

Mesoporous silica materials (M41S) have attracted tremendous interest due to their wide applications in the field of chemical Catalysis [1] and Biotechnology. They can be used as medical devices because of their high surface area, ordered mesopores and well defined structure [2-5]. Hence, they exhibit new possibilities for incorporating biological agents and controlling the drug from release from the matrix. A literature survey shows that they have found potential applications for encapsulating bioactive molecules [6-9].

---

* E-mail: aupatel_chem@yahoo.com.

A variety of materials have been reported as drug delivery systems such as polymers [10], polymer based composites, bioactive glasses or ceramics [11, 12]. As it is very difficult to get homogeneous distribution of drug through the matrix in these materials, they possess disadvantages of heterogeneity of samples and that can affect the release rate. To overcome these disadvantages, the matrix material should possess well-defined porosity and well-ordered structure. These requirements are clearly satisfied by mesoporous materials like MCM-41.

A literature survey shows various application of MCM-41 mesoporous silica as drug delivery and controlled system [13-22]. Vallet-Regi and coworkers studied the release of ibuprofen in MCM-41 mesoporous silica delivery systems in a simulated body fluid, and the results showed that the drug release behaviors depended much not only on the method for charging the drug in the MCM-41 but also on the structure of MCM-41 [13-14]. Andersson et al. reported the influences of materials on ibuprofen drug loading and release profiles from ordered micro- and mesoporous MCM-41, SBA-1, and SBA-3 silica materials [15]. Devoisselle and coworkers investigated the feasibility of loaded ibuprofen in MCM-41 material in various solvents [16]. Wang et al. synthesized the MCM-41 functionalized with polyelectrolyte multi-layer aptamer conjugate and studied the release of drug from this material as drug delivery system [23]. Synthesis of MCM-41 with different morphology through simple hydrothermal process and then functionalization of the same with luminescent YVO4:Eu3+ layer has been reported by Lina et al. they also studied released profile of ibuprofen [24]. Vallet-Regı and his group studied the release profile of bisphosphonate from MCM-41 and amino functionalized MCM-41 [25]. In vitro uptake and release of cytochrome C loaded on MCM-41 was examined by Lin et al. [26]. Shi Zhang Qiao et al. reported pH responsive drug carrier based on chitosan coated MCM-41 and studied the release profile of ibuprofen in SBF solution at PH 7.4 [27]. Lin and group studied the release profile of l-cysteine using organic functionalized mesoporous silica nano particle [28]. Delgado and group reported the effect of amine and carboxyl functionalized of MCM-41 on release of an anticancer drug, cisplatin. They suggest carboxy-functionalized MCM-41 sample as a superior carrier for controlled delivery of cisplatin because of higher percentage of drug release and favorable kinetics of the in vitro delivery process, which should facilitate drug delivery control over a significantly longer time period [29].

Similarly Popova and co-workers studied carboxylic modified mesoporous carriers MCM-41 and SBA-15 as drug delivery carriers for Sulfadiazine. They observed slower release rate of sulfadiazine from carboxylic modified MCM-41 and SBA-15 mesoporous particles compared to the non-modified ones. Also these silica materials demonstrated no cytotoxicity on Caco-2 cell line [30]. E. Santamaría and group studied hierarchically meso–macroporous materials as drug delivery systems, and ibuprofen as a model drug. They observed a two-step release process, consisting of an initial fast release and then a slower release. They also correlated drug release profiles with the textural properties of these materials. The more the macropore present in the material, the slower the release behavior was observed, as the ibuprofen adsorbed in the internal pores had to diffuse along the macropore channels up to the surface of the material [31]. Nunes and co-workers reported montmorillonite K10 and MCM-41 as drug delivery systems for nifedipine release, a potent calcium-channel blocking agent. They found the release profile from MCM-41 leads to higher drug solubility compared even with neat drug solubility, which is advantageous in a drug known for its low solubility [32]. Potrzebowski and group compared two methods of drug loading, incipient wetness and melting for the encapsulation of ibuprofen in the pores of MCM-41 through NMR spectroscopy. They showed that melting a mixture of ibuprofen and MCM is a much more efficient method of confining the drug in the pores compared to incipient wetness [33]. Recently Roik et al. described pH-controlled release ability of chemically modified silicas of MCM-41-type using biologically active amino acid, 4-aminobenzoic acid. MCM-41-type silica material with chemically anchored N-[N0-(N0- phenyl)-2-aminophenyl]-3-aminopropyl groups were obtained by combination of sol-gel synthesis and post synthetic modification in vapor and liquid phases [34].

All of these research suggest that the ordered mesoporous silica materials would be the appropriate candidates as the drug controlled released system, but many factors such as the solubility of the drug in the solvent, the diffusivity of the drug in the solvent and the structure characteristics of the pore materials can seriously affect the release behaviors of the drug molecule.

Considering all these properties of mesoporous materials, in the present work the potential of MCM-41, for hosting and delivery of drugs under appropriate conditions using a model molecule, was explored. Cysteine was selected as a model molecule.

Cysteine is a naturally occurring, sulfur-containing amino acid which has a thiol group and is found in most proteins. Because it is a sulfur-based amino acid, cysteine itself can act as an antioxidant in the body. Cysteine prodrugs are used to treat Schizophrenia and reduce drug cravings [35]. Cysteine is a limiting substrate in the production of glutathione in the body. The reduction of intracellular levels of glutathione contributes to chronic inflammatory conditions, which are associated with cancer, neurogenerative, cardiovascular and infertility diseases resulting in high demand of cysteine [36-38]. Current cysteine therapies are administration of different cysteine derivatives such as *N*-acetylcysteine. One of the major drawbacks of these therapies is high dosages that can provoke persistent damage and strong allergic reactions [39-42]. The mentioned drawbacks can be overcome by using Controlled drug delivery system.

The present work consists of non-hydrothermal synthesis of MCM-41. The obtained material was characterized by FT-IR, X-ray diffraction, N_2 adsorption- desorption, SEM and TEM. Cysteine was loaded over MCM-41 and the obtained material was characterized for various physicochemical techniques. The release profile was obtained for cysteine loaded samples in stimulated body fluid (SBF) at room temperature. The concentration of cysteine was determined by UV-visible spectrophotometer.

2. PROCEDURES

2.1. Materials

All chemicals used were of A.R. grade. Cetyl triethyl ammonium bromide (CTAB) (Loba chemie, Mumbai). Tetraethylorthosilicate (TEOS) was used as received from Merck.

2.2. Synthesis of Mesoporous Silica (MCM-41)

MCM-41 was synthesized using reported procedure [43] with slight modification. Surfactant (CTAB) was added to the very dilute solution of NaOH with stirring at 60^{0}C. When the solution became homogeneous, TEOS was added drop wise and the obtained gel was aged for 2hr. The resulting product was filtered, washed with distilled water, dried at room temperature. The obtained material was calcined at 555°C in air for 5 hr and designated as MCM-41.

2.3. Pro-Drug, (Cysteine), Loading (Cys-MCM-41)

Cysteine was loaded over MCM-41 using incipient wetness impregnation method. The advantage of using this method is precise amount of the drug is being loaded. Also time consuming equilibrium and filtration steps can be avoided as compared to other loading procedures. A sample containing 30% cysteine loading over MCM-41 was obtained by impregnation. 1 g of MCM-41 was impregnated with an aqueous solution of Cysteine (0.3/30 g/ml of double distilled water) and dried at 100°C for 5 h. The obtained materials were designated as Cys-MCM-41

2.4. Cysteine Release Studies

The release profile was obtained by soaking the samples in 150 ml of a simulated body fluid, SBF (1 mg of the cysteine sample per ml of fluid), and measuring the drug concentration in the fluid by means of a Uv-vis spectrophotometer at 570 nm. Simulated body fluid (SBF) has a composition very similar to the human plasma [44] (pmm: 142.0/5.0/2.5/1.5/147.8/4.2/1.0/0.5 Na$^+$/K$^+$/Ca^{2+}/Mg^{2+}/Cl$^-$/HCO$_3^-$/HPO$_4^{2-}$/SO$_4^{2-}$).

3. CHARACTERIZATION

Techniques	Conditions	Instrument specification
FT-IR	KBr wafer	Perkin-Elmer (Model-RX1)
Powder XRD	Cu-Kα radiation (1.5417 A°); scanning angle from 0° to 60°.	PHILIPS PW-1830
N$_2$ Adsorption – desorption isotherms	Surface area analyzer at -196°C using BET method	Micromatries ASAP 2010
Scanning Electron Microscopy (SEM)	With scanning electron electrode at 15 kV. Scanning was done at 1mm range and images taken at different magnifications.	JEOL SEM instrument (model-JSM-5610LV)
Transmission Electron Microscopy (TEM)	Dispersed in Ethanol and ultra-sonicated for 5-10 minutes	JEOL (JAPAN) TEM instrument (model-JEM 100CX II)

4. RESULTS AND DISCUSSION

4.1. Characterization of MCM-41

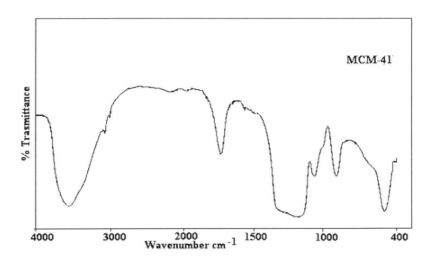

Figure 1. FT-IR spectra of MCM 41.

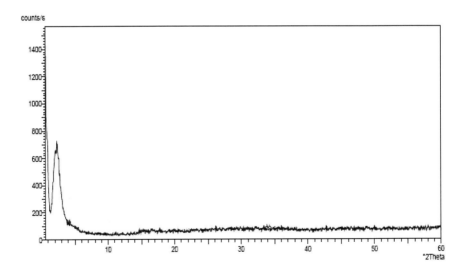

Figure 2. XRD of MCM 41.

FT-IR spectra (Figure 1) shows a broad band around 1300-1000 cm^{-1} corresponding to asymmetric stretching of Si-O-Si. The band at 801 and 458 cm^{-1} are due to symmetric stretching and bending vibration of Si-O-Si respectively. The band at 966 cm^{-1} corresponds to symmetric stretching vibration of Si-OH. The broad absorption band around 3448 cm^{-1} is the absorption of Si-OH on surface, which provides opportunities for forming the hydrogen bond.

The XRD patterns (Figure 2) showed a sharp peak around $2\theta=2°$ and few weak peaks in $2\theta=3\sim5°$, which indicated well-ordered hexagonal structure of MCM-41.

Figure 3. Shows the SEM images of MCM-41 3000x (Left) and 10,000x (Right).

SEM shows (Figure 3) the surface morphology of the synthesized material. The characteristic mesoporous hexagonal structure was observed in TEM.

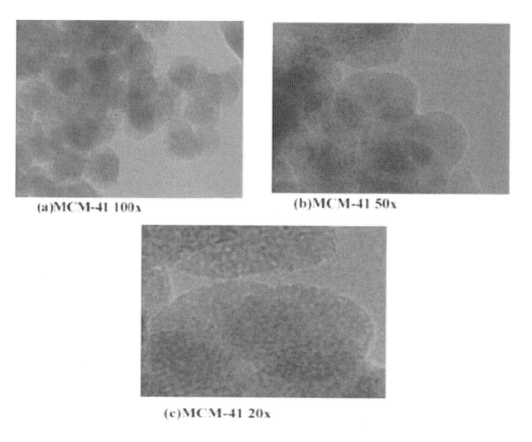

Figure 4. TEM images of MCM-41.

Figure 4 Clearly shows hexagonal mesopores in MCM-41. This reveals the presence of long range uniform hexagonal structure in the synthesized material..

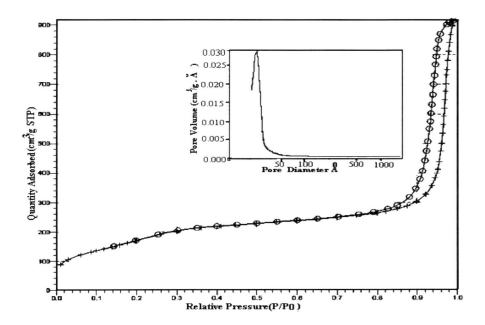

Figure 5. Nitrogen sorption isotherms and pore size distribution of MCM-41.

The N_2 adsorption–desorption isotherms (figure 5) are the type of IV in nature according to the IUPAC classification and exhibited an H1 hysteresis loop which is a characteristic of mesoporous solids [45]. The adsorption isotherm showed a sharp inflection, which means a typical capillary condensation within uniform pores. The position of the inflection point is clearly related to the diameter of the mesopore, and the sharpness of this step indicates the uniformity of the mesopore size distribution.

From N_2 adsorption–desorption isotherm the BET surface area for synthesized MCM-41 material was found to 636 cm^2g^{-1}, and average pore diameter of 47 Å.

4.2. Characterization of Cys- MCM-41

The FT-IR spectra for Cysteine and Cys- MCM-41 are shown in (Figure 6). The FT-IR spectra of cysteine shows bands at 1600 and 1390 cm^{-1} which corresponds to the asymmetric and symmetric stretching of COO–. Another characteristic band at 1520 cm–1 corresponds to N-H bend and the very broad band in the 3000–3500 cm^{-1} range was attributed to NH_3^+ stretching. Additionaly for thiol group, a very weak band near 2551 cm^{-1} is attributed to S-H stretching in the cysteine molecule. The observed FT-IR frequencies for Cysteine and MCM-41-Cys are presented in Table 1. It was clearly observed from the Table 1 that there was no considerable shift in carboxylic group -COOH frequency as well as in thiol group –SH frequency in MCM-41-Cys sample. But the broad band corresponding amine group –NH_2 from Cysteine seems to be merged with the broad absorption band around 3448 cm^{-1} which correcponds to the absorption of Si-OH on surface of MCM-41. The surface Si-OH groups of MCM-41 offers opportunities for forming the hydrogen bond as mentioned earlier, which reveals weak hydrogen bond interactions of –NH$_2$ with the surface silonol groups Si-OH of MCM-41 matrix.

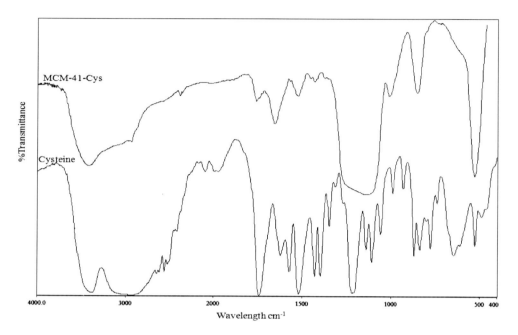

Figure 6. FTIR spectra of Cysteine and Cys-MCM-41.

Table 1. FT-IR absorption frequencies

Materials	FT-IR absorption frequencies(cm^{-1})			
	NH$_2$	SH	COOH	CH$_2$
Cysteine	3000-3400	2551	1695	1413
MCM-41-Cys	3448	2540	1680	1432

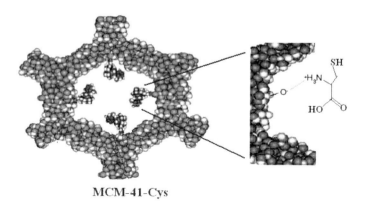

Scheme 1. Probable interaction of Cystiene with MCM-41.

Based on above observations, the probable interaction of cysteine with MCM-41 matrix is shown in scheme 1.

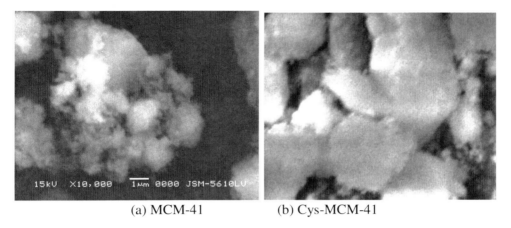

(a) MCM-41   (b) Cys-MCM-41

Figure 7. SEM images of (a) MCM-41 and (b) Cys-MCM-41.

Figure 7 (a) and (b) shows the SEM image of MCM 41 and Cys-MCM41. The surface morphology of Cysteine loaded sample is almost identical to that of pure MCM-41. Further no separate bulk phase of host that is cysteine was observed in SEM images of Cys-MCM41.

4.3. Cysteine Release from MCM-41

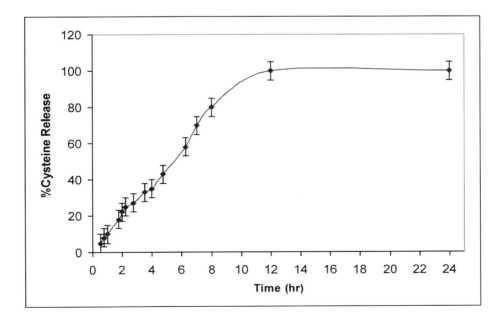

Figure 8. Cumulative release profile of cysteine over 24 hours for Cys-MCM-41.

It was observed that Cysteine loaded Cys-MCM-41(figure 8) did not show a sharp initial release burst during the first hours. The initial burst is due to the immediate dissolution and release of that portion of the drug located on and near the surface of matrix. This system shows a low release rate, followed by a constant rate over the subsequent hours. This fact is possibly related to an interaction between the host, Cysteine and the mesoporous silica, MCM-41 by hydrogen bonding between the functional groups amine, and carboxylic groups of cysteine and the silanol groups of MCM-41. 100% Cysteine release was observed in 12 h.

The release of cysteine from the MCM-41 matrix followed a diffusion-controlled model, where the quantity released per unit area is proportional to the square root of the time. A linear regression with the origin included among the data until 10 hours of release has been used to fit the data, with r =f 0.999. The mechanisms of release from this system include the leaching of the drug by the bathing fluid, which is able to enter the drug matrix phase through the pores. The drug is presumed to dissolve slowly into the liquid phase and diffuses from the system along the solvent-filled capillary channels.

4.4. Comparison with Traditional Drug Delivery

The important aim of designing any controlled drug delivery systems is to achieve a delivery profile that would yield a high blood level of the drug over a long period of time.

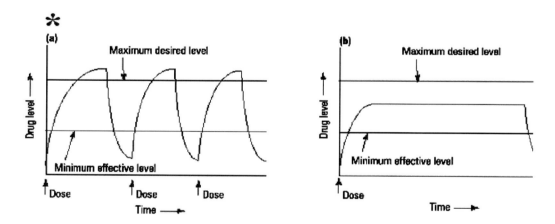

Figure 9. Drug levels in the blood with (a) traditional drug dosing and (b) controlled delivery dosing. (* Figure 8 was taken from reference [46]).

In case of traditional drug delivery systems, the drug level in the blood follows the profile shown in Figure 9a, in which the level rises after each administration of the drug and then decreases until the next administration. The major issue with traditional drug administration is that the blood level of the agent remains between a maximum value, in which it may be at a toxic level, and a minimum value, below which the drug is no longer effective [46].

In controlled drug delivery systems designed for long-term administration, the drug level in the blood follows the profile shown in Figure 9b, remaining constant, between the desired maximum and minimum, for an extended period of time.

In the present system cysteine was released over a long period of time, i.e., 12 h in a controlled manner. Similar results obtained by other authors with mesoporous materials as drug delivery carrier are summarized in table 2.

Table 2. Drug adsorption conditions and release rate in mesoporous silica materials

Mesoporous Silica	Textural properties SA cm^{2}g^{-1}	D Å	Drug molecule	solvent	% drug loading	%Release / time(hr)	Reference
MCM-41	636	47	Cysteine pro-drug	water	30	100/12	Present work
MCM-41A	1157	25	Ibuprofen	Hexane	30	80/72	[13]
MCM-41	928	37	Ibuprofen	Pentane	25	85/48	[14]
	341	28					
Amine-functionalized MCM-41	341	28	Ibuprofen	Pentane	30-36	97/150	[14]
MCM-41	1200	48	Ibuprofen	Hexane	41	~100/24	[15]
MCM-41	850	34	Ibuprofen	Ethanol	65 mg/ml	38/20*	[16]
Chitosan coated MCM-41-PO$_3^-$	1020	24	Ibuprofen	pH 5 acetate buffer	100mg	90/8	[27]
COOH-MCM-41	707	54	Cisplatin	water	14.7mg/g	37/40	[29]
MCM-41	1175	44	Sulfadiazine	Acetone: methanol (1:1)	39	100/4	[30]
MCM 41 COOH	412	37			49	100/25	
MCM-41	900	32	Nifedipine	dichloromethane	19.7	50/20	[32]

SA: Surface area, D: pore diameter
* release time in minutes

However, well-defined controlled drug release can be achieved by tailoring the textural properties such as surface area, surface functionalization, pore diameter and pore volume. Mesoporous materials are important as drug delivery carriers as they can accommodate poorly water-soluble drugs to enhance their dissolution and permeation behavior which is of great interest in the area pharmaceutical materials research [47,48].

CONCLUSION

Our studies show that cysteine was successfully loaded to MCM-41. Cysteine can be linked via surface Si-OH groups of MCM-41 and probable interaction of cysteine through amine groups was proposed on the basis of FT-IR. MCM-41 was found to be ideal drug delivery system for controlled release of Cysteine. 100% Cysteine release was observed in 12h. MCM-41 is able to release cysteine in a controlled manner and can exhibit better effect than traditional drug delivery systems. By an appropriate choice of the pore size of MCM-41 as well as the model molecule, a reliable drug delivery system can be designed.

ACKNOWLEDGMENTS

One of the authors, Ms. Varsha Brahmkhatri, is thankful to BRNS, Project. No.2007/37/20/BRNS/917, BARC, Mumbai, for the financial assistance. I am thankful to my students, Mr Soyeb Pathan, for assistance in preparation of chapter.

REFERENCES

[1] F. Zhang, G. Liu, W. He, H. Yin, X.Yang, H. Li, J. Zhu, H. Li, Y. Lu, *Adv Funct Mater*, 18, 3590(2008).

[2] M. Vallet-Regì, L. Ruiz-Gonzàlez, I. Izquierdo-Barba, JMJ. Gonzàlez-Calbet, *Mater. Chem*, 16, 26 (2006).

[3] M. Arruebo, M. Galan, N. Navascues, C. Tellez, C. Marquina, M. R. Ibarra, J. Santamaria, *Chem. Mater*,18, 1911 (2006).

[4] M. Arruebo, R. Fernandez-Pacheco, S. Irusta, J. Arbiol, M.R. Ibarra, J. Santamaria, *Nanotechnology*, 17, 4057 (2006)

[5] C. Y. Lai, B. G. Trewyn, D. M. Jeftinija, K. Jeftinija, S. Xu, S. Jeftinija, V. S. Y. Lin,. *J. Am. Chem. Soc*, 125, 4451 (2003).

[6] Y. Urabe, T. Shiomi, T. Itoh, A. Kawai, T. Tsunoda, F. Mizukami, K. Sakaguchi. *Chem Bio Chem*, 8, 668 (2007).

[7] F.Babonneau, L. Yeung, N. Steunou, C.Gervais, A. Ramila, M. Vallet-Regi, *J. Sol. Gel. Sci. Tech*, 31, 219 (2004).

[8] R. Mortera, J. Vivero-Escoto, I. Slowing, E. Garrone, B. Onida, V. Lin, *Chem. Comm*, 3219 (2009).

[9] C. Tourné-Péteilh, D. A. Lerner, C. Charnay, N. Nicole, S. Bégu, J. M. Devoisselle, *Chem Phys Chem*, 4, 281 (2003).

[10] H. Park, K. Park, *Pharm. Res*, 13, 1770 (1996).
[11] R. P. del Real, S. Padilla, M. Vallet-Regi, *J. Biomed. Mater. Res*, 52,1 (2000).
[12] M. Vallet-Regí, F. Balas, M. Colilla, M. Manzano, *Drug Metabol Lett*, 1, 37 (2007).
[13] M.Vallet-Regi, A. Ramila, R. P.del Real, J.P. Pariente, *Chem. Mater*, 13, 308 (2001).
[14] M. Manzano, V. Aina, C. O.Arean, F. Balas, V. Cauda, M. Colilla, M.R. Delgado, M.Vallet-Regi, M. *Chem. Eng. J*, 137,30 (2008).
[15] J. Andersson, J. Rosenholm, S. Areva, M. Linden, *Chem. Mater*, 16, 4160 (2004).
[16] C. Charnay, S.Begu, C. Tourne-Peteilh, L. Nicole, D. A. Lerner, J. M. Devoisselle, *Eur. J. Pharm. Biopharm.*, 57, 533 (2004).
[17] L.B. Fagundes, T.G.F.Sousa, A. Sousa, V.V.Silva, E.M.B. Sousa, *J. Non-Cryst. Solids*, 352, 3496 (2006).
[18] W. Xu, Y. Xu, D. Wu, Y. Sun, *Stud. Surface Sci. Catal*, 170, 861 (2007).
[19] B. Munoz, A. Ramila, J.P. Pariente, I. Diaz, M.V. Regi, *Chem. Mater*, 15, 500 (2003)
[20] Z. Wu, Y. Jiang, T. Kim, K. Lee, *J. Control. Release*, 119, 215 (2007)
[21] F. Qu, G. Zhu, S. Huang, S. Li, J. Sun, D. Zhang, S. Qiu, *Micropor. Mesopor. Mater*, 92, 1(2006).
[22] F. Qu, G. Zhu, H. Lin, W. Zhang, J. Sun, S. Li, S.Qiu, *J. Solid State Chem*, 179, 2027(2006).
[23] C-L. Zhu, X-Y. Song, W-H Zhou, H-H. Yang, Y-H. Wen, X-R. Wang, *J. Mater. Chem.*, 19, 7765, (2009).
[24] P. Yanga, Z. Quan, L. Lu, S. Huang, J. Lin, *Biomaterials*, 29, 692,(2008)
[25] F. Balas, M. Manzano, P. Horcajada, M. Vallet-Regı, *J. Am. Chem. Soc.*, 128, 8116, (2006).
[26] I. I. Slowing, B. G. Trewyn, V. S.-Y. Lin, *J. Am. Chem. Soc.*, 129, 8845, (2007).
[27] A. Popat, J. Liu, G. Qing (Max) Lua, S. Z. Qiao, *J. Mater. Chem.*, 22, 11173, (2012).
[28] R. Mortera, J. Vivero-Escoto, I. I. Slowing, E. Garrone, B. Onida, V. S.-Y. Lin, *Chem. Commun.*, 3219 (2009).
[29] C.O. Arean, M.J.Vesga, J.B.Parra, M.R.Delgado, *Ceramics International* 39,7407, (2013).
[30] M.D. Popova, A. Szegedi, I.N. Kolev, J. Mihaly, B.S. Tzankov, G.T. Momekov, N.G. Lambov, K.P. Yoncheva, *Int J Pharm* 436, 778,(2012)
[31] E. Santamaría, A. Maestro, M. Porras, J. M. Gutiérrez and C. González, *J. Solid. State Chem.* 210, 242,(2014)
[32] I.J. Marques, P.D. Vaz, A.C. Fernandes, C.D. Nunes, *Micropor. Mesopor. Mater*, 183,192, (2014) 192.
[33] E. Skorupska, A. Jeziorna, P. Paluch, M.J. Potrzebowski, *Mol. pharmaceutics*, 11,1512, (2014).
[34] N.V. Roik, L.A. Belyakova, *J. Solid State Chem*, 215, 284, (2014).
[35] J. Cook, A. Baker, W. Yin, *U S Patent application* 20090281109
[36] K. Senthil, S. Aranganathan, N. Nalini, *Clin Chim Acta*, 339, 27 (2004)
[37] Y. Gilgun-Sherki, E. Melamed, D. Offen, *Neuropharmacology*, 40, 959(2001).
[38] P. H. Black, L. D. Garbutt, *J. Psychosom. Res*, 52, 1 (2002).
[39] M. E. Anderson, A. Meister, *Methods Enzymol*, 143, 313 (1987)
[40] F. Santangelo, *Curr. Med. Chem*, 10, 2599 (2003)
[41] C. Brack, M. Labuhn, E. Bechter-Thüring, *X Cell. Mol. Life. Sci*, 53, 960 (1997).
[42] G. Auzinger, J. Wendon, *Curr Opin Crit Care*, 14, 179 (2008)

[43]] Q. Cai, Z. S. Luo, W. Q. Pang, Y. W. Fan, X. H. Chen, F. Z. Cui, *Chem. Mater,* 13, 258 (2001).
[44] T. Kokubo, H. Kushitani, S. Sakka, T. Kitsugi, T. Yamamuro, *J. Biomed. Mater. Res,* 24, 721 (1990).
[45] L. Yang, Y. Qi, X. Yuan, J. Shen, J. Kim, *J. Mol. Cat. A: Chem,* 229, 199 (2005).
[46] L. Brannon-Peppas, Polymers in Controlled Drug Delivery, Medical Plastics and Biomaterials Magazine, (1997) [Online] Available at: http://www.devicelink.com/mpb/archive/97/11/003.html
[47] W. Xu, J. Riikonen, V.P. Lehto, *Int. J. Pharm,* 453,181, (2013).
[48] F. Tang, L. Li, D. Chen, *Adv. Mater,* 24, 1504, (2012).

Chapter 11

MESOPOROUS MULTIVALENT TRANSITION METAL OXIDES (V, CR, MN, FE, AND CO) IN CATALYSIS

Altug S. Poyraz[1], Sourav Biswas[1], Eugene Kim[1], Yongtao Meng[1] and Steven L. Suib[1,2,]*

[1]Department of Chemistry University of Connecticut,
Storrs, Connecticut, US
[2]Department of Chemical Engineering,
Institute of Materials Science University of Connecticut,
Storrs, Connecticut, US

ABSTRACT

Multivalent first row transition metal oxides (V, Cr, Mn, Fe, and Co) can form various oxide structures with unique catalytic properties. The catalytic performance of a mesoporous, high surface area TM oxide is known to be better than its nonporous counterpart. However, the direct synthesis of multivalent mesoporous TM oxide materials with desired crystal structure and structural properties is still a challenge to date. The multivalent nature, lack of proper sources, weak inorganic-organic interactions, and poor control of reaction rates are problems for the direct synthesis of mesoporous multivalent TM oxides. We summarize here some recent results of synthesized mesoporous hybrid materials with these TMs and their catalytic performances.

INTRODUCTION

Twenty years after the seminal introduction of the first mesoporous materials by Mobil Company, the field of mesoporous materials has expanded enormously and almost 40,000 research articles have been published related to mesoporous materials during this period [1, 2]. The massive interest in the field of mesoporous materials is due to the control and fine

---

[*] Fax: (+1) (860)-486-2981; E-mail: steven.suib@uconn.edu.

tuning of structural properties such as pore size, pore volume, bi-modal pore distribution, unique morphologies, different mesostructures, and modifications for a wide range of specific applications. Such control of the structural parameters has been demonstrated to be very useful in numerous fields such as gas sensors, cathode materials in solar cells and lithium ion batteries, redox supercapacitors, adsorption and separation, catalysis and magnetic materials [3–10]. The synthesis of mesoporous materials and the fine tuning of the structural parameters are possible by controlling the soft-assembly (micellization of the surfactants), micelle-inorganic precursor interaction, and controlling the condensation of inorganic sols [11, 12]. Therefore, significant numbers of studies related to mesoporous materials focus on new synthetic approaches which involve the use of surfactant micelles as soft templates [13–18]. From a thermodynamic point of view, the crucial parameters are Inorganic-Inorganic (G_{I-I}), Inorganic-Surfactant (G_{I-S}) interactions [12, 19]. In order to obtain an ordered mesostructured material, controlled condensation of the inorganic component (G_{I-I}) and strong inorganic-surfactant interaction (G_{I-S}) are desired. The former is controlled by the sol-gel chemistry of the inorganic sols by proper pH adjustment and in alcoholic solutions [20, 21]. The later (inorganic-surfactant) interaction can be Coulombic such as S^+I^-; S^-I^+; $S^+X^-I^+$; $S^-X^+I^-$ where S is the surfactant, I is the inorganic sol and X is the mediator ion or can be a charge transfer interaction between empty d orbitals of transition metal and oxygen lone pair electrons of the surfactant [8, 11, 14, 15, 18, 21–24].

Known approaches have yielded the synthesis of numerous mesoporous materials from different parts of the periodic table. From the catalysis point of view the synthesis of mesoporous transition metal oxides (MTMO) and their use as heterogeneous catalysts have attracted substantial interest [3, 8, 9, 25–27]. Direct synthesis of MTMO is still a challenge to date. Despite the fact that the successful syntheses of numerous thermally stable MTMOs have been achieved, those efforts mostly focused on groups I-IV TMs such as Y, Ti, Hf, Zr, V, Nb, Ta, Cr, Mo, and W [14, 15, 28–33]. Some of the mesoporous groups I-IV TM oxides and inorganic-surfactant interactions (G_{I-S}) are summarized in Table 1. Moreover, compared to mesoporous silica materials, these MTMOs are not as well-ordered and not as porous as mesoporous silica and the control of structural parameters are also not an easy task. The difficulty of the synthesis of late MTMOs (groups V-X) arises from the lack of proper sol-gel chemistry, weak G_{I-S} (due to filled d orbitals), and easily interchangeable oxidation states with multiple different crystal structures.

Later MTMOs also suffer from having low thermal stability of the formed mesostructure, due to the low flexibility of the M-O-M bonds of the crystalline phases and therefore they cannot accommodate the curvature of the soft-templates (micelles) and increased ion mobility with high temperature heat treatments, especially when the temperatures gets close to the Tamman temperature [16, 34–36].

This chapter first focuses on the synthetic approaches for the synthesis of late MTMOs with various oxidation states and crystal structures of the same TM system (V, Cr, Mn, Fe, and Co).

Depending on the synthesis conditions, post treatments, and modifications, unique heterogeneous catalysts can be synthesized for specific reactions. The second part of the chapter summarizes the recent use of these MTMOs as heterogeneous catalysts. This part focuses on the advantage of having a mesostructure and mesoporosity in the MTMOS for numerous catalytic reactions.

The effects of structural parameters of the MTMOs on the catalytic results are also going to be discussed.

Table 1. Illustrative examples of some of the mesoporous group I-IV TM oxides, obtained crystal structures and inorganic-surfactant interactions (G_{I-S})

Transition Metal	Crystal Structure	G_{I-S} (inorganic-surfactant interactions)	Ref.
Yttrium	Amorphous	Coulombic	135
Titanium	TiO_2 (anatase)	Charge Transfer	18
Zirconium	ZrO_2 (tetragonal)	Charge Transfer	18
Hafnium	Amorphous	Charge Transfer	15, 17
Vanadium	Amorphous	Charge Transfer and Hydrogen Bonding	22
Niobium	Nb_2O_5 and Amorphous	Charge Transfer and Coulombic	18, 33
Tantalum	Ta_2O_5 and Amorphous	Charge Transfer and Coulombic	18, 28
Chromium	Cr_2O_3	Charge Transfer	29
Molybdenum	α-MoO_3	Charge Transfer	31
Tungsten	Amorphous	Charge Transfer	32

SYNTHETIC APPROACHES

The general synthetic approaches for the synthesis of MTMOs are given in Figure 1. The surfactant and inorganic self-assembly and true liquid crystal templating approaches are direct synthesis methods and are only applicable to certain transition metal systems. The methods were first developed for silica and then modified and applied to other metal oxide systems. However, only a few of these resulting materials have structural properties comparable to silica. In addition, for the later TMs, the number of successful efforts are very limited and control of neither structural properties nor the crystal structure is possible. Therefore, use of silica as either a hard template for nanocasting or as a support has been the main path for the design of later TM MTMO synthesis for catalytic reactions. Well known mesoporous silica syntheses with easily tunable structural properties, easiness of template removal by chemical etching in basic solutions, and high thermal and structural stabilities have widened the use of mesoporous silica. A different type of nanocasting can also be performed by packing nano-crystals and filling the interstitial voids with inorganic precursor.

a) Direct Synthesis Late MTMO by Soft Templating

Mesoporous materials with ordered mesostructure are typically characterized by low-angle diffraction and a type IV adsorption isotherm. Although there are numerous claims for mesoporosity in the literature, in this chapter we will only consider the contributions with regular mesoporosity. Chromium oxide is one of the most promising oxidation catalysts due to its mixed valent nature (2+, 3+, 5+, 6 oxidation states) [37].

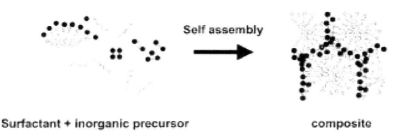

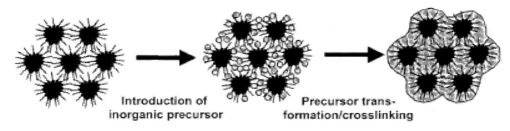

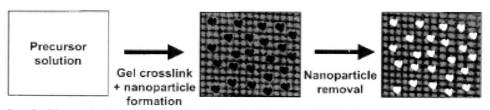

Reproduced with permission of the American Chemical Society (2001) ref. 15.

Figure 1. Schematic representation of general pathways leading to mesostructured and mesoporous non-siliceous materials.

The only known micelle templated mesoporous chromium oxide was synthesized by Sinha et al. [29,38]. The material has a cubic mesostructure and a Cr_2O_3 crystal phase which appeared after calcination. The synthesis requires long and multiple aging periods (7-14 days). Use of pluronic F127 surfactant gives thick mesopore walls (13 nm) which are probably the reason behind the thermal stability of the mesostructure. Despite the material preserving mesostructure during calcination steps, the surface area drastically decreases from 212 m^2/g to 78 m^2/g after calcination at 500°C. Moreover, XPS analyses suggests a multivalent nature of chromium but not a pure 3+ state (Cr_2O_3).

Another multivalent TM with various oxidation states and crystal structures is manganese. Different from the chromium oxides, manganese oxides can also form microporous tunnel structures known as octahedral molecular sieves (OMS) [26, 39]. Micelle templated mesoporous manganese oxides were synthesized first in 1997, but since then there has been no major progress for the synthesis of mesoporous manganese oxide [40–43]. The mesoporous materials obtained with the method of Tian et al. loses its mesostructure and mesoporosity mostly after calcination and the obtained surface areas are around ~50 m^2/g. This approach gave different crystal structures in different studies (γ-MnO_2, Mn_2O_3, Mn_5O_8, and Mn_3O_4) depending on heat treatment and post treatment conditions [41, 42]. Moreover, the XPS analyses of these materials show impurities in the structure and the preexistance of Mn^{2+}, Mn^{3+} and Mn^{4+} states in the same material [41–43]. In a typical synthesis procedure, manganese nitrate was added to a charged surfactant (CTABr) solution and the pH was adjusted to 8. The columbic interaction between cationic surfactant and negatively charged manganese hydroxide layers is the driving force for the formation of the mesostructure.

Other multivalent transition metals forming multiple crystal structures are vanadium, iron, and cobalt oxides. There is no reported thermally stable mesoporous oxide structures for these TMs so far. Despite this, there are several reports for the synthesis of mesostructured iron oxide, and no-thermal stability has been claimed so far [35, 44–46]. 2D hexagonal and 3D cubic mesostructured materials can be synthesized by evaporation induced self assembly but retaining porosity after heat treatment still remains as a challenge [35]. Moreover, certain crystal phases of transition metals are obtained by heating at high temperatures. Thermally unstable mesoporous TM oxides limit the use of these mesoporous materials in catalysis. From a catalytic point of view, heteregeneous catalysts are expected to be recyclable. Generally, to recycle the catalyst a regeneration step involves thermally removing adsorbed species from the surface. Therefore, thermal stability is an essential requirement. Similarly, for a gas phase catalytic reaction, a heteregeneous catalyst is expected to retain its catalytic activity for a substantial amount of time. The mesopore structure collapsed during stability tests due to high reaction temperatures after a period of time.

b) Reinforcement of Mesostructures

Thermally unstable mesostructures can be reinforced before heat treatment to ensure the mesoporosity after crystallization. The structures are reinforced by either carbon or silica. Figure 2 shows the schematic diagram of these approaches. Trimethylsiloxy-methyl silane (BTMS) and hexamethyldisilazane (HMDS) are the most common precursors for silica coating on the internal mesopore wall [8, 36, 47]. They react with surface hydroxyl groups to form silica coating and the coating supports the mesostructured material. The silica coating can easily be removed once the crystallization and stability of mesostructured TMs are established (path b in Figure 2). Carbon can also be a good structural support. The surfactant in the as synthesized mesostructured material or a diffferent carbon source introduced into the mesopores of the as synthesized material can be converted to carbon [8, 10, 36].

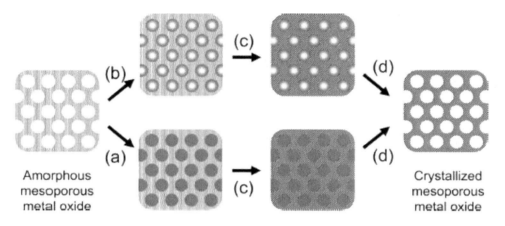

Reproduced with permission of the American Chemical Society (2008) ref 36.

Figure 2. Schematic illustration of the strategy for reinforced crystallization of mesoporous metal oxides: (a) back filling, (b) coating, (c) crystallization, and (d) removal of the reinforcement.

The organic compound is converted to carbon by heat treatment in either an inert atmosphere or under vacuum, after the crystallization of the mesostructure the carbon is removed by a second calcination step under air. Despite fact that there is no reported mesoporous late TM oxides in which the mesostructure is supported by either silica or carbon, the reinforcement of the mesostructure of these TMs is promising (path a in Figure 2).

c) The Use of Hard Templates

The use of a hard template is the most common approach for the synthesis of mesoporous late TM oxides because of the challenges in direct micelle templated syntheses. In this part of the chapter, we will give examples from manganese oxides for the illustration of the methods, since the methods are common to all TM oxides and one can find numerous examples for other TMs. The mesoporous hard template can be used either as a support to accomodate and stabilize the TM nano-particles or as a template for the synthesis of late mesoporous TM oxides by taking the replica of the existing pore structure of the template. The tunable mesostructure and the low bulk density of SiO_2 (~2.2 g.cm^{-3}) yield high specific surface area mesoporous silica materials, which makes them good candidates as support materials [25, 48–50]. However, homogeneous distribution of the TM oxides on the support is not quite possible and generally oxide clusters exist as scattered nano-particles on the support (Figure 3a) [51, 52].

The role of mesoporous silica as a hard template is more complicated and more parameters need to be considered. First, the filling of mesopores of the template is not always efficient and traditional impregnation methods generally result in formation of oxides both inside and outside the pore structure [51, 53–55]. Second, the stability of mesostructure upon template (silica) removal depends on the existence of intraconnecting micro or mesoporous tunnels, efficient fillling of these tunnels, and also the TM loading amount is important [8, 10, 25, 50, 56].

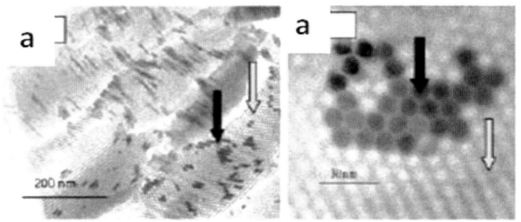

Reproduced with permission of the American Chemical Society (2004) ref 52.

Figure 3. (a)TEM images of Mn-loaded silica samples (SBA-15), prepared by adsorption with an aqueous solution of $MnCl_2$, $4H_2O$ (calcination at 5°C/min until 400°C annealing at this temperature for 3 h.

Therefore, SBA-15, SBA-16, and KIT-6 are the mesoporous templates which have been most commonly used due to the fact that they contain intraconnecting channels which ensure the stability of the mesostructure of TM oxides after template removal. New synthetic approaches such as, *in situ* surfactant oxidation, ion-exchange, two solvent methods, and chemical vapor deposition help in efficient pore filling and contribute to the stability of the mesostrucutre upon template removal [49, 50, 53–55, 57]. The materials prepared from these approaches with sufficient TM loading preserved mesoporosity upon template removal.

MESOPOROUS TMS (MN, CR, FE, V, AND CO) IN CATALYSIS

This part of the chapter focuses on the catalytic applications of mesoporous late TMs. The primary focus is on recent studies mostly conducted in the past decade. Tables 2 to 5 give details of the preparation methods, key structural properties and highlights from the catalytic reactions conducted. The selected illustrative examples demonstrate the advantage of mesoporosity and the nano-crystalline nature which comes with mesoporosity.

a) Mesoporous Manganese Oxide

Manganese oxide with easily exchangable multiple oxidation states and various crystal structures is known to be one of the best oxidation catalysts. Therefore, almost all catalytic reactions performed by mesoporous manganese oxide based catalysts utilize the redox properties of manganese oxide. The most common structures are Mn_2O_3, Mn_3O_4, and MnO_2 [39]. Moreover, in presence of cations these systems also form various microporous and layered structures known as octahedral molecular sieves (OMS) [26, 40]. This part of the chapter focuses on the catalytic applications of the mesoporous manganese oxides synthesized by either a soft-template, nanocasting, or manganese oxide clusters supported by mesoporous silica and further improvements in the catalytic performances by post treatments. Table 2 summarizes some of the recent catalytic studies conducted by mesoporous manganese oxides along with the synthetic approaches and key structural parameters.

The majority of the promising catalytic reactions done by mesoporous manganese oxides are gas phase oxidation reactions and volatile organic compound (VOCs) removal reactions (Entries 2, 4, 6, 8-10, 13, 14, and 16). The role of mesoporosity is dual in these studies. First, high surface areas and pore volumes make mesoporous materials very effective in removing these compounds from the gas stream, even at very low concentrations (10-500 ppm) [43, 58, 59]. For examples, Sinha et al. demonstrated in two different studies that gold deposited on mesoporous manganese oxides can remove n-hexane (77%, at 85°C), toluene (95%, at RT), and acetaldehyde (100%, at RT) effectively from a gas stream at low temperatures (entries 10 and 13) [42, 59]. The adsorbed compounds can either be oxidized afterwards or during the removal at high temperatures [43, 59–61]. Second, the oxidation power of nano-structured manganese oxide clusters on a support or nano-crystalline mesoporous manganese oxide are known to be better than bulk manganese oxides [43, 51, 62, 63]. Jiao et al. observed 8% manganese loaded mesoporous KIT-6 support is 26 times more active than non-porous manganese oxide (Mn_2O_3) for visible light driven O_2 evolution (Entry 11 in Table 2) [51]. The higher oxidation power of nano-crystalline mesoporous manganese oxide can be explained by easier reducibile nature. Oxidative power and better reducibility of mesoporous oxides studied by Deng et al. for the toluene complete oxidation with mesoporous manganese and cobalt oxides correlates with the easier reducibile nature (Entry 14 in Table 2) [63]. Mesoporous manganese and cobalt oxides show significantly higher performance for toluene oxidation (Figure 4a). The complete toluene oxidation (100% conversion) temperatures for mesoporous oxides are ~70°C lower than their non-porous counterparts. The high activity of mesoporous oxides can be explained by their reduction profiles. Figure 4b shows the H_2-TPR profiles of mesoporous and non-porous manganese and cobalt oxides. Better reducibility of mesoporous manganese and cobalt oxides was believed to be the reason for the high catalytic activity.

The catalytic performance of mesoporous manganese oxide can further be improved by post treatments. Post treatment can be done to change the crystal structure or to promote the catalyst surface by cations. For example, the different crystal structures (Mn_2O_3, Mn_3O_4, MnO_2 or K_xMnO_4) of mesoporous manganese oxide can be obtained by different heat treatments, acid treatments, and using different manganese precursors [42, 43, 51, 57–59, 62]. The activity of the crystal phase depends on the catalytic reactions studied. For example, Jiao et al. found that mesoporous Mn_2O_3 is the most active phase for visible-light driven O_2 evolution, however for VOC oxidation mesoporous the MnO_2 phase was found to be the most active phase in several studies [43, 58, 59, 64]. The typical way of preparing the MnO_2 phase is to treat manganese oxide with H_2SO_4. Sulfuric acid can convert manganese oxide structures with oxidation states of manganese being 2+ or 3+ (Mn_2O_3, Mn_3O_4) to the MnO_2 phase with 4+ oxidation state. The catalytic activity of manganese oxide structures can be further improved by introducing several ions (K$^+$, H$^+$, Cs$^+$, Ag$^+$, Na$^+$, Ca^{2+}, Mg^{2+}) in or on the surface of manganese oxide [26, 50, 65, 66]. In one of our recent studies, promoter ions (H$^+$, K$^+$, and Cs$^+$) were introduced on the surface of manganese oxide incorporated silicon oxide by an ion-exchange reaction between the cations and positively charged surfactant (CTABr) [50]. The ion promoted samples showed higher catalytic activity and selectivity for gas phase toluene oxidation (entry 16).

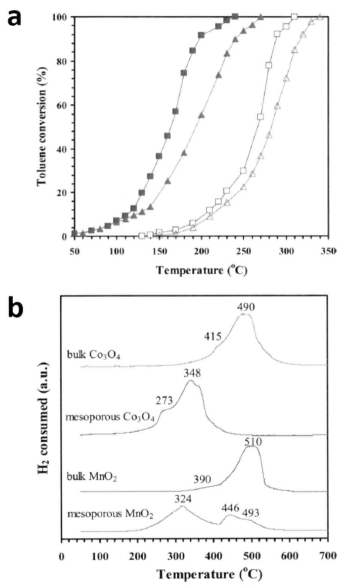

Reproduced with permission of the American Chemical Society (2010) ref 63.

Figure 4. (a) Toluene conversion as a function of reaction temperature over bulk MnO$_2$ (Δ), mesoporous MnO$_2$ (▲), bulk Co$_3$O$_4$ (□), and mesoporous Co$_3$O$_4$ (■) under the conditions of toluene concentration of 1000 ppm, toluene/O$_2$ molar ratio of 1/200, and space velocity of 20 000 h^{-1}. (b) H$_2$-TPR profiles of the bulk and ordered mesoporous MnO$_2$ and Co$_3$O$_4$ samples.

b) Mesoporous Cobalt Oxide

Mesoporous transition metal oxides are of particular interest in the field of catalysis due to the existence of easily convertible multiple oxidation states, large surface area, ordered mesopore structures, and the ability to extend d electrons in open shells [71]. Among other transition metal oxides, Co_3O_4 is one of the most studied oxide structures of cobalt in catalysis because of its diverse morphologies, for example zero (0D), one (1D) and two – dimensional (2D) hierarchical structures [72]. Extensive research has been done and is underway for the potential application of nano crystalline spinel type Co_3O_4 in various fields due to their structural richness, novel properties, and high available surface areas [73]. In size dependent catalytic chemistry, Co_3O_4 nanomaterials have a unique role because of their controlled morphology by different synthetic strategies. Like Mn, Co_3O_4 can be further modified by incorporation of different dopants to improve catalytic activity by several orders [74, 75]. This main focus of this part of the chapter is on the synthetic strategies to make mesoporous Co_3O_4 nanoparticles and the catalytic reactions performed by the Co_3O_4 nanoparticles. Table 3 summarizes some of the synthetic strategies with data for catalytic reactions.

The majority of the catalytic reactions performed by mesoporous Co_3O_4 are oxidation reactions either in the gas phase (Entries 5, 6, 8, 9, and 10) or in the liquid phase (Entry 4). The high catalytic activity is mainly attributed to the high surface area, uniform pore distribution, and the presence of a 3D mesoporous structure in Co_3O_4 materials. For example, Liu et al. performed low temperature CO oxidation with Co_3O_4 prepared by a soft templating method, where 50% CO conversion was achieved at -101°C (Entry 10) [76]. There is a correlation between the presence of spinel type Co_3O_4 on the mesoporous silica supported cobalt oxide and the catalytic activity, which is demonstrated by total oxidation of ethyl acetate under air (Entry 6) by Zhao et al. [77]. The controlled morphology of Co_3O_4 also has an important role in the catalytic reaction, as mentioned by Ma et al. (Entry 8), where the mesoporous Co_3O_4 nanoparticles prepared by impregnation performed superiorly to the Co_3O_4 nanoparticles prepared by precipitation. The preferential exposure of the catalytic active sites due to structural control is the main reason for the enhancement in catalytic activity. Higher catalytic activity is caused by the easily reducible nature of Co_3O_4. The higher activity in Fisher – Tropsch synthesis by fabricated Co_3O_4 nanoparticles on mesoporous carbon was attributed to enhanced reducibility and optimum particle size (6-10 nm).

An effective pathway to remove harmful volatile organic compounds (VOCs) is catalytic oxidation. Despite the fact that supported precious metals (Pt and Pd) can perform oxidation of VOCs at low temperatures, Co_3O_4 is a more promising catalyst for VOC oxidation due to its availability and cost efficiency [78, 79]. 3D ordered mesoporous cubic Co_3O_4 showed much higher catalytic activity compared to its nonporous counterparts (Entry 5) [80]. The high concentration of adsorbed O_2 species in the mesoporous Co_3O_4 is one of the main reasons for its higher activity.

With respect to clean and renewable energy, regenerative solid fuel cells are a very attractive approach to replacing fossil fuels [81, 82]. The evolution of O_2 by oxidation of water (OER) is one of the more important reactions to providing a large scale production of solid fuels [71].

Table 2. Synthetic approaches, important structural parameters and catalytic performances of mesoporous manganese oxide materials

Entry	Method of Preparation	Key Structural Properties	Catalytic Reaction	The Results and Conclusions	Ref.
1	Impregnation of MCM-41 by Mn(NO$_3$)$_2$ Mn/Si : 0.008-0.04 550°C (6h)	Amorphous Mn content w.t.% (0.09-3.76) S.A.: 502-846 m^2/g Pore size 3.15-2.73 nm	Ozonation of nitrobenzene in water (@ 288K)	Hydroxyl radicals are the active species and they are created by reaction of ozone on MnOx surface. A max conversion of 93.3% was obtained by 1.15% Mn loaded MCM-41	67
2	Impregnation of SBA-15 by KMnO$_4$ Multiple impregnation cycles 550°C (6h)	Cryptomelane (K$_x$MnO$_4$) Mn/Si: 8.3-13.8 AOS: 3.68-3.96 S.A.: 200-225 m^2/g	Gas phase ethyl acetate oxidation (VOC removal)	Total oxidation (100% conversion) achieved for all catalysts <350°C. Catalyst with the lowest Mn loading (8.3%) and highest AOS is the most active catalyst (t$_{50}$= 227°C).	60
3	Template free synthesis by reduction of KMnO4 with maleic acid 300°C (4h) and impregnated by Cu(NO$_3$)$_2$	Amorphous (MnO$_2$ after impregnation of Cu) Cu content 4.06-4.56 w.t% 206-132 m^2/g	NO reduction by CO NO+CO→CO$_2$+1/2N$_2$	The activity of mesoporous MnOx increases with T upto 340°C (0.011 mmol g cat^{-1} min^{-1}). The activity decrease is attributed to reduction of MnO$_2$ to Mn$_3$O$_4$ Cu loaded samples show higher activity (0.014 mmol g cat^{-1} min^{-1} @ 430°C)	68
4	Impregnation of SBA-16 and Kit-6 by Mn(NO$_3$)$_2$ 460°C (3h) followed by template removal by NaOH	MnO$_2$ S.A.:184 m^2/g (SBA-16) and 142 m^2/g (KIT-6) Pore sizes: 4.9 nm (SBA-16) and 15.8 nm (KIT-6)	Gas phase toluene removal (toluene/O$_2$ 1/400) CO oxidation (toluene/O$_2$ 1/1)	Catalyst Toluene (T$_{100\%}$) CO (T$_{100\%}$) MnO$_2$-SBA-16 230°C 200°C MnO$_2$-KIT-6 230°C 180°C MnO$_2$ (commercial) 340°C 290°C	61
5	Two-solvent impregnation of Kit-6 by Mn(NO$_3$)$_2$ 460°C (3h) followed by template removal by NaOH	Mixed phase mesoporous MnO$_2$ and Mn$_2$O$_3$ S.A.:73.3 m^2/g	Tetraethylated rhodamine (RhB) oxidation in water at various pH and T(°C) (30-100°C)	The best decolorization of RhB was achieved at pH 1.8 and at 100°C. Highly pH and T dependent decolorization. 91% dye removal with mesoporous MnOx catalyst vs. 12% with non porous manganese oxide at pH1.8 and at 30°C.	62

Table 2. (Continued)

Entry	Method of Preparation	Key Structural Properties	Catalytic Reaction	The Results and Conclusions	Ref.
6	Impregnation of SBA-16 and Kit-6 by $Mn(NO_3)_2 \cdot 4H_2O$, $Mn(CH_3COO)_2 \cdot 4H_2O$, and $KMnO_4$ 500°C (6h)	Precursor dependent crystal structure Nitrate (MnO_2), acetate (amorphous), permanganate (cryptomelane) S.A. (m^2/g) 579 (nitrate), 451 (acetate), and 259 (permanganate)	Ethyl acetate oxidation (VOC removal) 315 ppm ethyl acetate in air (500 ml/min)	*Catalyst* EA (T_{50} K) SBA-15-nitrate 490 SBA-15-acetate 498 SBA-15-permanganate 530 *Presence of Mn^{4+}/Mn^{3+} is responsible for high activity*	58
7	Manganosilicate synthesized with TEOS, $KMnO_4$, hexadecyl amine and maleic acid 550°C (5h)	Amorphous material with cubic mesostructure S.A.:552 m^2/g with 3.4 nm pore size Mn(II) immpobilized in silica matrix	Oxidation of alcohols in water using ammonium peroxydisulfate as oxidazing agent	High conversion and selectivity for oxidation of heterocyclic and alicyclic alcohols in water. High reusability, 4 recycle with no perfomance loss.	69
8	Direct synthesis by precipitation in basic aqueous solution by using CTABr and $Mn(NO_3)_2 \cdot 6H_2O$ 500°C (6h)	Hausmanite (Mn_3O_4) MnO_2 phase was also obtained by H_2SO_4 treatment S.A.:20 m^2/g (MnSc) and 90 m^2/g (MnSc-At) Sc:surfactant and At: acid treated	Formaldehyde total oxidation under air 500 ppm of HCOH under 20% O_2 balanced by He	Mesoporous manganese oxides perform better than non-porous manganese oxides. T_{100} is 145°C and 160°C for MnSc-At and MnSc respectively. T_{100} values for non-porous manganese are 160°C and 200°C. The catalytic activity correlates with reducibility obtained from TPR studies. MnO_2 phase is found to be more active than Mn_3O_4.	43
9	Impregnation of SBA-15 by In-Situ surfactant oxidation with $KMnO_4$ 500oC (6h)	Amorphous Dissapearance of mesostructure upon pore filling.	CO oxidation 50mLmin^{-1} flow rate (CO/O_2 0.2)	$T50 = 221.6$°C and $T100 = 300$°C The catalyst was recycled 4 time without any performance loss	53
		Mn/Si: 0.12-0.96 S.A.:223.6– 275.2 m2/g			

Entry	Method of Preparation	Key Structural Properties	Catalytic Reaction	The Results and Conclusions	Ref.
10	Direct synthesis by precipitation in basic aqueous solution by using CTABr and Mn(NO$_3$)$_2$.6H$_2$O 550°C (6h) Followed by a acid treatment with H$_2$SO$_4$ and Au deposition by VALA	Mn$_3$O$_4$ (assynthesized material). γ-MnO$_2$ (after H$_2$SO$_4$ treatment) S.A.:72.6 m^2/g for Mn$_3$O$_4$ and 315.8 m^2/g for mesoporous γ-MnO$_2$. Au loading is 1.6- 2.8%	VOC removal of a mixture of 19.5 ppm of toluene, 27 ppm of acetaldehyde, and 50 ppm of hexane in 20%O$_2$ flow.	Mesoporous γ-MnO$_2$ can eliminate all acetaldehyde rapidly at room temperature in 1 h. However it can only remove hexene <5% (@80°C). It can also reach 90% toluene removal at 85°C (vs. 60% at RT). Au deposition dramatically enhances toluene and hexane removal. 80% hexane removal (at 85°C) and 93% toluene removal (at RT)	59
11	Impregnation of KIT-6 by Mn(NO$_3$)$_2$ 400°C-900°C (3h)	Amorphous (<700°C) α-Mn$_2$O$_3$ (>700°C) 8%Mn loading Mn^{+4} (<500°C) by XANES Mn^{+3} (600°C) by XANES Mn$^{+2.85}$ (900°C) by XANES analyses	Visible-light driven O$_2$ evolution in aqueous solution	TOF= 1630 S^{-1} (400°C), 1210 S^{-1} (500°C), 3330 S^{-1} (600°C), 1260 S^{-1} (700°C), 1590 S^{-1} (800°C), and 1830 S^{-1} (900°C) The sample calcined at 600°C is the most active which also have the highest amount of Mn$_2$O$_3$ phase. Mesoporous Mn$_2$O$_3$ is 26 times more active than commercial non-porous Mn$_2$O$_3$.	51
12	Manganese oxide implantation in to MCM-41 by molecular organic chemical vapor deposition (MOCVD) 300°C (2h)	<9% Mn loading Amorphous S.A.:1052- 1216 m^2/g More reactive Mn-O species than impregnated samples (TPR).	Trans-stilbene epoxidation by using TBHP as oxidant. Oxidation of diphenylmethane using O$_2$ in air as oxidant.	Trans-stilbene epoxidation: Implanted MCM-41 samples perform better (18.3% yield) than impregnated samples (9.6% yield) at 55°C (72 h). TOF numbers are 365 and 138 for Mn implanted and impregnated samples respectively. Diphenylmethane oxidation: Benzophenone yields are almost identical for Mn implanted (35.6% yield) and Mn impregnated samples (36.1% yield).	57
13	Direct synthesis by precipitation in basic aqueous solution by using CTABr and Mn(NO$_3$)$_2$.6H$_2$O 500°C (4h) Followed by a acid treatment with H$_2$SO$_4$ and Au deposition by VALA	γ-MnO$_2$. S.A.:316 m^2/g with 3.6 nm pore size Partially positively charged Au (δ+) 2.8 wt% Au loading	VOC removal of a mixture of toluene (19.5 ppm), acetaldehyde (27 ppm), and n-hexane (50 ppm).	VOC elimination by adsorption significantly enhanced at higer temperatures and in presence of gold. γ-MnO$_2$ can remove 60% of tolune at RT (93% at85°C) vs. γ-MnO$_2$/Au 95% at RT. γ-MnO$_2$ can remove 2% of n-hexane at at 85°C vs. γ-MnO$_2$/Au 30% at RT (77% at 85°C)	42

Table 2. (Continued)

Entry	Method of Preparation	Key Structural Properties	Catalytic Reaction	The Results and Conclusions	Ref.
14	Ultrasound assisted impregnation of SBA-16 by Mn(NO$_3$)$_2$·4H$_2$O and Co(NO$_3$)$_2$ 260°C (1.5 h)– 450°C (2h) template removal by NaOH	MnO$_2$ S.A.:266 m^2/g with 6.4 nm pore size Co$_3$O$_4$ S.A.:313 m^2/g with 6.1 nm pore size	Complete gas phase toluene oxidation (1000ppm) Toluene/O$_2$ 1/200	Mesoporous MnO$_2$ and Co$_3$O$_4$ shows significantly higher performance than non-porous oxides. Catalyst / T_{100} °C Meso- MnO$_2$ / 270 Bulk- MnO$_2$ / 340 Meso- Co$_3$O$_4$ / 240 Bulk- Co$_3$O$_4$ / 310	63
15	Impregnation of SBA-15 by M(II)acetyl acetonate in acetone 550°C (5h)	Mn$_3$O$_4$ 13wt% Mn loading S.A.:299 m^2/g with 6.3 nm pore size	Ethanol oxidation in aqueous solution using H$_2$O$_2$ as oxidant at various different pHs.	The maximum conversion of ~45% was obtained with 20mL ethanol, pH of 6.7, 1 ml of 30%H$_2$O$_2$, at 70°C and with 5 mg of catalyst. The high conversion was attributed to faster peroxide decomposition rate in the presence of mesoporous manganese oxide catalyst.	64,70
16	Bi-Modification of mesoporous silicon oxide by a coupled in-situ oxidation at interfaces (IOI) and ion-exchange by KMnO$_4$ Cs$^+$, K$^+$, H$^+$ ions used for ion exchange with CTABr. 500°C (5h)	Amorphous S.A.:535-605 m^2/g Pore sizes: 3.4- 4.3 nm Mn loading is 7% (w.r.t. Si)	Gas phase toluene oxidation under air flow. Flow rate is 20 sccm	Bi-modified samples show higher selectivity for total oxidation of toluene than MPS-Mn samples. Cs as a promoter ion was found to be the most active promoter. The highest TOF (h^{-1}) was obtained for MPS-Mn-Cs sample which is 8h^{-1} at 425 °C. Promoted samples show higher performance in catalytic stability tests. MPS-Mn-K sample shows 100% selectivity for toluene oxidation for 50h with no performance lost.	50

Table 3. Synthetic approaches, important structural paramaters and catalytic performances of mesoporous cobalt oxide materials

Entry	Method of Preparation	Key Structural Properties	Catalytic Reaction	The Results and Conclusions	Ref.
1	Impregnation of KIT-6 by Co(NO$_3$)$_2$.6H$_2$O, and Mg(NO$_3$)$_2$.6H$_2$O. Followed by selective leaching of Mg and template (Kit 6) removal by NaOH 450°C (3h)	Mesoporous Mg –Co$_3$O$_4$ and Co$_3$O$_4$ spinel. S.A. : 250 m^2/g Pore size 3.7 nm	Chemical water oxidation using Ce^{4+} (at 550 mV)	Substitution of Mg in both tetrahedral and octahedral sites of Co$_3$O$_4$ affected the O$_2$ evolution activity. Leaching of Mg due to strong acidic environment created defect in the meaterial, responsible for higher activity. TOF = 1.6 × 10^{-4} s−1 per Co atom for Mg- Co$_3$O$_4$ TOF = 2.2 × 10^{-3} s−1 per Co atom for Mg leached- Co$_3$O$_4$	71
2	Impregnation of KIT-6 and SBA-15 by Co(NO$_3$)$_2$.6H$_2$O. 550°C (5h)	Cubic mesostructured Co$_3$O$_4$, formed inside the channel of KIT-6 and SBA-15. S.A. : 58.9 – 104.2 m^2/g Pore size 1.9 – 10.5 nm	Investigating the electrochemical properties in Li – O$_2$ cell.	Additional hydrophilic channels for oxygen diffusion to the air electrodes and accessibility of the active sites due to large specific surface area are the reasons for the improved performance. Round trip frequency of 81.42% and specific capacity of 2250 mA h g^{-1} for the catalyst with highest surface area and pore size.	86
3	Impregnation of KIT-6 by Co(NO$_3$)$_2$.6H$_2$O 450°C (6h) Template removal by NaOH Multiple impregnation cycles	FCC mesostructure Pore size 3.8 – 21.9 nm S.A. : 114 – 135 m^2/g Higer oxidation state of Co than bulk Co$_3$O$_4$ (XANES)	Bifunctional elctrocatalysis O$_2$ evolution reaction (OER), O$_2$ reduction reaction (ORR)	The mesoporous Co$_3$O$_4$ shows high activity and stabilty in OER in alkaline medium, compare to Co$_3$O$_4$ nanoparticles and Pt/C catalyst. It shows promising activity in ORR reaction with superior methanol tolerance compare to the Pt/C catalyst. The total overpotential is 1.03 V for OER and ORR.	87
4	Selective growth of Co$_3$O$_4$ nanopaticles inside SBA -15 using Co(NO$_3$)$_2$.6H$_2$O and two-solvent method.	Crystalline spinel Co$_3$O$_4$ phase S.A. : 347 – 620 m^2/g Pore size 5.7 – 5.8 nm Controlled location of nano-particles	Liquid phase oxidation of cyclohexanol to cyclohexanone using TBHP as oxidant.	Catalyst with lowest Co contain exhibited high catalytic activity. Inaccessiblty of the reacting species is the reason for low catalytic activity for catalyst with high Co content.	88

Table 3. (Continued)

Entry	Method of Preparation	Key Structural Properties	Catalytic Reaction	The Results and Conclusions	Ref.
5	Vacuum aided Impregnation of KIT-6 and SBA-15 by Co(NO$_3$)$_2$·6H$_2$O. Template removal by aqueous HF solution. 400°C (3h)	Co$_3$O$_4$ Bimodal mesopore formation S.A.: 118 – 121 m^2/g Pore size 6.2 – 6.5 nm Co3+/Co2+ ratio : 1.46 - 1.54, indicates higher surface O$_2$ vacancy than bulk Co$_3$O$_4$	Gas phase VOC (toluene and methanol) oxidation	90% toluene and methanol conversions were achieved at 180 and 139°C. Large surface area, low reducibility, high concentration of adsorbed O$_2$ species and presence of 3D mesoporous structure are the reason for high activity.	80
6	Impregnation of KIT-6 and SBA – 15 by Co(NO$_3$)$_2$·6H$_2$O 500°C (2 h) template removal by Ethanol/HCl 120°C - 550°C (2h)	Co$_3$O$_4$ spinel phase S.A.: 283 – 490 m^2/g Pore size 4.5 – 9.8 nm Co/Si ratio : 0.062 – 0.087	Total oxidation of ethyl acetate (1.21 mol%) under air (WHSV 100h^{-1})	Corelation between catalytic activity and with the presence of spinel type Co$_3$O$_4$. Interconnectivity of pores and uniformity of channel dimension are the key parameters yield high activity. High CO$_2$ formation from ethyl acetate at 250°C – 350°C Low selectivity at low temperatures due to partial oxidation	77
7	Electrodeposition of Co(OH)$_2$ on the Ni foam and then thermally (250 °C) transformed in to Co$_3$O$_4$	Co$_3$O$_4$ Uniform Co$_3$O$_4$ nanosheets with high density on skeleton of Ni foams S.A.: 118 m^2/g Pore size 2 - 5 nm	High performance eletrode for electrochemical capacitors	Exceptional capacitive performance with ultrahigh specific capacitance in the range of 2735–1471 F g^{-1} and excellent cycling stability upto 3000 cycles. Excellent structural stabilities, large electroactive surface area and endowment of fast ions and electron transportation are the reasons for high performance as a electrochemical capacitor.	89
8	Impregnation of KIT-6 by Co(NO$_3$)$_2$·6H$_2$O and HAuCl$_4$·3H$_2$O 300°C (3 h) Multiple impregnation cycles Template removal by NaOH solution.	High order 3D mesopore structures. S.A.: 84-100 m^2/g Pore size : 3.4/11.5 nm. Gold particles enter pores, embed into pore walls, and drill through the Co$_3$O$_4$ walls.	Low temperature ethylene (50 ppm) oxidation under air.	76% ethylene conversion at 0°C for 2.5% Au/Co$_3$O$_4$ compared to 30% for Co$_3$O$_4$ and 0% for mesoporous Co$_3$O$_4$ Structural control of Co$_3$O$_4$ allows the preferential exposure of catalytically active sites. Crystal facet {110} of Co3O4 played an essential role. Au particle promoted the catalytic reaction by producing surface active O$_2$ species.	90

Entry	Method of Preparation	Key Structural Properties	Catalytic Reaction	The Results and Conclusions	Ref.
9	Impregnation of mesoporous caron materials (OMC, CMK-3 and MSU-F-C) by Co(NO$_3$)$_2$·6H$_2$O 300°C (5h) Multiple impregnation cycles	Co$_3$O$_4$ nanoparticles formed in the mesopore structures. particle size is 6-16 nm	Fisher – Tropsch synthesis (FTS) under the conditions of: 220°C, 2.0 MPa, and H$_2$/CO = 2.	FT activity of 3.22×10^{-5} mol CO g^{-1} Co s^{-1} and C$_{5+}$ selectivity of 78.0% for Co – CMK -3 and FT activity of 2.02×10^{-5} mol CO g^{-1} Co s^{-1} and C$_{5+}$ selectivity of 73.0% for Co – MSU –F – C. High reducibility of Co particles and optimum average particle size of Co for FT reaction is 6-10 nm	91
10	Soft reating grinding based on mechanical activation. Grinding of a mixture of 2CoCO$_3$·3Co(OH)$_2$·H$_2$O, oxalic acid and polyethelelglycol (PEG-20000) 300°C (4h)	Co$_3$O$_4$ Crystallize size: 10 nm. S.A.: 124 m^2/g Pore size: 3.8 nm	Oxidation of CO 50mL min^{-1} flow rate (CO/O$_2$: 1/20)	The light off temperature T50 (temperature at 50% conversion) is - 101°C. Enhanced surface area of the material attributed to the superior activity.	76
11	Triblock polymer/ ice crystal double templated sol-gel synthesis of Co$_3$O$_4$ cryogel.	Co$_3$O$_4$ Interconnected of nanocrystals introduce mesoporous structure S.A.: 52 – 82 m^2/g Pore size: 10 – 20 nm	Electrochemical energy strogae. 3 electrode system, 2M KOH, using potentiostat.	Superior electrochemical performance with a specific capacitance of 742.3 F g–1 (potential window of 0.5 V) The capacity retention was 86.2% after 2000 cycles. Intimate electrolyte contact with Co$_3$O$_4$, due to unique channels in hierarchical pores network facilitate electrolyte diffusion	84

Mesoporous Co_3O_4 is a transition metal oxide that can perform OER. For example, Rosen et al. demonstrated fabricated mesoporous Co_3O_4 through Mg substitution, followed by a Mg-selective leaching process which showed superior catalytic activity in OER using the Ce^{4+}/Ce^{3+} system (Entry 1). Bifunctional electrocatalysis was performed by mesoporous Co_3O_4 as mentioned by Sa et al. where the catalyst can perform both OER and ORR (O_2 reduction reaction) with a superior methanol tolerance for ORR (Entry 3).

In modern electronics and power systems, supercapacitors have received attention as an energy storage device due to their high power densities, long cycle life, and energy backup [83]. Mesoporous Co_3O_4 materials are efficient supercapacitors because of their well defined redox properties, high theoretical capacitance, and high morphological diversity [84]. Additionally, mesoporous Co_3O_4 possesses large specific surface area with small crystal size, both of which are major criteria for categorizing a material as electronegative [85]. Mesoporous Co_3O_4 nanomaterials demonstrated superior electrochemical performance with excellent reversibility and recycle stability (Entry 7, 11). The accessibility of active sites due to large specific surface area and unique channels in hierarchical pore networks are the main reasons for improved performance.

c) Mesoporous Vanadium Oxide

The sol-gel based synthesis of mesoporous vanadium oxide is one of the most challenging TM systems due to the limited number of soluble vanadium sources, multiple easily exchangeable oxidation states of the sols, acid-base pairs of the vanadium sols, and multiple crystal structures [92–97]. Apart from the costly anhydrous chloride salt and oxo-alkoxide sources, other sols as precursors are generally prepared by digesting vanadium oxide or ammonium vanadate by peroxide and organic acids containing carboxyl groups [92, 98–102]. The sols are either used in direct synthesis or supported structures. Despite the fact that there are few reported mesoporous vanadium oxide materials, these materials have different structural properties from traditional mesoporous materials. Li et al. prepared mesoporous vanadium oxide materials which consisted of nano-particle aggregates and mesopores which were the intra-particle voids. [92]. The crystal structure of the materials can be either VO_2, V_6O_{13}, or V_2O_5 depending on the post treatment conditions. The overall approach used a freeze-drying process on nano-fibers synthesized by a sol-gel approach. Xue et al. synthesized mesoporous vanadium oxides that have a slit-like pore structure, which is quite unusual for a typical sol-gel method [95]. The synthesized materials have high surface areas (182 m^2/g) with the V_2O_5 crystal structure. However, the majority of the reported vanadium containing mesoporous materials focus on the dispersion of nano vanadium oxide clusters on a mesoporous support or the doping of the clusters in mesoporous inorganic walls.

From a catalysis point of view, the focus has been mostly on supported mesoporous vanadium oxides. Vanadium can form a wide variety of catalytically active oxide structures on these supports such as monomeric Td (VO_4), oligomeric Td, oligomeric Oh, alternated VO_5 square pyramids, microcrystalline V_2O_5, isolated species (V^{+4}, V^{+5} Td), distorted Oh, mixed valent species, and hydrated species [93, 99, 101–108]. The nature of supported VOx species has been mostly investigated spectroscopically using FTIR, DR-UV-Vis, Raman, XPS, and EPR due to their unique redox, electronic and coordination properties [92, 93, 103, 105, 107, 109, 110]. Silica, titania, and aluminosilicates are the most studied mesoporous templates for supporting vanadium species (VOx). The supported VOx species create Lewis acid-base pairs, Brønsted acid sites, enhanced redox properties which are the main motivations for their use in catalysis [93, 99, 103, 104, 110, 111]. Table 4 summarizes some of the recent catalytic studies conducted by supported mesoporous vanadium oxides along with the synthetic approaches and key structural parameters.

As a multivalent transition metal oxide, supported VOx species are good oxidizing agents similar to other first row TMs such as Mn, Fe, and Co. Therefore, supported mesoporous vanadium oxide materials have been widely used in oxidation, epoxidation, and hydroxilation reactions (Table 4 entries 1, 2, 3, 5, and 7). In these studies, polymeric VOx species (multilayer) and microcrystalline V_2O_5 were found to be the most active vanadium species supported on titania and silica for oxidation reactions. The formation of desired polymeric VOx species and microcrystalline V_2O_5 depends highly on pH, V amount, and heat treatment temperatures [100, 101, 106]. Zhao et al. found pH 2 was the optimum pH for the formation of redox active VOx species in direct hydrothermally synthesized vanadium doped mesoporous silica (entry 2) [106].

Nano-clusters of vanadium oxides supported on a mesoporous template offer more than enhanced redox properties. The vanadium oxide nano-clusters also create acid-base pairs on the surface. Different from other supported mesoporous TM oxides, vanadium oxide incorporated mesoporous silica samples were found to be promising catalysts for oxidative de-hydrogenation and hydrosulfurization reactions due to their oxidative and acidic properties (Table 4 entries 4, 6, 8, and 9). The rate determining step for oxidative dehydrogenation (ODH) of low molecular weight hydrocarbons is the C-H bond activation by lattice oxygen on the heterogeneous catalysts (Mars and van Krevelen mechanism) followed by a dehydration step on the acidic surfaces [112–114]. The C-H bond activation is the rate determining step and has a high activation energy which causes low selectivity to alkenes. However the use of mesoporous materials decreases the activation energy due to the strong adsorption of alkanes and increases selectivity by reducing the reaction temperature [112]. Therefore, mesoporous catalysts perform better than their non-porous counterparts (Table 4 entry 9). Moreover, by controlling the nature of nano-VOx clusters it is possible to manipulate conversions and selectivity. Micro V_2O_5 clusters are found to be very active for conversion and the isolated V species embedded in the silica matrix are found to increase selectivity to alkenes (Table 4 entry 6).

d) Mesoporous Iron Oxide

Mesoporous iron oxides are another type of most widely pursued porous nanomaterials. They have found applications in catalysis, energy storage and conversion, and magnetic devices [35, 115–118]. The last one is of particular interest, due to its unique magnetic properties related to particle sizes and phases. The research by Alam et al. is a representative example of tuning the magnetic properties by control the mesopore size of the silica SBA-15, the smaller the pore size, the stronger the magnitization; and the sample with lowest iron dopant has the highest magnetization [115]. The mesoporous iron oxide was found to have ten times higher magnetization than the pure iron oxide prepared without a silica matrix (Table 5, entry 3).

Table 4. Synthetic approaches, important structural paramaters and catalytic performances of mesoporous vanadium oxide materials

Entry	Method of Preparation	Key Structural Properties	Catalytic Reaction	Results and Conclusions	Ref.
1	Impregnation of mesoporous titania and Direct incorporation of Vanadium in mesoporous titania 250 oC (2h)	Impregnation (V5+) Ti:V 100:5 Doping (V4+) Ti:V 100:10 S.A. (m2/g) 228-1030 Pore Size 2.61-4.2 nm	Gas Phase Propene Oxidation (227 oC) Propene → CO → CO2	V/Meso TiO2 > V-meso TiO2 > V/Nonporous TiO2 5 wt% Vanadium Loading Rates of Formation (V/Meso TiO2) CO2 r (min-1) = 0.279 CO r (min-1) = 0.091 H2O r (min-1) = 0.039	102
2	Direct hydrothermal synthesis of V-SBA-16 550 oC (4h) pH 0.5-2.5	Td VO4 species in silica matrix V Content= 0-1.6 wt.% S.A. (m2/g) 685.9-997.6 Pore Size 4.35-7.83 nm	Hydroxylation of benzene to phenol by O2	The best catalytic performance obtained for 1.67 V wt% at pH 2. Phenol yield 30.4% with 90% selectivity and TON of 105	106
3	Direct hydrothermal synthesis of V-SBA-15 500 oC (6h)	Wormhole-like pore morphology V Content= 0-4.8 wt.% S.A. (m2/g) 630-820 Pore Size 3.28-3.57 nm	Aerobic Dichloromethane Oxidation (Cl-VOC) 1000-3000 ppm 200-500 oC	Optimum vanadium loading is 2.2 wt%. Conversion rises from 7 to 68 with the selectivity increasing from 9 to 50 in a temperature range of 300 to 500 oC. Activity is attributed to the existence of polymeric V species on the catalysts	101
4	Impregnation of MCM-41, Al-MCM and Nb-MCM-41 by NH4VO3 and (CH3COO)3Sb 650 oC (96h)	Td coordinated V (IV) SbandV content= 5 wt.% S.A. (m2/g) 483-966 Pore Size 2.1-2.8 nm	Methanol hydrosulphurisation at 450 oC	The conversion follows the order of SbV/NbMCM-41 (60%) > SbV/AlMCM-41 (48%) > SbV/MCM-41(18%). The yield of MeSH follows the order of SbV/NbMCM-41 (98%) > SbV/MCM-41(92%) > SbV/AlMCM-41 (51%). The high selectivity attributed to the presence of Lewis acid-base pairs.	111
5	Atomic Layer Deposition (ALD) of SBA-15 and FDU-15 by VO(iPrO)3 550 oC(6h)	Mono-triple layer coverage of Vanadium S.A. (m2/g) 598-731 Pore Size 6.39-9.30 nm	Liquid phase epoxidation of cyclohexene using tert-butyl hydroperoxide at 80 oC under Ar	The highest conversion for monolayer coverage is 55.2% with a selectivity of 46.3%. The highest conversions for double and triple layer coverage are enhanced catalytic performance with the conversion reaching 63.2% and selectivity being 53.7%. SBA-15 was found to be a better support than FDU-15.	100

Entry	Method of Preparation	Key Structural Properties	Catalytic Reaction	Results and Conclusions	Ref.
6	Direct hydrothermal synthesis of V-SBA-15 and V-MCF Impregnation of SBA-15 and MCF by NH4VO3 600 oC (5h)	Vandium content= 2.4-2.8wt.% S.A. (m2/g) 530-925 Pore Size 3.5-23 nm	Oxidative De-hydrogenation of propane	Highest conversion (9.9%) was obtained for V/SBA-15. Highest selectivity was obtained for V-MCF (67.7%). Isolated V species (doping) increase selectivity. however microcrystalline V2O5 increases conversion.	99
7	Solid state incorporation of V2O5 in MCM-41, SBA-15, and MCM-48 Ball milled (.5h) 500 oC (1h)	V2O5 content= 10 wt.% S.A. (m2/g) 103-919 Pore Size 3.54-9.42 nm Grafted V2O5 on silica	Gas phase ethylacetate (1.21 mol%) oxidation in air WHSV-100 h-1	The conversion follows the order of V/SBA-15 (48%) > V/MCM-41 (32%) > V/MCM-48 (30%) ~ 400 oC. The catalytic was increased by pre-reduction of the catalyst and the conversion can be as high as 69%. However, pre-reduction does not affect selectivity to CO2 ~25%.	104
8	Direct Synthesis of mesoporous vanadosilicate using TEOS and NH4VO3 550 oC(4h)	Si/V ratio: 10-100 S.A. (m2/g) 697-819 Pore Size 14.8-15.1 nm	Oxidative de-hydrogenation of Propane. propane/O2/N2 with a molar ratio of 1/1/8 450-650 oC	Higer activity than V impregnated mesoporous silica for ODH of propane. 34.2% conversion with a yield of 21.6% with Si/V ratio of 10 at 575oC. Vs. 217% conversion and 12.4% yield for impregnated sample	110
9	Impregnation of MCM,SBA, HMF. MCF by ammonium vandate NH4VO3 500 oC (1h)	V content=0.7-8.0 wt.% S.A. (m2/g) 770-966 Pore Size 2.6-6.9 nm	Oxidative de-hydrogenation of propane. Prapane:O2 is 1:1 and flow rate of 105 ccm 425-600 oC	The use of mesopores increases the conversion from 3.3 to 20% with a selectivity of 50.8%. The highest yield obtained is 19% at 600oC with mesoporous support.	95

Table 5. Synthetic approaches, important structural paramaters and catalytic performances of mesoporous iron oxide materials

Entry	Method of Preperation	Key Structural Properties	Catalytic Reaction	The Results and Conclusions	Ref
1	Impregnation of SBA-16 thin films on indium-tin oxide(ITO) by electodeposition (1.4-1.5 V for 1h) Electrolyte:0.1 M FeSO4. Template removal by aqueous HF solution	γ-Fe2O3 nanorodes, i.d.= 15nm Pore size: 10 nm	Synthesis of carbon nanotube arrays in situ by CVD. Fe work as a catalyst to produce	After template removal with HF, Fe can work as catalyst to produce well aliglined CNTs arrays by catalytic decomposition of acelylene at 700°C.	118
2	2D-mesoporous iron oxides(2DMIO) using Fe(III) ethoxide (3mmol), ethanol, decylamine (3mmol) as surfactent, 40 °C (2h) + 40 °C (24h) + 80 °C (6h); For 3DMIO: additional aging at 150 °C	Bimodal pore structure (Microporous and mesoporous) S.A.: 2D(340 m2/g), 3D(610 m2/g). Pore size: 5.4 nm Wall thickness: 5.1nm	Superparamagnetic Mesostructure is stable up to 250 oC	Micro- and mesoporous structure was formed simultaneously, first report of alkyl amine was used as bifunctional surfactants.	35
3	Impregnation of SBA-15 with Fe(NO3)3. 9H2O 300 °C (4 h) under O2 flow	Fe2O3 particles grown from SBA-15 with tunable pore sizes. Particle size: 6.5-9.0 nm	Particle size dependent Maganetization. Controllable by ajusting pore size of the template	The smaller the particle size of Fe2O3, the stonger the magnetization. The less dopant amount, the higher the saturation magnetic moment (Ms),7.5% Fe has the highest Ms= 25 emu/g, whereas the one of 50% Fe has a Ms = 6 emu/g. Non-porous Fe$_2$O$_3$ has a Ms=2.1 emu/g.	115
4	Impreganation of KIT-6 by Fe(NO3)3. 9H2O 600 °C (6h) Further reduction at 350 °C (1 h) under (H2/Ar=5/95) mixture	γ-Fe2O3 and Fe3O4 after reduction S.A. of γ-Fe2O3: 86 m2/g Pore size distribution:centered at 3.6 nm. Crystalline wall (a.u. 7 nm thickness)	Retention of magnetic order at higher temperature for mesoporous structure iron oxide (>340 K)	The retention of magnetic order for mesoporous iron oxides at higher temerpature may find usfull applications comparing with general nanoparticles which lose magnetic order at much lower termpature (when particle size < 8 nm)	119
5	Sonochemical synthesis (20 KHz, 100 W/cm2 at 65% efficiency, air, 3h) Useing Iron(III) ethoxideand CTABr, at pH of 10.6. 250-500 °C (2-5 h)	γ-Fe$_2$O$_3$ (250°C heating) Particle size: 100-200 nm; Pore size: 3-5nm; S.A.:274 m^2/g (after solvent extraction)	cyclohexane oxidation using 1 atm O$_2$ and 70 °C; Magnetic study	35.6% converion; > 80% selectivity to alcohol Calcined material above 250 °C forms γ-Fe$_2$O$_3$, shows superparamagnetic properties.	122

Entry	Method of Preperation	Key Structural Properties	Catalytic Reaction	The Results and Conclusions	Ref
6	Reduced graphene used (RG) as substrate and iron(II) acetylacetonate as iron source. Nucleation at 170 °C (0.5-5h), thermal annealing at 250 -400 °C.	Fe_3O_4 and α-Fe_2O_3 phases were obtained Ribbons-like morphololy, Pore size: 2-20 nm, S.A. 115 m2/g	Lithium ion battery Lithium insertion-extraction	Discharge capacity of 1426 mAh/g Reversible charge capacity remained for 130 cycles. 250 °C is the premium annealing temperature	123
7	Surfactant templated mesoporous iron oxide incoparated silica film. Condensation under supercritic conditions with CO_2 400 °C (6h).	Film thickness: 200-300 nm. 1% wt -10% wt Fe loading	Magnetic study Zero field cooled and Field cooled magnetization were measured	1% and 5% wt ferritin containg smaple shows supermagnetic above 225 K, Whereas 10% wt sample only shows the property over than 300 K.	120
8	Impregnation of Al-SBA-15 by $FeCl_2.4H_2O$ using Ball milling and Microwave assisted. 400 °C, 4h;	< 0.1% Fe loading by ball milling; >0.5% Fe loading by microwave Increased acidity	Alkylation of toluene with benzyl chloride and benzyl alcohol under MW	Small loading mount of Fe (0.04%) achieved >99% and 47% conversion in both reactions, respectively. Comparable with high loading ones, presumably due to the presenc eof samll and homogeneously distributed nano particles.	116
9	Mesoporous iron oxide hybridized with layered titanate. calcine up to 400 °C	Pore Size.: 3 nm-6 nm S.A.:190-230 m2/g Bind gap: 2.3 eV Loosely packed FeO_6 octahedra	Photocatalytic degradation of methylene blue and dichloroacetic under visible light	Visible light dye degradation is achieved by a mesoporous hybrid material incoporated titanium and iron oxides comparing with the nonactive titanate and α-Fe_2O_3; increased chemical stability of iron oxide by encapsulation.	124

The magnetic properties of mesoporous iron are particularly related to heteregeneous catalysis since such processes offeres an easy seperation from the reaction media by a magnet. The synthesis is the key to desired properties for these applications, and thus has been explored in many ways.

The most widely used synthetic approach is using hard templates, such as SBA-16, KIT-6, and MCM-41. After a successful mesopore filling and conversion to an oxide structure, the template was removed to obtain a pure iron oxide phase with the porous structure of the template replicated, with similar particle sizes close to the preserved pore sizes [115, 117–120]. The pore sizes and mesostructure can be tuned by choosing the right hard templates and the phases of iron oxides can be adjusted by applying different heat treatments under various conditions. For example, Jiao et al. obtained the α-Fe_2O_3 phase under calcination at 600 °C, further heat treatment under reducing atmosphere (H_2/Ar) forms the Fe_3O_4 phase, and a continuous heat treatment at 150 °C for 2 hour forms the γ-Fe_2O_3 phase [119]. Nevertheless, the mesoporosity remained intact during the phase changes under heat treatments, which was ascbribed to the flexibility afforded by the reduced wall thickness (~7 nm). In addition, apart from the powder samples, mesoporous films can also be a hard template for the synthesis of mesostructured iron oxide materials [118]. Shi et al. used mesoporous silica films as hard templates to synthesize mesostructured iron and the resulting material was then used as a catalyst for the carbon nano-tube synthesis (Table 5, entry 1). The hard template can be either removed or preserved depending on the specific application desired. The typical way of template removal is treating with either aqueous HF and NaOH solutions. However, in some applications, the SiO_2 is desired to exist as a structural support or protection shell, for example, when iron oxide is applied for bio-imaging, drug delivery and other bio-medical applications.

As discussed in several studies, SiO_2 plays an important role as a host and protection layer preventing the direct contact between iron oxide and the human body [117, 121]. Lots of variations in impregnation approach methods can be done. Some of them are wet impregnation, electrodeposition, solide state and microwave assisted impregnation (Table 5, *Method of Preperation* colum). The iron precursors being most used are nitrates and acetylacetonates. The selection is based on easiness of counter ion removal by heat treatment and low cost. In addition, a biological iron-containing material called ferritin could also be used as iron source.

Direct synthesis of mesostructured iron oxide is also possible with a micellar soft-template, however the material loses its mesostructure (and mesoporosity) upon template removal (surfactant) [35, 122]. The direct synthesis method uses iron ethoxides as iron sources and decylamine or cetyltrimethylammonium bromide (CTAB) as surfactant. With this direct synthesis approach, both 2D and 3D structures are formed, with the thin wall (5.1 nm) promising ideal superparamagnetic properties (Table 5, entry 2). Srivastava et al. synthesized amorpous iron oxides having surface areas of 274 m^2/g after solvent extraction with a sonochemical technique.

The catalytic activity of this mesoporous iron oxide was tested for cyclohexane oxidation using oxgen: 35.6% conversion, and 80% selectivity of alcohol was achieved (Table 5, entry 5).

However, the nonporous couterparts showed less than 16% conversion with a 60% selectivity. A simple heat treatment of the amorphous material under 250 °C, γ-Fe_2O_3 phase was obtained with superparamagnetic properties.

e) Mesoporous Chromium Oxide

Mesoporous chromium oxide is the least studied late TM system due to its known toxicity [125, 126]. Despite the fact that there are only a limited number of studies about mesoporous chromium because of environmental concerns, the Cr_2O_3 phase is known to be a very active redox catalyst. Mesoporous chromium oxide with the Cr_2O_3 phase can be used in several catalytic processes such as oxidative dehydrogenation, epoxidation, and VOC removal [38, 127, 128]. The generic methods of preparation include nanocasting a hard template such as SBA-15, mesoporous carbon, mesoporous metal CeO_2 and soft templates like P-123 and F-127 with chromium nitrates as precursors [38, 127–130]. The counter ion is generally removed by a heat treatment once the oxide structure is established. Another way is loading chromium oxide on mesoporous silicates, which can be done by either the well established *in-situ* addition before the crystallization step, or a versatile impregnation method [131, 132]. Cr-containing mesoporous silica molecular sieves (Cr-HMS) were reported to be photoactive under both UV and visible light, due to the highly dispersed chromium oxide on the mesoporous silica, which was tetrahedral bonded in the silica framework [131].

Photocatalytic decomposition of NO to N_2, N_2O and O_2 was observed over the Cr-HMS material at room temperature. The reaction rate under visible light was lower than the one under UV light irradiation, but showed higher selectivity for N_2 (97%) than the latter (45%). Nevertheless, photocatalytic oxidation of propane towards partial oxygen-containing hydrocarbons revealed that the selectivity was higher for oxidation reactions under visible light. In some other applications, mesoporous Cr_2O_3 has been incorporated with promoters and the resulting material was used as a catalyst for complete volatile organic compound (VOCs) removal [127]. In this study, the authors used poly 4-vinylpyridine (P4VP) as a template, combining with chromium nitrates through a hydrothermal synthesis at 180°C. They obtained a crystalline Cr_2O_3 material having a surface area of 38 m^2/g. Further modification by other transition metals (i.e., Co, Cu, and Mn) tuned both porosity and catalytic activity in a trend of reduction of the pore size and volume after impregnation. The highest activity was obtained in the presence of both Mn and Ce: complete oxidation of benzene was achieved at 250°C [127]. The mesoporosity can generally enhance the catalytic activity due to the improved diffusion of substrates. Mesoporous Cr_2O_3 can be used as a premium catalyst support. A Ru supported Cr_2O_3 showed catalytic activities in ammonia decomposition for production of hydrogen [133]. The mesoporous Cr_2O_3 support has a high surface area (270.7 m^2/g) and the mesopores were formed by thermal decomposition of citric acid [133]. Complete ammonia decomposition was observed at 600°C with a high space velocity (30,000 mL / h·g). The high activity was attributed to the interactions between the mesoporous support and Ru metal. Mesoporous Cr_2O_3 can also be modified to make a mesoporous composite material with polyoxometalates (POMs) and SBA-15 as a hard template [134]. The resulting mesoporous catalyst was found to be a highly effective catalyst for oxidation of 1-phenylethanol with H_2O_2. The synthesized mesoporous materials have high surface areas (>100 m^2/g) and regular mesoporosity with a pore diameter of 4 nm. The composites show enhanced activity >80%, turnover frequency (TOF) =105 h^{-1} as compared to the unmodified meso-Cr_2O_3 showing 10% conversion with a TOF of 11 h^{-1}. All the above dta indicate that meso-Cr_2O_3 is a promising catalyst and catalyst support, and with its unique incorporation with other metal oxides, rare earth metal, zeolites, and polyoxometalates, open up various avenues for applications in heterogeneous catalysis.

REFERENCES

[1] Kresge, C. T.; Leonowicz, M. E.; Roth, W. J.; Vartuli, J. C.; Beck, J. S. *Nature*, 1992, 359, 710–712.
[2] Web of Knowledge http://apps.webofknowledge.com/ summary.do?SID= 4ETXCXJipj1aPEw9JFyandproduct=UAandqid=1andsearch_mode=GeneralSearch (accessed Septermber, 20, 2013).
[3] Debecker, D. P.; Hulea, V.; Mutin, P. H. *Appl. Catal. A: Gen.*, 2013, 451, 192–206.
[4] Walcarius, A. *Chem. Soc. Rev.*, 2013, 42, 4098.
[5] Bibby, A.; Mercier, L. *Green Chem.*, 2003, 5, 15–19.
[6] Wu, Z.; Zhao, D. *Chem. Commun.*, 2011, 47, 3332–3338.
[7] Wagner, T.; Haffer, S.; Weinberger, C.; Klaus, D.; Tiemann, M. *Chem. Soc. Rev.*, 2013, 42, 4036.
[8] Ren, Y.; Ma, Z.; Bruce, P. G. *Chem. Soc. Rev.*, 2012, 41, 4909.
[9] Taguchi, A.; Schüth, F. *Microporous Mesoporous Mater.*, 2005, 77, 1–45.
[10] Vos, D. E. D.; Dams, M.; Sels, B. F.; Jacobs, P. A. *Chem. Rev.*, 2002, 102, 3615–3640.
[11] Grosso, D.; Cagnol, F.; Soler-Illia, G.; Crepaldi, E.; Amenitsch, H.; Brunet-Bruneau, A.; Bourgeois, A.; Sanchez, C. *Adv. Funct. Mater.*, 2004, 14, 309–322.
[12] Huo, Q.; Margolese, D. I.; Ciesla, U.; Demuth, D. G.; Feng, P.; Gier, T. E.; Sieger, P.; Firouz, A.; Chmelka, B. F.; Schüth, F.; Stucky, G. D. *Chem. Mater.*, 1994, 6, 1176–1191.
[13] Tian, B.; Liu, X.; Tu, B.; Yu, C.; Fan, J.; Wang, L.; Xie, S.; Stucky, G. D.; Zhao, D. *Nat. Mater.*, 2003, 2, 159–163.
[14] Boettcher, S. W.; Fan, J.; Tsung, C.-K.; Shi, Q.; Stucky, G. D. *Accounts Chem. Res.*, 2007, 40, 784–792.
[15] Schüth, F. *Chem. Mater.*, 2001, 13, 3184–3195.
[16] Lee, J.; Orilall, M. C.; Warren, S. C.; Kamperman, M.; DiSalvo, F. J.; Wiesner, U. *Nat. Mater.*, 2008, 7, 222–228.
[17] Yang, P.; Zhao, D.; Margolese, D. I.; Chmelka, B. F.; Stucky, G. D. *Nature*, 1998, 396, 152–155.
[18] Fan, J.; Boettcher, S. W.; Stucky, G. D. *Chem. Mater.*, 2006, 18, 6391–6396.
[19] Fan, J.; Boettcher, S. W.; Tsung, C.-K.; Shi, Q.; Schierhorn, M.; Stucky, G. D. *Chem. Mater.*, 2008, 20, 909–921.
[20] Brinker, C. J.; Scherer, G. W. Sol-gel Science: The Physics and Chemistry of Sol-gel Processing; Academic Press, 1990.
[21] Soler-Illia, G. J. de A. A.; Crepaldi, E. L.; Grosso, D.; Sanchez, C. *Curr. Opin. Colloid and Interface Sci.*, 2003, 8, 109–126.
[22] De AA Soler–Illia, G. J.; Sanchez, C. *New J. Chem.*, 2000, 24, 493–499.
[23] Grosso, D.; Boissière, C.; Smarsly, B.; Brezesinski, T.; Pinna, N.; Albouy, P. A.; Amenitsch, H.; Antonietti, M.; Sanchez, C. *Nat. Mater.*, 2004, 3, 787–792.
[24] Soler-Illia, G. J.; Azzaroni, O. *Chem. Soc. Rev.*, 2011, 40, 1107–1150.
[25] Arends, I. W. C. E.; Sheldon, R. A. *Appl. Catal. A: Gen.*, 2001, 212, 175–187.
[26] Brock, S. L.; Duan, N.; Tian, Z. R.; Giraldo, O.; Zhou, H.; Suib, S. L. *Chem. Mater.*, 1998, 10, 2619–2628.

[27] Choudhary, T. V.; Banerjee, S.; Choudhary, V. R. *Appl. Catal. A: Gen.*, 2002, 234, 1–23.
[28] Antonelli, D. M.; Ying, J. Y. *Chem. Mater.*, 1996, 8, 874–881.
[29] Sinha, A. K.; Suzuki, K. *Angew. Chem.*, 2005, 117, 275–277.
[30] Bruce, D. W.; O'Hare, D.; Walton, R. I. Porous Materials; Wiley, 2011.
[31] Brezesinski, T.; Wang, J.; Tolbert, S. H.; Dunn, B. *Nat. Mater.*, 2010, 9, 146–151.
[32] Yuan, J.; Zhang, Y.; Le, J.; Song, L.; Hu, X. *Mater. Lett.*, 2007, 61, 1114–1117.
[33] Vettraino, M.; Trudeau, M. L.; Antonelli, D. M. *Adv. Mater.*, 2000, 5, 337–341.
[34] Carreon, M. A.; Guliants, V. V. *Eur. J. Inorg. Chem.*, 2005, 2005, 27–43.
[35] Jiao, F.; Bruce, P. G. *Angew. Chem.*, 2004, 116, 6084–6087.
[36] Kondo, J. N.; Domen, K. *Chem. Mater.*, 2008, 20, 835–847.
[37] Weckhuysen, B. M.; Wachs, I. E.; Schoonheydt, R. A. *Chem. Rev.*, 1996, 96, 3327–3350.
[38] Sinha, A. K.; Suzuki, K. *Appl. Catal. B: Environ.*, 2007, 70, 417–422.
[39] Suib, S. L. *J. Mater. Chem.*, 2008, 18, 1623–1631.
[40] Suib, S. *Chem. Commun.*, 1997, 1031–1032.
[41] Tian, Z.; Wang, J.; Duan, N.; Krishnan V. V.; Suib, S. L. *Science*, 1997, 276, 926–930.
[42] Sinha, A. K.; Suzuki, K.; Takahara, M.; Azuma, H.; Nonaka, T.; Fukumoto, K. *Angew. Chem. Int. Ed.*, 2007, 46, 2891–2894.
[43] Torres, J. Q.; Giraudon, J.-M.; Lamonier, J.-F. *Catal. Today*, 2011, 176, 277–280.
[44] Brezesinski, T.; Groenewolt, M.; Antonietti, M.; Smarsly, B. *Angew. Chem. Int. Ed.*, 2006, 45, 781–784.
[45] Jiao, F.; Harrison, A.; Jumas, J.-C.; Chadwick, A. V.; Kockelmann, W.; Bruce, P. G. *J. Am. Chem. Soc.*, 2006, 128, 5468–5474.
[46] Bruce, A. P.; Harrison, A. A.; Chadwick, A. A.; Jumas, A. J.; Jiao, A. F. *J. A. Chem. Soc.*, 2008, 128, 12905-12909.
[47] Shirokura, N.; Nakajima, K.; Nakabayashi, A.; Lu, D.; Hara, M.; Domen, K.; Tatsumi, T.; Kondo, J. N. *Chem. Commun.*, 2006, 2188–2190.
[48] Lide, D. R. CRC Handbook of Chemistry and Physics: A Ready-reference Book of Chemical and Physical Data; CRC Press, 2004.
[49] Gómez, S.; Giraldo, O.; Garcés, L. J.; Villegas, J.; Suib, S. L. *Chem. Mater.*, 2004, 16, 2411–2417.
[50] Poyraz, A. S.; Biswas, S.; Genuino, H. C.; Dharmarathna, S.; Kuo, C.-H.; Suib, S. L. *Chem. Cat. Chem.*, 2012.
[51] Jiao, F.; Frei, H. *Chem. Commun.*, 2010, 46, 2920.
[52] Imperor-Clerc, M.; Bazin, D.; Appay, M.-D.; Beaunier, P.; Davidson, A. *Chem. Mater.*, 2004, 16, 1813–1821.
[53] Dong, X.; Shen, W.; Zhu, Y.; Xiong, L.; Shi, J. *Adv. Funct. Mater.*, 2005, 15, 955–960.
[54] Kumar, G. S.; Palanichamy, M.; Hartmann, M.; Murugesan, V. *Microporous Mesoporous Mater.*, 2008, 112, 53–60.
[55] Jiao, F.; Harrison, A.; Hill, A. H.; Bruce, P. G. *Adv. Mater.*, 2007, 19, 4063–4066.
[56] Ryoo, R.; Joo, S. H.; Kruk, M.; Jaroniec, M. *Adv. Mater.*, 2001, 13, 677–681.
[57] Caps, V.; Tsang, S. C. *Catal. Today*, 2000, 61, 19–27.
[58] Pérez, H.; Navarro, P.; Delgado, J. J.; Montes, M. *Appl. Catal. A: Gen.*, 2011, 400, 238–248.

[59] Sinha, A. K.; Suzuki, K.; Takahara, M.; Azuma, H.; Nonaka, T.; Suzuki, N.; Takahashi, N. *J. Phys. Chem. C*, 2008, 112, 16028–16035.

[60] Pérez, H.; Navarro, P.; Torres, G.; Sanz, O.; Montes, M. *Catal. Today*, 2013, 212, 149–156.

[61] Du, Y.; Meng, Q.; Wang, J.; Yan, J.; Fan, H.; Liu, Y.; Dai, H. *Microporous Mesoporous Mater.*, 2012, 162, 199–206.

[62] Sun, S.; Wang, W.; Shang, M.; Ren, J.; Zhang, L. *J. Mol. Catal. A: Chem.*, 2010, 320, 72–78.

[63] Deng, J.; Zhang, L.; Dai, H.; Xia, Y.; Jiang, H.; Zhang, H.; He, H. *J. Phys. Chem. C*, 2010, 114, 2694–2700.

[64] Han, Y.-F.; Chen, F.; Ramesh, K.; Zhong, Z.; Widjaja, E.; Chen, L. *Appl. Catal. B: Environ.*, 2007, 76, 227–234.

[65] Kim, S. C.; Shim, W. G. *Appl. Catal. B: Environ.*, 2010, 98, 180–185.

[66] Suib, S. L. *Curr. Opin. Solid State Mater. Sci.*, 1998, 3, 63–70.

[67] Sui, M.; Liu, J.; Sheng, L. *Appl. Catal. B: Environ.*, 2011, 106, 195–203.

[68] Patel, A.; Shukla, P.; Chen, J.; Rufford, T. E.; Rudolph, V.; Zhu, Z. *Catal. Today*, 2013, 212, 38–44.

[69] Manyar, H. G.; Chaure, G. S.; Kumar, A. *Green Chem.*, 2006, 8, 344.

[70] Han, Y.-F.; Chen, F.; Zhong, Z.; Ramesh, K.; Chen, L.; Widjaja, E. *J. Phys. Chem. B*, 2006, 110, 24450–24456.

[71] Rosen, J.; Hutchings, G. S.; Jiao, F. *J. Am. Chem. Soc.*, 2013, 135, 4516–4521.

[72] Xie, X.; Shen, W. Nanoscale 2009, 1, 50.

[73] Wang, X.; Tian, W.; Zhai, T.; Zhi, C.; Bando, Y.; Golberg, D. *J. Mater. Chem.*, 2012, 22, 23310.

[74] El-Shobaky, G. A.; El-Molla, S. A.; Zahran, A. A. *Mater. Lett.*, 2003, 53, 1612–1623.

[75] Gasparotto, A.; Barreca, D.; Bekermann, D.; Devi, A.; Fischer, R. A.; Fornasiero, P.; Gombac, V.; Lebedev, O. I.; Maccato, C.; Montini, T.; Tendeloo, G. V.; Tondello, E. *J. Am. Chem. Soc.*, 2011, 133, 19362–19365.

[76] Liu, Q.; Liu, C.-X.; Nie, X.-L.; Bai, L.; Wen, S.-H. *Mater. Lett.*, 2012, 72, 101–103.

[77] Tsoncheva, T.; Ivanova, L.; Rosenholm, J.; Linden, M. *Appl. Catal. B: Environ.*, 2009, 89, 365–374.

[78] Łojewska, J.; Kołodziej, A.; Łojewski, T.; Kapica, R.; Tyczkowski, J. *Appl. Catal. A: Gen.*, 2009, 366, 206–211.

[79] Okumura, K.; Kobayashi, T.; Tanaka, H.; Niwa, M. *Appl. Catal. B: Environ.*, 2003, 44, 325–331.

[80] Xia, Y.; Dai, H.; Jiang, H.; Zhang, L. *Catal. Commun.*, 2010, 11, 1171–1175.

[81] Grätzel, M. *Nature*, 2001, 414, 223–344.

[82] Dahl, S.; Chorkendorff, I. *Nat. Mater.*, 2012, 11, 100–101.

[83] Wei, T.-Y.; Chen, C.-H.; Chien, H.-C.; Lu, S.-Y.; Hu, C.-C. *Adv. Mater.*, 2010, 22, 347–351.

[84] Wang, X.; Sumboja, A.; Khoo, E.; Yan, C.; Lee, P. S. *J. Phys. Chem. C*, 2012, 116, 4930–4935.

[85] Long, J. W.; Dunn, B.; Rolison, D. R.; White, H. S. *Chem. Rev.*, 2004, 104, 4463–4492.

[86] Cui, Y.; Wen, Z.; Sun, S.; Lu, Y.; Jin, J. *Solid State Ionics*, 2012, 225, 598–603.

[87] Sa, Y. J.; Kwon, K.; Cheon, J. Y.; Kleitz, F.; Joo, S. H. *J. Mater. Chem.*, 2013, 1, 9992.

[88] Taghavimoghaddam, J.; Knowles, G. P.; Chaffee, A. L. *J. Mol. Catal. A: Chem.*, 2012, 358, 79–88.
[89] Yuan, C.; Yang, L.; Hou, L.; Shen, L.; Zhang, X.; Lou, X. W. (David) *Energy and Environ. Sci.*, 2012, 5, 7883.
[90] Ma, C. Y.; Mu, Z.; Li, J. J.; Jin, Y. G.; Cheng, J.; Lu, G. Q.; Hao, Z. P.; Qiao, S. Z. *J. Am. Chem. Soc.*, 2010, 132, 2608–2613.
[91] Kwak, G.; Hwang, J.; Cheon, J.-Y.; Woo, M. H.; Jun, K.-W.; Lee, J.; Ha, K.-S. *J. Phys. Chem. C*, 2013, 117, 1773–1779.
[92] Li, H.; He, P.; Wang, Y.; Hosono, E.; Zhou, H. *J. Mater. Chem.*, 2011, 21, 10999.
[93] Venkov, T. V.; Hess, C.; Jentoft, F. C. *Langmuir*, 2007, 23, 1768–1777.
[94] Krins, N.; Bass, J. D.; Grosso, D.; Henrist, C.; Delaigle, R.; Gaigneaux, E. M.; Cloots, R.; Vertruyen, B.; Sanchez, C. *Chem. Mater.*, 2011, 23, 4124–4131.
[95] Karakoulia, S. A.; Triantafyllidis, K. S.; Lemonidou, A. A. *Microporous Mesoporous Mater.*, 2008, 110, 157–166.
[96] Shyue, J.-J.; Guire, M. R. D. *J. Am. Chem. Soc.*, 2005, 127, 12736–12742.
[97] Zhang, W.; Zhang, B.; Wolfram, T.; Shao, L.; Schlögl, R.; Su, D. S. *J. Phys. Chem. C*, 2011, 115, 20550–20554.
[98] Xue, M.; Chen, H.; Ge, J.; Shen, J. *Microporous Mesoporous Mater.*, 2010, 131, 37–44.
[99] Piumetti, M.; Armandi, M.; Garrone, E.; Bonelli, B. *Microporous Mesoporous Mater.*, 2012, 164, 111–119.
[100] Muylaert, I.; Musschoot, J.; Leus, K.; Dendooven, J.; Detavernier, C.; Voort, P. V. D. *Eur. J. Inorg. Chem.*, 2012, 2012, 251–260.
[101] Piumetti, M.; Bonelli, B.; Armandi, M.; Gaberova, L.; Casale, S.; Massiani, P.; Garrone, E. *Microporous Mesoporous Mater.*, 2010, 133, 36–44.
[102] Yoshitake, H.; Tatsumi, T. *Chem. Mater.*, 2003, 15, 1695–1702.
[103] Gao, F.; Zhang, Y.; Wan, H.; Kong, Y.; Wu, X.; Dong, L.; Li, B.; Chen, Y. *Microporous Mesoporous Mater.*, 2008, 110, 508–516.
[104] Tsoncheva, T.; Ivanova, L.; Dimitrova, R.; Rosenholm, J. *J. Colloid Interface Sci.*, 2008, 321, 342–349.
[105] Piumetti, M.; Bonelli, B.; Massiani, P.; Millot, Y.; Dzwigaj, S.; Gaberova, L.; Armandi, M.; Garrone, E. *Microporous Mesoporous Mater.*, 2011, 142, 45–54.
[106] Zhao, L.; Dong, Y.; Zhan, X.; Cheng, Y.; Zhu, Y.; Yuan, F.; Fu, H. *Catal. Lett.*, 2012, 142, 619–626.
[107] Bulánek, R.; Čapek, L.; Setnička, M.; Čičmanec, P. *J. Phys. Chem. C*, 2011, 115, 12430–12438.
[108] Shylesh, S.; Singh, A. *J. Catal.*, 2004, 228, 333–346.
[109] Borah, P.; Ma, X.; Nguyen, K. T.; Zhao, Y. *Angew. Chem. Int. Ed.*, 2012, 51, 7756–7761.
[110] Liu, Y.-M.; Xie, S.-H.; Cao, Y.; He, H.-Y.; Fan, K.-N. *J. Phys. Chem. C*, 2010, 114, 5941–5946.
[111] Golinska, H.; Decyk, P.; Ziolek, M.; Kujawa, J.; Filipek, E. *Catal. Today*, 2009, 142, 175–180.
[112] Chen, K.; Bell, A. T.; Iglesia, E. *J. Phys. Chem. B*, 2000, 104, 1292–1299.
[113] Silberova, B.; Fathi, M.; Holmen, A. *Appl. Catal. A: Gen.*, 2004, 276, 17–28.
[114] Heracleous, E.; Lemonidou, A. A. *Appl. Catal. A: Gen.*, 2004, 269, 123–135.

[115] Alam, S.; Anand, C.; Ariga, K.; Mori, T.; Vinu, A. *Angew. Chem.*, 2009, 121, 7494–7497.
[116] Pineda, A.; Balu, A. M.; Campelo, J. M.; Luque, R.; Romero, A. A.; Serrano-Ruiz, J. C. *Catal. Today*, 2012, 187, 65–69.
[117] Parma, A.; Freris, I.; Riello, P.; Cristofori, D.; Fernández, C. de J.; Amendola, V.; Meneghetti, M.; Benedetti, A. *J. Mater. Chem.*, 2012, 22, 19276.
[118] Shi, K.; Chi, Y.; Yu, H.; Xin, B.; Fu, H. *J. Phys. Chem. B*, 2005, 109, 2546–2551.
[119] Jiao, F.; Jumas, J.-C.; Womes, M.; Chadwick, A. V.; Harrison, A.; Bruce, P. G. *J. Am. Chem. Soc.*, 2006, 128, 12905–12909.
[120] Hess, D. M.; Naik, R. R.; Rinaldi, C.; Tomczak, M. M.; Watkins, J. *J. Chem. Mater.*, 2009, 21, 2125–2129.
[121] Taylor, K. M. L.; Kim, J. S.; Rieter, W. J.; An, H.; Lin, W.; Lin, W. *J. Am. Chem. Soc.*, 2008, 130, 2154–2155.
[122] Srivastava, D. N.; Perkas, N.; Gedanken, A.; Felner, I. *J. Phys. Chem. B*, 2002, 106, 1878–1883.
[123] Yang, S.; Sun, Y.; Chen, L.; Hernandez, Y.; Feng, X.; Müllen, K. *Sci. Reports*, 2012, 2.
[124] Kim, T. W.; Ha, H.-W.; Paek, M.-J.; Hyun, S.-H.; Baek, I.-H.; Choy, J.-H.; Hwang, S.-J. *J. Phys. Chem. C*, 2008, 112, 14853–14862.
[125] Barnhart, J. *Soil Sediment Contam.*, 1997, 6, 561–568.
[126] Fendorf, S. E. *Geoderma*, 1995, 67, 55–71.
[127] Xia, P.; Zuo, S.; Liu, F.; Qi, C. *Catal. Commun.*, 2013, 41, 91–95.
[128] Liu, F.; Zuo, S.; Xia, X.; Sun, J.; Zou, Y.; Wang, L.; Li, C.; Qi, C. *J. Mater. Chem.*, 2013, 1, 4089.
[129] Sinha, A. K.; Suzuki, K. *Angew. Chem. Int. Ed.*, 2005, 44, 271–273.
[130] Yan, Z. F.; Lu, G. Q.; Zhu, Z. H. with Li J. *Phys. Chem. B*, 2006, 110, 178–183.
[131] Yamashita, H.; Yoshizawa, K.; Ariyuki, M.; Higashimoto, S.; Anpo, M.; Che, M. *Chem. Commun.*, 2001, 435–436.
[132] Yamashita, H.; Anpo, M. *Curr. Opin. Solid State Mater. Sci.*, 2003, 7, 471–481.
[133] Li, L.; Wang, Y.; Xu, Z. P.; Zhu, Z. *Appl. Catal. A: Gen.*, 2013, 467, 246–252.
[134] Tamiolakis, I.; Lykakis, I. N.; Katsoulidis, A. P.; Stratakis, M.; Armatas, G. S. *Chem. Mater.*, 2011, 23, 4204–4211.
[135] Yada, M.; Kitamura, H.; Machida, M.; Kijima, T. *Inorg. Chem.*, 1998, 37, 6470-6475.

In: Comprehensive Guide for Mesoporous Materials. Volume 3 ISBN: 978-1-63463-318-5
Editor: Mahmood Aliofkhazraei © 2015 Nova Science Publishers, Inc.

Chapter 12

NANOCOMPOSITES EMBEDDED BY MESOPOROUS MATERIALS

Chunfang Du, Yiguo Su and Zhiliang Liu
College of Chemistry and Chemical Engineering, Inner Mongolia University, Hohhot, Inner Mongolia, P. R. China

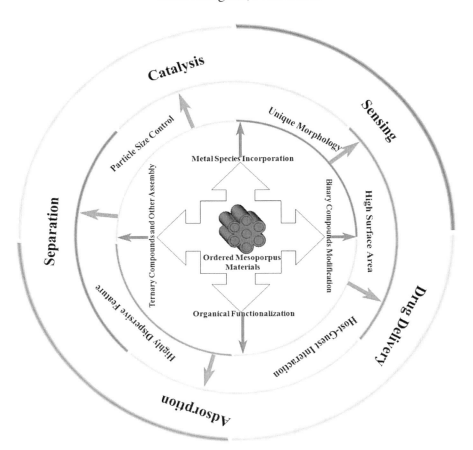

Abstract

"Host-guest chemistry" is a potential subject, attracting a great deal of attention from various research areas, including large-molecule catalysis, separation techniques, adsorption procedure and drug delivery. Mesoporous materials with huge specific surface area, large pore volume and uniform pore size distribution could act as a kind of "micro-reactor" for synthesis of functional nanoparticles. The interaction of host species-mesoporous materials and guest nanoparitcles as well as the restriction effect of framework of host species endow the nanocomposites embedded by mesoporous materials unique and enhanced properties that facilitate future applications. Undoubtedly, the combination of various functional species and mesoporous matrices would open a realm of new possibilities for exploring of novel materials with unexpected functions.

1. Introduction

Since their discovery in 1992 [1,2], ordered mesoporous materials have been extensively investigated due to their high surface area, big pore volume, uniform pore size distribution, tunable pore structure and ease of functionalization. This kind of material has been applied in various fields, including catalysis [3], separation [4], adsorption [5], ion exchange [6], sensing [7], drug delivery [8] and so on. In comparison to microporous materials such as zeolites, the exceptional mesostructure makes mesoporous materials appreciable supports for guest molecules. With the introduction of various functional groups into the mesopores, unexpected and excellent properties become available. Until now, a large number of studies have been reported, referring constructing nanoparticles or attaching organic groups using the channels of mesoporous materials as micro-reactors. For example, organic functional groups, metal species, metal oxides, sulfides, phosphates, tungstates and other macromolecules have been successfully incorporated into the channels of mesoporous materials for desirable physical and chemical properties (Table 1). The high surface area, large pore volume as well as uniform pore size distribution of ordered mesoporous materials bring the guest species with particle size reduction, unique morphology and non-agglomeration. Furthermore, the abundant hydroxyl groups on the pore surface contribute to effective interaction with the guest molecules [21]. Associated with the synergetic effects of guest species and mesoporous structures is the superior physicochemical properties of the final products that differ from those of their individual, isolated components.

2. Organically Functionalized Mesoporous Materials

From the viewpoint of material science, the construction of inorganic and organic building blocks within a single material has attracted great attention because their combined properties may open the realm of new possibilities to combine the enormous functional variation of organic chemistry with the advantages of a thermally stable and robust inorganic substrate. Therefore, functionalization of organic moieties into mesoporous matrices becomes a major topic research because it offers a further possibility of tailoring the physical and chemical properties of the porous materials.

Table 1. Reports of various functionalities in the channels of mesoporous materials

Functionalities	Examples of functional groups	Mesoporous Materials	Refs
organic functional groups	amino groups	MCM-41, mesoporous thin films	[9, 10]
	disulfide-containing organotrimethoxysilanes with different anionic groups	MCM-41	[11]
	photoisomerizable 2-hydroxychalcones	MCM-41, SBA-15	[12]
	saccharides	Mesoporous UVM-7	[13]
	mercaptopropyl	MCM-41, SBA-15	[14]
metal complexes	Eu(III) complexes	mesoporous SiO_2, SBA-15, MCM-41	[15-18]
	Pt(II) complexes	MCM-41, MCM-48, SBA-15	[19, 20]
	V(IV) complexes	MCM-41	[21, 22]
	Cu(II) complex	MCM-41, SBA-15	[23]
	Ru(II) complexes	mesoporous silica	[24, 25]
	Mn(II) complexes	MCM-41	[22, 26]
	Ir(III) complexes	sodium aluminosilicate glasses, MCM-41, MCM-48, SBA-15	[27, 28]
	Er(III), Nd(III), Yb(III), Sm(III), Pr(III) complexes	MCM-41	[29]
substituted metals	Al	SBA-15, MCM-41, MCM-48	[30-32]
	Ti	MCM-41	[33]
	V	SBA-15	[34]
	Fe	SBA-15, MCM-41	[35, 36]
	Zr	MCM-41	[37]
	Cr	MCM-41	[38]
	Al-Cr	MCM-41	[39]
deposited metals	Au, Ag, Pt, Pd, Fe nanoparticles	SBA-15, FSM-16, HMM-1, mesoporous carbon-silica nanocomposites	[40-42]
	Pt, Pd nanowires	MCM-41, FSM-16, HMM-1	[41, 43]
	Au, Ag clusters	MCM-41, mesoporous TiO_2 film	[44, 45]
metal oxides	Al_2O_3	SBA-15	[46]
	Fe_2O_3	MCM-48	[47]
	ZnO	CMI-1	[48]
	TiO_2	SBA-15	[49, 50]
	CuO	mesoporous alumina	[51]
	Eu_2O_3	SBA-15	[52]
	SnO_2, SnO_2:Eu^{3+}	SBA-15, Al-MCM-41	[53, 54]
	mixed La-Co oxides	SBA-15	[55]
sulfides	CdS	MCM-41, SBA-15	[56, 57]
	ZnS	MCM-41	[58]
	PbS	SBA-15	[59]
	MnS	mesoporous SiO_2	[60]
	Ag_2S	Zeolite A	[61]
quantum dots	CdSe	MCM-41	[62]
	GaN	MCM-41	[63]
ternary Compounds	$Ti_3(PO_4)_4$	SBA-15	[64]
	$CaWO_4$:Eu^{3+}	SBA-15	[65]
	$GdVO_4$:Eu^{3+}	mesoporous silica nanoparticles	[66]
other macromolecules	polyelectrolytes	mesoporous silica	[67]
	carboxymethylated polyethyleneimine	MCM-48	[68]

Table 2. Examples of grafting and co-condensation methods

Methods	Reaction stages	Examples of organic groups	Properties	Refs
grafting (post-synthesis)	after the synthesis of pure mesoporous materials	Methylcyclopentadien-yl and pentamethylcy-clopentadienyl	excellent catalysts in the transesterification of dimethyl oxalate (DMO) with phenol	[69]
		coumarin derivatives	uptake, storage, and release of guest molecules under different UV wavelength	[70]
		mono-, di-, and triaminosilane	absorb chromate and arsenate	[71]
		poly (N-isopropyl acrylamide)	absorb and transport molecular species controlled by temperature	[72]
co-condensation (direct synthesis)	introduce the organic groups in the existence of the structure-directing agent	N,N'-bis(salicylidene)-thiocarbohydrazide (BSTC)	show dominant blue emission	[73]
		aminopropyl	chemisorptions of carbon dioxide and deposit gold nanoparticles for catalytic application	[74]
		mercaptopropyl	adsorption of Hg^{2+}	[14]

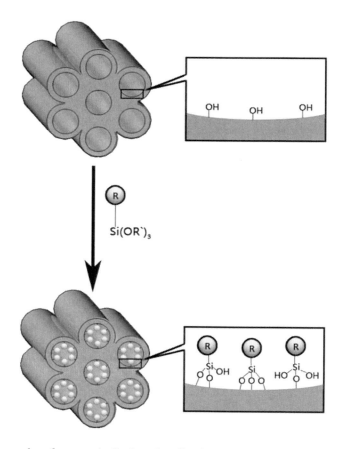

Figure 1. Grafting procedure for organically functionalized mesoporous materials.

There are two strategies to anchor functionalities onto the mesochannels of mesoporous materials: (1) subsequent introduction of organic components onto a pure mesoporous matrix (grafting or post-synthesis procedure); (2) co-condensation of siloxane and organosiloxane precursors in the existence of the structure-directing agent (direct synthesis or one-pot synthesis) (Table 2).

2.1. Grafting

Grafting procedures are based on the subsequent modification of the inner surfaces of mesoporous matrix with organic groups through silylation reactions occurring on isolated ($\equiv$Si-OH) and germinal (=Si(OH)$_2$) silanol groups (Figure 1). This process refers to the organosilanes of the type (R'O)$_3$SiR, chlorosilanes ClSiR$_3$, or silazanes HN(SiR$_3$)$_3$. The alkoxide bonds are hydrolyzed upon reaction with the surface silanols and/or adsorbed water, ultimately resulting in surface attachment of the organosilane *via* condensation with the silanols. Generally, functionalization with a variety of organic groups can be realized in this way by variation of the organic residue R [75]. There are both advantages and disadvantages of this method: retain the original mesostructure of the starting mesoporous phase and hydrolytically more stable than the co-condensation method, and nonhomogeneous distribution of the organic groups within the pores and a lower degree of occupation or pore blocking.

Two kinds of organic groups concerning methylcyclopentadienyl (CP') and pentamethylcyclopentadienyl (CP'') grafting MCM-41 were prepared, denoted as CP'-MCM-41 (Figure 2) and CP''-MCM-41, respectively [69]. These two functionalized materials showed active catalysis in the transesterification of dimethyl oxalate (DMO) with phenol. It is noted that the material CP''-MCM-41 displayed higher conversion of DMO and selectivity of diphenyl oxalate (DPO) than CP'-MCM-41, which was possibly attributed to the relatively strong Lewis basicity of CP'' group. Coumarin derivative-modified MCM-41, for the purpose of uptake, storage, and release of guest molecules, were regulated through the photoresponsive reversible intermolecular dimerization of coumarin derivatives attached preferentially on the pore outlets [70]. Under irradiation of UV light (λ>310 nm), guest molecules such as phenanthrene neither could enter nor escape from the mesopores of MCM-41, which was related to the photodimerization of coumarin to close the pore outlet with cyclobutane dimer under such UV light. However, the guest molecules could be released when the modified MCM-41 was irradiated under UV light around 250 nm (Figure 3). It was found that selective functionalization was highly essential for these materials with photoresponsive reversible properties. The utilization of as-prepared MCM-41without removing the template, short-time grafting procedure, and one-dimensional, isolated, individual pore structures of MCM-41 are responsible for the selective functionalization.

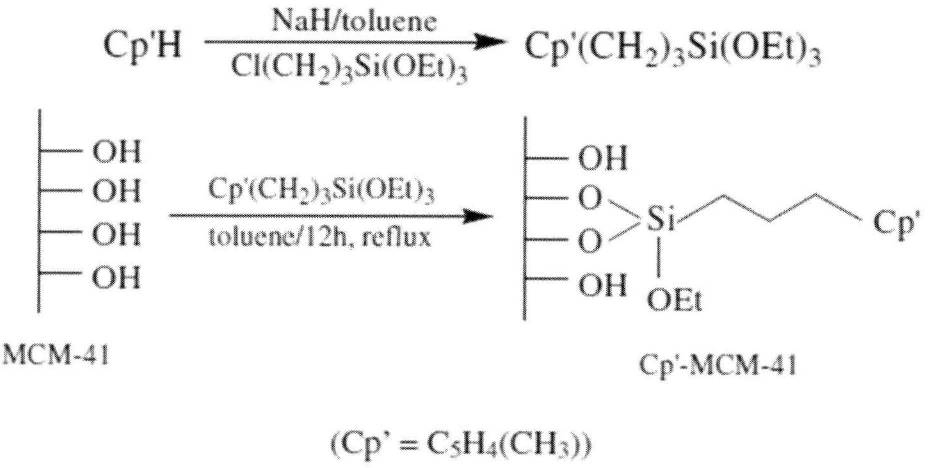

Figure 2. Synthesis of cyclopentadienyl-functionalized MCM-41 material (Reprinted from *Catal. Commun. 9,* Liu, Y.; Zhao, G.; Liu, G.; Wu, S.; Chen, G.; Zhang, W.; Sun, H.; Jia, M., Cyclopentadienyl-functionalized mesoporous MCM-41 catalysts for the transesterification of dimethyl oxalate with phenol, p 2022, Copyright (2008), with permission from Elsevier).

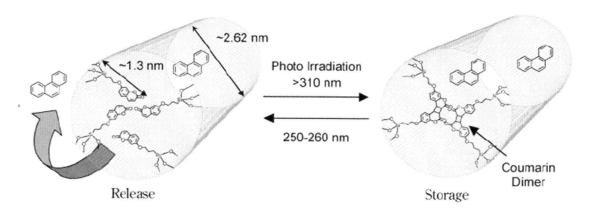

Figure 3. Conceptual scheme of photo-switched storage-release controlled release by coumarin-modified MCM-41 (Reprinted with permission from Mal, N. K.; Fujiwara, M.; Tanaka, Y.; Taguchi, T.; Matsukata, M. *Chem. Mater.* 2003, *15*, 3385. Copyright (2003) American Chemical Society).

In order to investigate the adsorption properties of oxyanions in acidic conditions, aminosilane-grafted mesoporous materials were prepared [71]. The influence of amino group density and mesoporous structures on the adsorption properties were mainly studied through employing mono-, di-, and triaminosilane as function groups, and SBA-1 and MCM-41 as mesoporous hosts, respectively. It was demonstrated that the triaminosilane-grafted mesoporous silica adsorbed more chromate and arsenate than the other two adsorbents functionalized by mono- and diaminosilane, respectively. Furthermore, functionalized SBA-1 presented larger adsorption capacities than MCM-41 derivatives. Modulating the adsorption and transport of molecular species in mesoporous structures was realized by grafting poly (*N*-isopropyl acrylamide) (PNIPAAm), a stimuli responsive polymer, onto mesoporous materials

[72]. The uptake and release of molecular species from porous structures was greatly dependant on temperature. At low temperature, PNIPAAm inhibited the transport of solutes due to its hydrophilicity and extension. However, it allowed the solute diffusion at higher temperature because of its hydrophobicity and being collapsed within the pore networks.

2.2. Co-Condensation Reaction

An alternative strategy for incorporation of functional groups onto inner walls of mesoporous matrix is co-condensation method. In this approach, functional groups are introduced in the initial stages for the synthesis of the mesoporous hosts, i.e., in the presence of the structure directing agent. Mesostructured silica phases are possibly synthesized by the co-condensation of tetraalkoxysilanes [(RO)$_4$Si (TEOS or TMOS)] with terminal trialkoxyorganosilanes of the type (R'O)$_3$SiR in the presence of structure-directing agents leading to materials with organic residues anchored covalently to the pore walls [75](Figure 4). One coin has two sides. In contrast to grafting methods, co-condensation procedure could not only provide a uniform surface coverage with functional groups and better control over the amount of organic groups [76], but also could solve the problem of pore blocking which seriously exists in grafting method. However, the order degree of the final products synthesized by co-condensation method decreases with increasing the amount of organic groups and the general limited content for organic functionalities is 40% [75]. Moreover, it is should be cautious to the approaches for removing the surfactants that involved in the synthesis of mesostructured matrix. Therefore, extractive method other than calcination is commonly adopted to remove the surfactants for protecting the functional groups as much as possible.

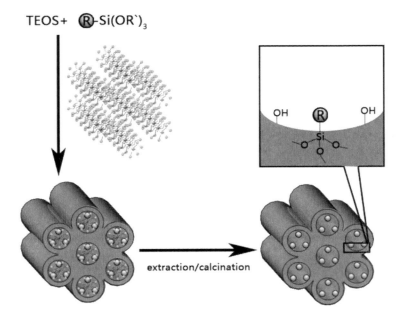

Figure 4. Co-condensation procedure for the organically functionalized mesoporous materials.

A novel organic-inorganic luminescent hybrid product functionalized by *N,N'*-bis(salicylidene)-thiocarbohydrazide (BSTC) grafted to the coupling agent 3-(triethoxysilyl)-propyl isocyanate (TESPIC) was prepared *via* a co-condensation of tetraethyl orthosilicate (TEOS) and the organosilane in the existence of P123 template, which was denoted as BSTC-SBA-15 [73]. For comparison, SBA-15 doped with BSTC was also synthesized, denoted as BSTC/SBA-15. Both of the two functionalized materials showed regular uniform microstructures and no phase separation was observed. However, these two materials displayed different luminescent properties: BSTC-SBA-15 showed dominant blue emission, while BSTC/SBA-15 presented strong dominant green luminescence.

Aminopropyl-functionalized mesoporous KIT-5 was prepared *via* a co-condensation method using different surfactants as templates [74]. The KIT-5 synthesized with carboxyl-terminated Pluronic F127 as a structure-directing agent showed higher aminopropyl content than the counterpart synthesized with hydroxyl-terminated copolymers. The aminopropyl functionalized KIT-5 could not only make contribution to chemisorptions of carbon dioxide, but also was used to deposit gold nanoparticles for catalytic application.

Acid and base co-functionalized mesoporous silicas were synthesized by grafting and co-condensation methods, and the structural characteristics and cooperative catalytic properties were studied [77]. It was shown that a weaker Brønsted acid silanol contributed more in cooperative catalysis of the aldol condensation in combination with an amine base than the stronger carboxylic acid for all samples. Thus, the removal of silanols and introduction of the carboxylic acid did not make a beneficial effect on catalytic activity.

3. METAL SPECIES EMBEDDED BY MESOPOROUS MATERIALS

Metal elements play an important role in the development of catalysis. In practical catalytic process, metal species are often supported by inorganic matrices or organic polymer supports for the purpose of low-cost and easy-recovery. As supports, inorganic matrices, especially mesoporous materials have several pivotal advantages over other supports: chemical stability, mechanical stability and thermal stability. The combination of metal elements and mesoporous materials, mostly silica-based mesoporous materials also could improve the activity and acidity of the mesoprous matrices, which is essential for catalysis. Metal species are introduced into mesoporous materials mainly through three ways: metal complexes functionalized, metal-substituted and metal-deposited approaches (Table 3).

3.1. Metal Complexes Functionalized Mesoporous Materials

There are many strategies for immobilization of metal complexes on mesoporous materials: Incipient-wetness impregnation, ion exchange, tethering approach and so on. The tethering approach is an acceptable way that it could control the homogeneous distribution of the metal complexes [95]. A functionalized ligand was firstly covalently bound to the surface of the mesoporous matrices and then the metal specie was directly added to this ligand, forming the complex directly inside the mesoporous materials.

Table 3. Examples of various metal species embedded in mesoporous materials

Metal species	Application fields	Functional groups	Properties	Refs
metal complexes	catalysis	oxovanadium and manganese complexes	active catalysts for the oxidation of *cis*-cyclooctene using tert-butylhydroperoxide as oxygen donor	[22]
		Pd complex	hydrocarboxylation of aryl olefins and alcohols	[78]
		copper (II) schiff base complex	showed excellent catalytic activities in styrene epoxidation reaction	[79]
	luminescence	lanthanide (Eu, Er, Nd, Yb, Sm, Pr, Tb) complexes	phosphors with photostability, thermal stability and mechanical property	[29, 80, 81]
	sensing	Eu complex	oxygen sensing	[82]
substituted metals	catalysis	Al	enhanced thermal and hydrothermal stability, adsorption and the catalytic decomposition of isopropyl alcohol and nitrosamine, catalyst in the reaction of 1-butene to 2-butene	[32, 83, 84]
		Ti	decomposition of NO into N_2 and O_2 under UV irradiation, higher conversion of cyclohexene	[33, 85]
		V	oxidation of cyclododecane with H_2O_2, hydroxylation of benzene to phenol using molecular O_2 as the oxidant	[34, 86]
		Fe	benzylation of benzene,	[35, 87]
		Au-Ti	selective epoxidation of styrene	[88]
		Al-Zn	selective production of *p*-cymene by isopropylation of toluene	[89]
	adsorption	Cu, Co and Fe	adsorption of dibenzothiophene from hexane	[90]
		Fe	adsorption of arsenic from water	[91]
	luminescence	Eu	two europium species existed in zeolite-1	[92]
	drug delivery	Al	delivery of diflunisal, naproxen and ibuprofen with loading (wt%) 8.7, 7.3 and 6.4, respectively	[93]
	sensing	Eu	humidity sensing	[94]
deposited metals	catalysis	Pd nanoparticles and nanowires	CO oxidation	[41]
	luminescence	Ag cluster	enhanced luminescent property	[45]

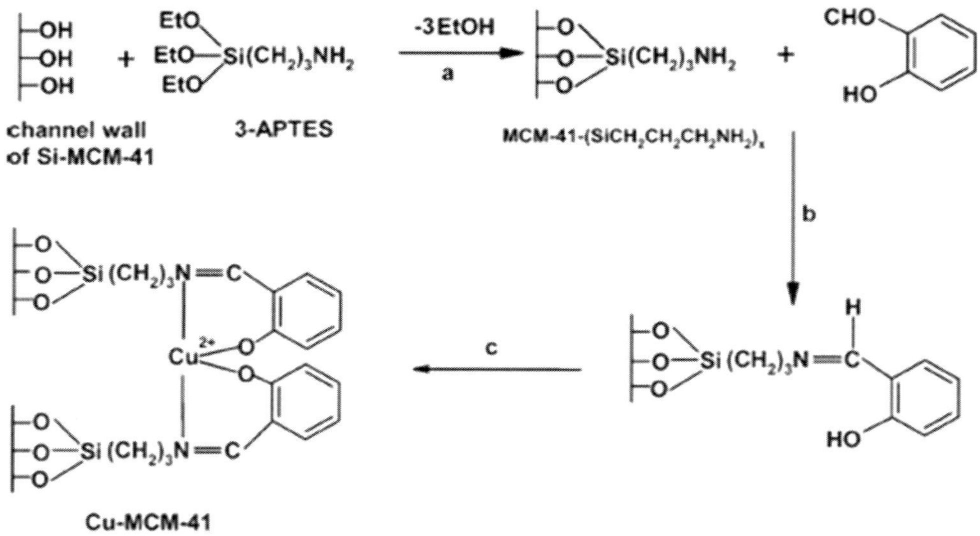

Figure 5. Modification of Si-MCM-41 channel wall: APTES/CHCl₃ (a); condensation with salicylaldehyde in methanol (b); metal complex formation: Cu(NO₃)₂·3H₂O/MeOH (Reprinted with permission from Jana, S.; Dutta, B.; Bera, R.; Koner, S. *Langmuir* 2007, *23*, 2492. Copyright (2007) American Chemical Society).

Most metal complexes supported on mesoporous materials are employed in catalysis fields. The oxovanadium and manganese complexes bearing the ligand with triethoxysilyl groups Ph-DAB-(CH₂)₃Si(OEt)₃ were anchored onto ordered MCM-41 through a grafting method [22]. The supported materials are active catalysts for the oxidation of *cis*-cyclooctene using tert-butylhydroperoxide as oxygen donor, with the 1,2-epoxycyclooctane as the main reaction product and 1,2-cyclooctanediol as a by-product. In comparison, the vanadium catalysts showed more active and selective epoxidation property than the corresponding Mn materials. Mukhopadhyay [78] reported a kind of novel heterogeneous catalysts containing a Pd complex anchored in mesoporous supports such as MCM-41 and MCM-48 for hydrocarboxylation of aryl olefins and alcohols with high regioselectivity, activity, and recyclability without the leaching of Pd complex from the supports. Another commonly focused transition metal is Cu [96]. Jana [79] reported the copper (II) Schiff base complex has been successfully anchored into Si-MCM-41 matrix *via* covalent bond (Figure 5). The catalyst Cu-MCM-41 showed excellent catalytic activities (a high conversion of 97% and selectivity of 89%) in styrene epoxidation reaction.

Another important property of metal complexes supported on mesoprous materials is luminescence. It is known that lanthanide complexes give sharp, intense emission lines upon ultraviolet irradiation, which are due to effective intramolecular energy transfers from the coordinated ligands to the luminescent central lanthanide ions [97]. Therefore, they are potential promising luminescent components for the preparation of hybrid phosphors and other optical sources. The luminescence properties of lanthanide complexes embedded in mesoporous matrices have attracted intensive attention in recent years due to controllable properties by the interaction of the phosphors with the host structure. The photostability, thermal stability and mechanical property would be improved after lanthanide (Eu, Er, Nd, Yb, Sm, Pr, Tb) complexes were incorporated into mesoporous materials [29, 80, 81].

Among lanthanide elements, Eu complexes supported on mesoporous materials were mostly fascinating. A novel organic-inorganic mesoporous hybrid titania material, denoted as Eu(Ti-MAB-S15)$_2$(NTA)$_3$, was synthesized by introducing the Eu(NTA)$_3 \cdot$2H$_2$O (NTA=1-(2-naphthoyl)-3,3,3-trifluoroacetonate) complex into the hybrid material Ti-MAB-S15 *via* a ligand exchange reaction, which Ti-MAB-S15 (MAB=*meta*-aminobenzoic acid) was a modification product of SBA-15 with organic reagents [98]. Compared to the pure complex Eu(NTA)$_3 \cdot$2H$_2$O, the mesoporous hybrid titania material Eu(Ti-MAB-S15)$_2$(NTA)$_3$ exhibited longer luminescent lifetime and higher quantum efficiency. Wang [99] reported another luminescent mesoporous hybrid material synthesized through the coordination reaction between europium nitrate and chelated quinoline-amide type ligands immobilized in mesoporous materials (MCM-41 and SBA-15). The results demonstrated that the SBA-15-type hybrid materials, with pore sizes twice that of MCM-41 counterpart, showed more efficient emission due to the spatial confinement of the nanochannels of the mesoporous matrix. Zuo [82] synthesized a novel Eu^{3+} complex of Eu(DPIQ)(TTA)$_3$ (DPIQ=10H-dipyrido [*f,h*] indolo [3,2-*b*] quinoxaline, TTA=2-thenoyltrifluoroacetonate) and incorporated it into the mesoporous MCM-41, in order to explore the oxygen-sensing property of the composite. The results indicated that the final product exhibited the characteristic emission of Eu^{3+} ion and the fluorescence intensity of $^5D_0$-$^7F_2$ obviously decreased with increasing oxygen concentrations.

3.2. Metal-Substituted Mesoporous Mateirals

It is known that mesoporous materials is a kind of proper candidate for catalysis application due to huge surface area, big pore volume, controlled pore structure and uniform pore size distribution. The catalytic study with mesoporous materials firstly focused on the metal-substituted type. Al is the most commonly used metal element to incorporate into the mesostructure, such as MCM-41, MCM-48, SBA-3 and SBA-15, for the purpose of enhancing the acidity and thermal stability of the mesoporous materials [100-103]. Aluminosilicate MCM-48 materials with Si/Al ratio between 4~50 were prepared and the hydrothermal stability in boiling water and steam was investigated [83]. The results demonstrated that the stability of Al-SBA-15 with various Si/Al ratios in boiling water was higher than that of pure silica material, which was extensively dependent on the Al content and reached the optimum hydrothermal stability in the Si/Al range 8~15. The relationship between Al content and stability in steam was contrary to that observed for boiling water. Wu [84] reported a successful synthesis of Al-SBA-15 by direct synthesis procedure in an aqueous solution that containing triblock copolymer templates, nitrates, and silica sources but without using mineral acid. The obtained Al-SBA-15 samples showed excellent performances in the adsorption and the catalytic decomposition of isopropyl alcohol and nitrosamine.

The introduction of transition metal elements such as Ti [104, 105], V [106, 107], Cr [108], Zr [109, 110] and Fe [111] into the framework of mesoporous materials is also worthy to be mentioned for the purpose of synthesis for mesoporous catalysts with redox properties. Bérubé [85] reported an efficient and stable Ti-SBA-15 epoxidation catalysts synthesized by a postgrafting method. It was shown that the hydrothermal treatment and calcination temperatures of the silica supports greatly influence the catalytic properties of the resulting mesoporous Ti-SBA-15. Higher conversion of cyclohexene was obtained with catalyst prepared using SBA-15 calcined at 550°C prior to the Ti grafting step. Framework Ti-substituted, three-dimensional, mesoporous titanosilicates, Ti-SBA-12 and Ti-SBA-16 were prepared by directly hydrothermal synthesis method [112]. Cubic Ti-SBA-16 with interconnected cage-like mesopore structure was more active than hexagonal Ti-SBA-12. An epoxide selectivity of 100% and olefin conversion greater than 92% were obtained, higher than the hitherto known titaosilicates for the oxidation of cyclic olefins with tert.-butyl hydroperoxide. V-SBA-16 catalysts with ordered cubic mesoporous structure, exhibiting highly dispersed framework V species and relative high vanadium content, were synthesized by one-pot hydrothermal method [86]. The catalytic activities of V-SBA-16 catalysts were evaluated for the hydroxylation of benzene using molecular O_2 as the oxidant, and the highest phenol yield of 30.4% with a selectivity of 90% and turnover number of 105 were obtained. Large-pore hexagonal SBA-15 partially substituted with iron (III) were obtained in highly acidic media [87]. Under optimized reaction conditions, the catalyst FeSBA15 displayed a superior catalytic property in the benzylation of benzene and other aromatics using benzyl chloride, with a clean conversion of benzyl chloride to the monoalkylated product (100% selectivity) with a very high rate constant in comparison to other mesoporous materials, such as AlSBA-15 and FeHMS.

Besides single metal-substituted mesoporous materials, multimetal-substituted mesoporous materials have also been paid attention. Wang [113] reported a Ni, Ce-substituted Al mesoporous materials prepared *via* a one-pot route: evaporation-induced self-assembly (EISA) strategy. The catalyst exhibited the highest catalytic activity (with CO_2 and CH_4 initial conversions being 70% and 68% at 700°C, respectively) and remained stable in a methane dry reforming reaction. Selvaraj [89] compared the physicochemical properties of Zn-Al-MCM-41(x) and Al-MCM-41(y) with various Si/(Zn+Al)(x) and Si/Al(y) ratios synthesized under hydrothermal conditions. The results indicated that all the synthesized samples showed good Zn-O-Si and Al-O-Si frameworks, high surface area, good hydrothermal stability, and more Brønsted acid sites in the production of *p*-cymene by isopropylation of toluene. Zn-Al-MCM-41(75) was found to be more effective for the selectivity of *p*-cymene (84%) under vapor-phase reaction conditions: 2-propanol/tulune molar ration of 1/2, reaction temperature of 275°C and WHSV (weight hourly space velocity)=0.5 h^{-1}. Highly ordered mesoporous Au-Ti-SBA-15 with larger pore diameter, pore volume and uniform pore size distribution were successfully prepared by traditional hydrothermal procedure [88]. Highly selective epoxidation of styrene in 10 min with high turnover frequency (TOF) of 4.75×10^3 min^{-1} over Au-Ti-SBA-15 was obtained.

There are many approaches to synthesize metal-substituted mesoporous materials. Jana [114] prepared Al containing MCM-41 by four different method: sol-gel, hydrothermal, template cation exchange and grafting. The samples prepared by sol-gel, grafting and template cation exchange methods are effective for the incorporation of large amounts of aluminum into the framework of MCM-41. The Al containing MCM-41 catalysts prepared by different methods behaved differently in acting as acidic catalysts: the catalyst obtained by sol-gel method showed higher cracking activity, whereas that prepared by template cation exchange method showed higher dehydration activity. Gallo [103] synthesized Al-SBA-15 through three different direct synthesis methods and one postsynthesis procedure. Based on the different synthetic procedures, the final samples exhibited various structural, textural, and surface characteristics, especially in terms of Brønsted and Lewis acid sites content. The samples synthesized by the pH-adjusting method showed the highest Brønsted/Lewis ratio of 3.49.

Besides the metal-substituted mesoporous materials, there are many examples of metal-doped mesoporous materials. A series of Zr-doped titania photocatalysts were prepared through the acid catalyzed sol-gel process [115]. The degradation experiments of the quinalphos solution demonstrated that the mesoporous catalyst Zr-TiO$_2$ with 0.5% Zr composition showed the best catalytic property. The substituted TiO$_2$ mesostructured materials have narrowed band gap, smaller crystallite size, adequate surface area and greater thermal stability, which are desirable features for catalysts, support materials, semiconductors, and electrodes in dye-sensitized solar cells. Gastro [116] reported the mesostructured rare earth (Eu^{3+}, Sm^{3+} and Er^{3+})-doped Y$_2$O$_3$ films using the evaporation induced self-assemble (EISA). The luminescence properties of the coatings displayed the characteristic features of each rare earth ion, which depended on the nature of the doping rare earth ion, being sensitive to the temperature and doping concentration.

3.3. Metal-deposited Mesoporous Materials

Different from metal framework structures in metal-substituted mesoporous materials, metal elements generally exist in the mesochannels of mesoporous matrices for metal deposited mesoporous materials. Among the metal species deposited in mesoporous materials, nanospheres and nanorods are the most common architectures. The former are commonly obtained either by impregnation or grafting of the precursor salts on to the mesoporous matrices [117]. The latter are mostly synthesized through pore filling procedure by wet or vapor infiltration of high concentration of metal precursor into the channels, which is subsequently reduced by kinds of methods to form nanorods [118]. Different morphologies of platinum nanoparticles (spheroids, nanorugby balls and nanorods) (Figure 6) were obtained from very low concentration of precursors and could be isolated exclusively inside the mesochannels of SBA-15 [119]. This was achieved by dispersing a platinum precursor in surfactant modified polymer to different extents and using these composite materials as template for the formation of mesoporous silica.

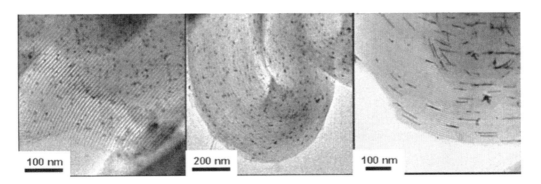

Figure 6. HRTEM images of Pt/SBA-15 with different morphologies: spheroid (left), rugby balls (middle) and nanorods (right) (Reprinted with permission from Prashar, A. K.; Hodgkins, R. P.; Chandran, J. N.; Rajamohanan, P. R.; Devi, R. N. *Chem. Mater.* 2010, *22*, 1633. Copyright (2010) American Chemical Society).

Zhu [40] synthesized mesoporous structure SBA-15 containing Au, Ag, Pt nanoparticles in the 2-20 nm range. HRTEM images indicated that the metal nanoparticles in the 2-10 nm range could be successfully incorporated into the ordered mesoporous matrix SBA-15. However, metal nanoparticles with particle size of 20 nm could not be inserted because of their larger diameter than the SBA-15 pore size. In the case of 5 and 10 nm diameter nanoparticles, regardless of metal kinds, the inclusion is controlled by the larger size in the population (Figure 7).

Nanowires and nanoparticles of palladium were synthesized in mesoporous silicas FSM-16 and HMM-1 [41]. Under UV-irradiation in the presence of water and methanol vapors, the Pd nanowires with the diameter less than 3 nm and the length up to 300 nm were prepared by subsequent H_2-reduction. In contrast, one-step H_2-reduction gave Pd nanoparticles in the mesoporous structure of matrices. Platinum nanowires contained in the MCM-41 channels were reported and unsupported Pt nanowires were also obtained by removing the silica framework from the Pt/MCM-41 samples in the literature [43]. According to TEM observations, most of the Pt clusters were synthesized in the form of nanowires, which followed the bending of the channels, although some Pt particles were observed on the external surfaces of the MCM-41. After removing of the silica matrix, pure Pt nanowires with diameter of 3.0 nm were obtained, which was corresponding to the mesochannels of the MCM-41. The Pt nanowires have good crystallinity, each can be regarded as a single crystal, and the wires have fairly smooth surfaces with specific indices.

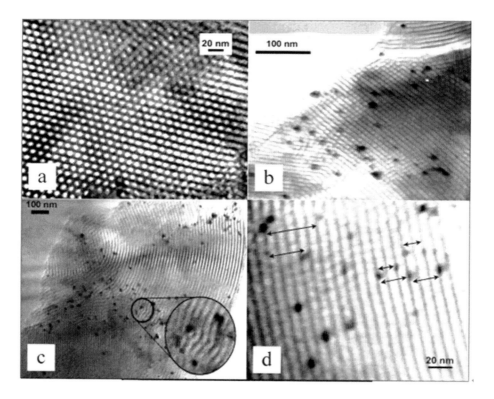

Figure 7. TEM pictures of SBA-15 mesoporous silica containing 10 nm gold nanoparticles (Reprinted with permission from Zhu, J.; Kónya, Z.; Puntes, V. F.; Kiricsi, I.; Miao, C. X.; Ager, J. W.; Alivisatos, A. P.; Somorjai, G. A. *Langmuir* 2003, *19*, 4396. Copyright (2003) American Chemical Society).

The application of metal-deposited mesoporous materials are mainly focused on the catalytic area. Juan [120] reported a novel catalyst Au/SBA-16 with highly dispersed Au nanoparticles (2-3 nm) in the nanocage of SBA-16. The developed catalyst is active in the direct oxidation esterification of alcohols, even for straight chain alcohols. The highest activity was observed for the Au loading of 5 wt%, which did not decrease significantly throughout the consecutive eight reaction cycles. A functionalized mesoporous aluminosilicate supporting Au-Ag bimetallic nanocatalyst Au-Ag@APTS-MCM (APTS, $H_2N(CH_2)_3$-Si(OMe)$_3$) was prepared by a two-step synthesis procedure [121]. The catalysts exhibited higher activity in catalysis for low-temperature CO oxidation, up to 100% for the sample with Au/Ag=8/1. Furthermore, the catalytic properties of the products showed high stability, remaining 90% CO oxidation after 1 year of storage under room conditions. Wang [122] reported a catalyst made of Pd nanoparticles supported on a mesoporous graphitic carbon nitride, Pd@mpg-C_3N_4, which displayed highly activity and promoted selective formation of cyclohexanone under atmospheric pressure of hydrogen in aqueous media without additives. The catalytic properties of this composite could be improved at higher temperature, but 99% conversion and 96% selectivity could still be obtained even at room temperature.

Other applications for metal-deposited mesoporous materials are also attractive. Gu [123] reported a facile strategy to synthesize highly dispersed Au nanoparticles within the pore channels of mesoporous thin films (MTFs). The prepared composite films demonstrated an ultrafast femtosecond-scale nonlinear optical (NLO) response and enhanced third-order NLO susceptibility due to high content, uniform size, and high dispersion of the incorporated gold nanoparticles and local-field effects. Low-density ordered mesoporous carbon-silica nanocomposites with various Fe content was prepared by a facile solvent-evaporation-induced self-assembly approach, followed by carbothermal reduction [42]. The Fe-containing nanocomposites exhibited dual-frequency absorption characteristics and the bandwidth lower than -10 dB was found to exceed 5.0 GHz for an absorber thickness of 2 mm. The composites also presented a lower infrared emissivity in the wavelength from 8 to 14 μm than that of Fe-free powder. Luminescence of silver nanoparticles photodeposited on titan dioxide nanoparticles of mesoporous film was studied [45]. The results indicated that the Ag/TiO_2 mesoporous films have high concentration of bright luminescence spots revealing stability to degradation under long illumination, which suggests potential application in single molecule spectroscopy and biological objects visualization.

4. BINARY COMPOUNDS EMBEDDED BY MESOPOROUS MATEIALS

The constrained growth of binary compound, especially metal oxides and sulfides within the nanoscaled channels of mesoporous materials have gained extensively attention due to their wide application in various fields. The introduction of metal oxides and sulfides into mesoporous materials can endow regular nanoparticles with a well-controlled particle size and distribution, which would facilitate the enhanced properties of the final composites.

4.1. Metal Oxides

Metal oxides loaded on mesoporous materials is mainly based on impregnating mesoporous matrices in the metal salt solution, and transforming to metal oxides after subsequent calcinations. Uniform and highly dispersed γ-Fe$_2$O$_3$ nanoparticles loaded on C/SBA-15 and CMK-5 carbons composites were successfully synthesized by conversional impregnation and thermal treatment [124]. γ-Fe$_2$O$_3$/CMK-5 composites were obtained from γ-Fe$_2$O$_3$/CSBA-15 and subsequent calcination under argon atmosphere to remove the thin carbon layer inside of the channels of SBA-15. This kind of catalyst showed complete ammonia conversion at 600°C at a space velocity of 7500 cm^{3}g$_{cat}^{-1}$h^{-1}, which conversion maintained for 16 h without any deactivation. Mesoporous composite ZnO@MCM-48 was prepared through solution, liquid and gas-phase infiltration under a non aqueous condition using diethyl zinc as organometallic precursor [125]. It was found that the gas phase loading of MCM-48 with ZnEt$_2$ at reduced pressure was the most effective way to highly loaded MCM-48. A new CO$_2$ sorbents CaO/C composites were synthesized with nanocrystalline calcium oxides confined in ordered mesoporous carbon [126]. The catalysts showed excellent performance for CO$_2$ physisorption with very competitive capacities (up to 7 mmol·g^{-1}), which was highly dependent on the surface area, temperature and pressure. Furthermore, the CaO/C composites displayed not only good adsorption selectivity for CO$_2$ over N$_2$, but also presented highly active for CO$_2$ chemisorption, with high initial conversion (~100%) and fast reaction-based kinetics at a low temperature of ~450°C.

Multi-component material is a kind of potential candidate in the field of catalyst design. With the aim of creating enhanced and multi-functional composites, mixed oxides were loaded on mesoporous materials. Liu [127] reported a series of catalysts concerning dual metal oxides co-supported on mesoporous materials La$_2$O$_3$-V$_2$O$_5$/MCM-41, which prepared by an incipient wetness impregnation procedure. The catalytic properties for the dehydrogenation of ethylbenzene to styrene were influenced by the La/V atomic ratio. At 600°C, the conversion of ethylbenzene and selectivity to styrene using CO$_2$ as oxidant over 10La15V/MCM-41 after an on-stream time of 4h was about 86.5% and 91.0%, respectively. The addition of lanthanum oxide could promote the removal of surface carbon resulted by catalyst deactivation and hence brought stability of catalyst. Mixed metal oxides of CuO and Nb$_2$O$_5$ supported on MCM-41 were prepared by wet impregnation route with a 1:1 mass ratio (CuO:Nb$_2$O$_3$) and a wide mixed oxides concentration range (2-25wt%) [128]. The mixed oxides formed well-dispersed small crystallites and an amorphous phase with lower loading weight, and displayed bigger particles with the higher oxide loadings. The catalytic properties of the composites were tested in the oxidation of diesel soot particulates. The most attractive catalytic property over CuO/Nb$_2$O$_5$/MCM-41 with 25 wt% loadings presented an onset temperature of 388°C and excellent activity at temperatures as low as 450°C without the addition of any chemical promoter.

4.2. Sulfides

Besides high dispersion and particle size confinement, there exists another two favorable factors for incorporating of sulfides into the mesoporous materials: (1) facilitating the separation of electrons and holes [129]; (2) preventing sulfides from undergoing photocorrosion due to less exposed surface area [130].

CdS supported cubic MCM-48 mesoporous composites were prepared by a post-impregnation method for the photocatalytic evolution of hydrogen from water [131]. All the samples displayed photocatalytic activity under visible light (λ>400 nm) for production of hydrogen from splitting of water in the absence of any activators, such as Pt. The highest hydrogen evolution rate was 1.81 mmol·h^{-1}·g$_{CdS}^{-1}$ and the apparent quantum yield was estimated to be 16.6%. Karakassides [58] reported the synthesis of quantum dots ZnS through a metal exchange reaction route inside the mesopores of MCM-41 and subsequent reaction with H$_2$S. The final composite has both porous and semiconducting properties, showing the low-dimensional character of the ZnS semiconductor particles. Semiconductor PbS nanocrystals and nanowires with uniformity were synthesized inside the channels of SBA-15 by a new flexible approach [59]. The key of the morphology transformation for PbS from uniform nanocrystals (5 nm) to nanowires (6 nm in diameter and several hundred nanometers in length) was to increase the amount of the Si-OH group on the channel surface of SBA-15, which was available by ethanol extraction method for removing the block copolymer template. The photoluminescence spectra demonstrated a massive blue shift for the PbS, which is obviously related to the quantum size effects of the PbS nanoparticls and nanowires.

4.3. Other Binary Compounds

It is demonstrated that transition-metal silicides are conceivably stable and sulfur-resistant in the catalytic reaction with sulfur-containing compounds. Cobalt silicide nanoparticles supported on SBA-15 were synthesized by metal-organic chemical vapor deposition of a single-source precursor Co(SiCl)$_3$(CO)$_4$ at atmosphere pressure and moderate temperature [132]. It was proved that the CoSi with particle size of 2-4 nm was highly dispersed in the mesochannels of SBA-15. The loading amount was massively dependant on the amount of precursor and the hydroxyl groups in the mesoporous silica. The sample CoSi-SBA-15 with the CoSi/SBA-15 ratio of 8.3% displayed a lower conversion and higher selectivity to tetralin (about 100%) in the naphthalene hydrogenation, while the 16.1% CoSi-SBA-15 showed a higher catalytic activity up to 87% and a lower selectivity to tetralin (35%).

Incorporating semiconductor in mesoporous materials has been a general concept to create novel mesoscopic systems. CdSe is a quite essential semiconductor due to its potential in the fabrication of optoelectronic devices such as light-emitting diodes and quantum dot lasers [62]. Liu prepared CdSe nanocrystals incorporated into mesoporous silica [133]. A blue shift was observed in photoluminescence emission accompanied by a broadened linewidth, which was attributed to a combination of phonon confinement and compressive strain effects. Gallium nitride confined in mesoporous silica MCM-41 was prepared by Fischer [63]. Similar to CdSe, a blue shift was also observed in luminescence spectra at 290 K for the mesoporous composite GaN/MCM-41.

5. TERNARY COMPOUNDS AND OTHERS EMBEDDED BY MESOPOROUS MATEIALS

5.1. Ternary Compounds

Different from metals or metal oxides, two or three different types of inorganic species should be involved in synthesis of binary or ternary compounds embedded by mesoporous materials. The difficulty in introducing multi-component species into mesochannels of mesoporous materials makes this kind of composites very rare.

Metal phosphates have great potential in applications of catalysis, separation and proton conduction. Zhang [64] adopted two approaches to graft liquid-phase titanium phosphate onto mesoporous silica SBA-15. These two methods referred: (1) alternate grafting of Ti(OPr^i)$_4$ and then POCl$_3$ (Figure 8); (2) one-pot grafting of titanium phosphate formed in situ by using Ti(OPr^i)$_4$ (a base) and POCl$_3$ (an acid) as an acceptable "acid-base pair" (Figure 9). The content of titanium phosphate could be adjusted by repeating the "layer-by-layer" technique. The obtained composites Ti$_3$(PO$_4$)$_4$/SBA-15 showed higher catalytic activity in isopropanol dehydration and cumene cracking as well as better ability in metal ion adsorption than pure mesoporous material and titanium phosphate.

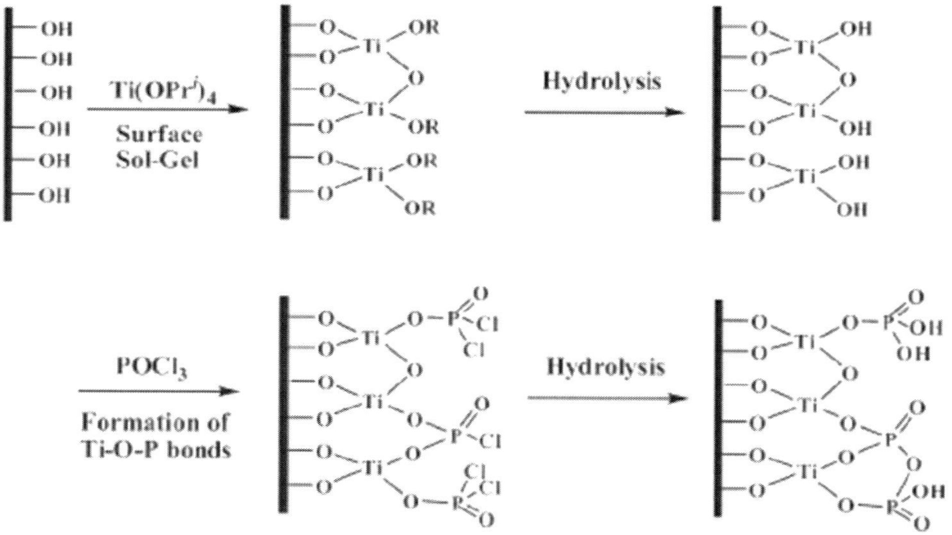

Figure 8. Modification of SBA-15 surfaces by alternate grafting with Ti(OPr^i)$_4$ and POCl$_3$ (Method (1)) (Reprinted with permission from Zhang, J.; Ma, Z.; Jiao, J.; Yin, H.; Yan, W.; Hagaman, E. W.; Yu, J.; Dai, S. *Langmuir* 2009, 25, 12541. Copyright (2009) American Chemical Society).

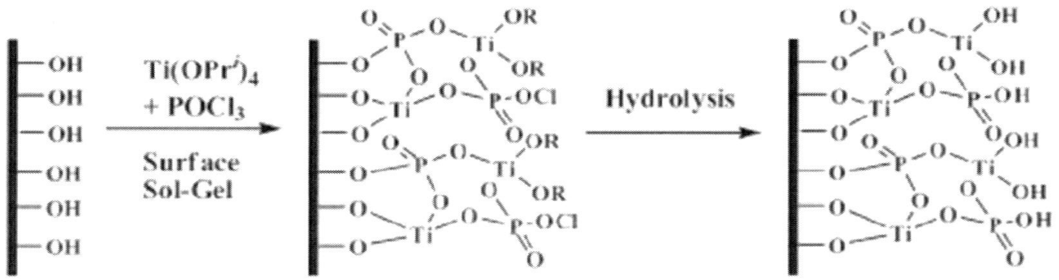

Figure 9. Modification of SBA-15 surfaces by one-pot grafting of titanium phosphate formed in situ (method (2)) (Reprinted with permission from Zhang, J.; Ma, Z.; Jiao, J.; Yin, H.; Yan, W.; Hagaman, E. W.; Yu, J.; Dai, S. *Langmuir* 2009, *25*, 12541. Copyright (2009) American Chemical Society).

Calcium tungstate (CaWO$_4$) with scheelite structure has been confirmed as a highly functional material and have found profound potential applications in various fields, such as photoluminescence, microwave applications, optical fibers, scintillator materials, and so on [134]. The luminescence properties of this classic phosphor could be further improved by doping with rare earth (RE) ions. The composites of terbium doped calcium tungstate loaded on mesoporous matrix SBA-15 with various regular morphologies were synthesized by a Pechini-type process [135]. These composites displayed strong green emission under UV lamp irradiation (254 nm) and showed various lysozyme adsorption capacities in buffer solution with different pH values. The adsorption behavior was highly influenced by the morphology, particle size, mesopore structure of the adsorbent, and the pH value of the buffer solution. The sample CaWO$_4$:Tb^{3+}/SBA-15 (morphology of rod-like) represented a maximum equilibrium adsorption amount at solution pH 7.0 and 10.0. The composites of europium doped calcium tungstate supported on mesoporous material SBA-15 (CaWO$_4$:Eu^{3+}/SBA-15) were also prepared [65]. The photoluminescence spectra indicated that the resultant CaWO$_4$:Eu^{3+}/SBA-15 exhibited enhanced-luminescent properties (nearly 2 times) than that of pure CaWO$_4$:Eu^{3+} nanoparticles, which was mainly related to the dopant concentration of Eu^{3+}, passivation of OH groups and defects on the surface of composites through the interaction of calcium tungstate and mesoporous matrix.

Luminescent GdVO$_4$:Eu^{3+} functionalized mesoporous silica nanoparticles (GdVO$_4$:Eu^{3+}@MSN) with diameters in the range of 80-120 nm were synthesized by a Pechini sol-gel approach [66]. It was demonstrated by a series of tests that the samples had favorable biocompatibility and could be effectively taken up by SKOV3 ovarian cancer cells and A549 lung adenocarcinoma cells. Furthermore, the composites showed a red emission under UV irradiation, and the photoluminescence intensity was closely relating to the pH-dependant cumulative release of doxorubicin. It was also indicated that the composites could act as T$_1$ contrast agents for magnetic resonance imaging. Thus, all the results suggested that the composites GdVO$_4$:Eu^{3+}@MSN can potentially act as a multifunctional drug carrier system with luminescent tagging, magnetic resonance imaging and pH-controlled release property for doxorubicin.

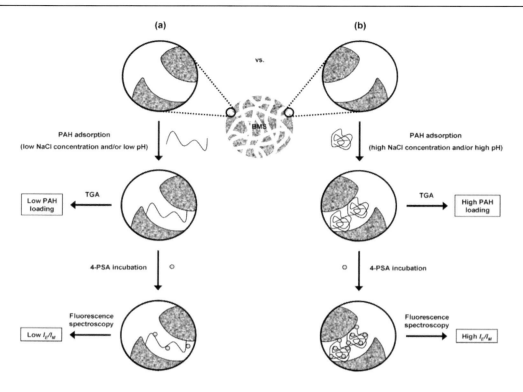

Figure 10. Schematic illustration of two extreme cases: (a) low NaCl concentration and/or low pH during PAH adsorption, resulting in limited infiltration of more linear PAH molecules into the nanopores, and hence a relatively low PAH loading and low I_E/I_M; and (b) high NaCl concentration and/or high pH during PAH adsorption, resulting in significant infiltration of more coiled PAH molecules into the nanopores, and hence a relatively high PAH loading and high I_E/I_M (Reprinted with permission from Angelatos, A. S.; Wang, Y.; Caruso, F. *Langmuir* 2008, *24*, 4224. Copyright (2008) American Chemical Society).

5.2. Others

Except for various species mentioned above embedded by mesoporous materials, polymers or macromolecules should also be referred due to their derivatives showed non-negligible applications in luminescence, sensing, catalysis and adsorption.

In order to follow the trail of conformation process of a macromolecule, poly(allylamine hydrochloride) (PAH) in the bimodal mesoporous silica (BMS) particles, Angelatos [67] reported a fluorescence-base approach involving monitoring the fluorescent properties of the probe, 1,3,6,8-pyrenetetrasulfonic acid tetrasodium salt (4-PSA) through electrostatic binding to PAH molecules adsorbed in the nanopores of the BMS particles. The ratio of the excimer to monomer emission intensity (I_E/I_M) was adopted to discern differences in the PAH conformation in the BMS particles. There were many parameters influencing the PAH conformation and the greatest two factors were NaCl concentration and pH value, which were shown in Figure 10. This study shed light on macromolecule infiltration and conformation in mesoporous materials, which are essential for the application of mesoporous matrix in various fields, such as sensing, separation, catalysis and adsorption/immobilization.

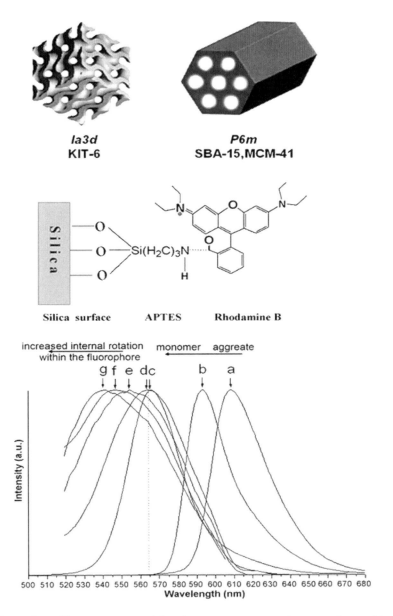

Figure 11. The illustration of host-guest material and pore structure are shown on the upside. Photoluminescence spectra of liquid Rh B in ethanol with the concentration of (a) 1×10^{-3} mol L^{-1}, (b) 1×10^{-4} mol L^{-1}, (c) 1×10^{-6} mol L^{-1}, solid PL of Rh B in (d) none porous silica, (e) SBA-15, (f) MCM-41, (g) KIT-6 are shown on the underside (Reprinted from *Mater. Chem. Phys. 118*, Tu, J.; Li, N.; Chi, Y.; Qu, S.; Wang, C.; Yuan, Q.; Li, X.; Qiu, S., The study of photoluminescence properties of rhodamine B encapsulated in mesoporous silica, p 273, Copyright (2009), with permission from Elsevier).

A Chelating polymer carboxymethylated polyethyleneimine (CMPEI) was functionalized on mesoporous silica (MCM-48) for the purpose of adsorbing Eu^{3+} ions, which is a representative trivalent lanthanide and especially used as a surrogate for Am^{3+} due to its similar chemical behaviors to those of Am^{3+} [68]. The results demonstrated that the CMPEI/MCM-48 showed much higher adsorption capacity at pH 3.0~5.0 compared to that of unfunctionalized MCM-48, which indicated the potential application of this material in the recovering of trivalent actinide species.

Rhodamine B contained by various mesoporous materials were prepared by postgrafting method [136]. It was found that the composites showed significant blue-shift in photoluminescence spectra. Furthermore, the extent of shift was influenced by the mesoporous silica kinds, which was attributed to the topological structure of mesopores (Figure 11).

SUMMARY AND OUTLOOK

Mesoporous materials are well known for their spread applications in catalysis, adsorption, separation, ion exchange and drug delivery. Drawbacks such as low acidity, poor thermal or hydrothermal stability inhibit the further development of mesoporous materials. The combination of various functional species and mesoporous matrices induces a thrilling beginning for exploring of new materials and unexpected functions. The appreciable properties of the composites are so amazing because they are not only the simple combination of respective properties of functional species and mesoporous hosts, but the products of constriction effect of mesoprous structure and cooperative effect between active functional groups and hosts. Various species, including metals, metal complexes, metal oxides, sulfides, phosphates, tungstates and others has been attached, incorporated, or contained in various mesoporous materials. It could be expected that more and more advanced functional composites based on mesoporous materials would emerge and be applied in various fields in human society.

REFERENCES

[1] Kresge, C. T.; Leonowicz, M. E.; Roth, W. J.; Vartuli, J. C.; Beck, J. S. *Nature* 1992, *359*, 710.
[2] Beck, J. S.; Vartuli, J. C.; Roth, W. J.; Lionowicz, M. E.; Kresge, C. T.; Schmitt, K. D.; Chu, C. T. W.; Olson, D. H.; Sheppard, E. W.; McCullen, S. B.; Higgins, J. B.; Schlenker, J. L. *J. Am. Chem. Soc.* 1992, *114*, 10834.
[3] Liu, Y.; Zhao, G.; Liu, G.; Wu, S.; Chen, G.; Zhang, W.; Sun, H.; Jia, M. *Catal. Commun.* 2008, *9*, 2022.
[4] Zornoza, B.; Irusta, S.; Téllez, C.; Coronas, J. *Langmuir* 2009, *25*, 5903.
[5] Acatrinei, A. I.; Hartl, M. A.; Eckert, J.; Falcao, E. H. L.; Chertkov, G.; Daemen, L. L. *J. Phys. Chem. C* 2009, *113*, 15634.
[6] Mokaya, M. *Adv. Mater.* 2000, *12*, 1681.
[7] Wada, A.; Tamaru, S.; Ikeda, M.; Hamachi, I. *J. Am. Chem. Soc.* 2009, *131*, 5321.
[8] Wang, S. *Micropor. Mesopor. Mater.* 2009, *117*, 1.
[9] Descalzo, A. B.; Jimenez, D.; Marcos, M. D.; Martínez-Máñez, R.; Soto, J.; Haskouri, J. E.; Guillém, C.; Beltrán, D.; Amorós, P.; Borrachero, V. *Adv. Mater.* 2002, *14*, 966.
[10] Calvo, A.; Joselevich, M.; Soler-Illia, G. J. A. A.; Williams, F. J. *Micropor. Mesopor. Mater.* 2009, *121*, 67.

[11] Radu, D. R.; Lai, C. Y.; Huang, J.; Shu, X.; Lin, V. S.-Y. *Chem. Commun.* 2005, *10*, 1264.
[12] Gago, S.; Fonseca, I. M.; Parola, A. J. *Micropor. Mesopor. Mater.* 2013, *180*, 40.
[13] Rodríguez-López, G.; Marcos, M. D.; Martínez-Máñez, R.; Sancenón, F.; Soto, J.; Villaescusa, L. A.; Beltrán, D.; Amorós, P. *Chem. Commun.* 2004, *19*, 2198.
[14] Walcarius, A.; Delacôte, C. *Chem. Mater.* 2003, *15*, 4181.
[15] Duarte, A. P.; Gressier, M.; Menu, M.-J.; Dexpert-Ghys, J.; Caiut, J. M. A.; Ribeiro, S. J. L. *J. Phys. Chem. C* 2012, *116*, 505.
[16] Peng, C.; Zhang, H.; Yu, J.; Meng, Q.; Fu, L.; Li, H.; Sun, L.; Guo, X. *J. Phys. Chem. B* 2005, *109*, 15278.
[17] Gago, S.; Fernandes, J. A.; Rainho, J. P.; Ferreira, R. A. S.; Pillinger, M.; Valente, A. A.; Santos, T. M.; Carlos, L. D.; Ribeiro-Claro, P. J.A.; Goncalves, I. S. *Chem. Mater.* 2005, *17*, 5077.
[18] Xu, Q.; Li, L.; Liu, X.; Xu, R. *Chem. Mater.* 2002, *14*, 549.
[19] Mori, K.; Watanabe, K.; Kawashima, M.; Che, M.; Yamashita, H. *J. Phys. Chem. C* 2011, *115*, 1044.
[20] Feng, K.; Zhang, R.-Y.; Wu, L.-Z.; Tu, B.; Peng, M.-L.; Zhang, L.-P.; Zhao, D.-Y.; Tung, C.-H. *J. Am. Chem. Soc.* 2006, *128*, 14685.
[21] Bhunia, S.; Saha, D.; Koner, S. *Langmuir* 2011, *27*, 15322.
[22] Fernandes, T. A.; Nunes, C. D.; Vaz, P.D.; Calhorda, M. J.; Brandão, P.; Rocha, J.; Goncalves, I. S.; Valente, A. A.; Ferreira, L. P.; Godinho, M.; Ferreira, P. *Micropor. Mesopor. Mater.* 2008, *112*, 14.
[23] Yoshitake, H.; Otsuka, R. *Langmuir* 2013, *29*, 10513.
[24] Frasconi, M.; Liu, Z.; Lei, J. *J. Am. Chem. Soc.* 2013, *135*, 11603
[25] Lei, B.; Li, B.; Zhang, H.; Zhang, L.; Li, W. *J. Phys. Chem. C* 2007, *111*, 11291.
[26] Luan, Z.; Xu, J.; Kevan, L. *Chem. Mater.* 1998, *10*, 3699
[27] Queiroz, T. B.; Botelho, M. B. S.; Fernández-Hernández, J. M.; Eckert, H.; Albuquerque, R. Q.; Camargo, A. S. S. *J. Phys. Chem. C* 2013, *117*, 2966.
[28] Mori, K.; Tottori, M.; Watanabe, K.; Che, M.; Yamashita, H. *J. Phys. Chem. C* 2011, *115*, 21358.
[29] Sun, L.-N.; Yu, J.-B.; Zhang, H.-J.; Meng, Q.-G.; Ma, E.; Peng, C.-Y.; Yang, K.-Y. *Micropor. Mesopor. Mater.* 2007, *98*, 156.
[30] Luan, Z.; Hartmann, M.; Zhao, D.; Zhou, W.; Keyan, L. *Chem. Mater.* 1999, *11*, 1621.
[31] Shen, S.-C.; Kawi, S. *J. Phys. Chem. B* 1999, *103*, 8870.
[32] Russo, P. A.; Carrott, M. M. L.R.; Carrott, P. J. M.; Lopes, J. M.; Ribeiro, F. R.; Rocha, J. *Micropor. Mesopor. Mater.* 2008, *114*, 293.
[33] Hu, Y.; Martra, G.; Zhang J.; Higashimoto, S.; Coluccia, S.; Anpo, M. *J. Phys. Chem. B* 2006, *110*, 1680.
[34] Selvaraj, M.; Park, D. W. *Appl. Cata. A: Gen.* 2010, *388*, 22.
[35] Liu, Y.-M.; Xu, J.; He, L.; Cao, Y.; He, H.-Y.; Zhao, D.-Y.; Zhuang, J.-H.; Fan, K.-N. *J. Phys. Chem. C* 2008, *112*, 16575.
[36] Samanta, S.; Giri, S.; Sastry, P. U.; Mal, N. K.; Manna, A.; Bhaumik, A. *Ind. Eng. Chem. Res.* 2003, *42*, 3012.
[37] Occelli, M. L.; Biz, S.; Auroux, A. *Appl. Cata. A: Gen.* 1999, *183*, 231.
[38] Zhu, Z.; Chang, Z.; Kevan, L. *J. Phys. Chem. B* 1999, *103*, 2680.

[39] Lezanska, M.; Szymanski, G. S.; Pietrzyk, P.; Sojka, Z.; Lercher, J. A. *J. Phys. Chem. C* 2007, *111*, 1830.
[40] Zhu, J.; Kónya, Z.; Puntes, V. F.; Kiricsi, I.; Miao, C. X.; Ager, J. W.; Alivisatos, A. P.; Somorjai, G. A. *Langmuir* 2003, *19*, 4396.
[41] Fukuoka, A.; Araki, H.; Sakamoto, Y.; Inagaki, S.; Fukushima, Y.; Ichikawa, M. *Inorg. Chim. Acta* 2003, *350*, 371.
[42] Zhou, J.; He, J.; Li, G.; Wang, T.; Sun, D.; Ding, X.; Zhao, J.; Wu, S. *J. Phys. Chem. C* 2010, *114*, 7611.
[43] Liu, Z.; Sakamoto, Y.; Ohsuna, T.; Kenji, H.; Terasaki, O.; Ko, C. H.; Shin, H. J.; Ryoo, R. *Angew. Chem. Int. Ed.* 2000, 39, 3107.
[44] Wojtaszek, A.; Sobczak, I.; Ziolek, M.; Tielens, F. *J. Phys. Chem. C* 2010, *114*, 9002.
[45] Aiboushev, A. V.; Astafiev, A. A.; Lozovik, Y. E.; Merkulova, S. P.; Nadtochenko, V. A.; Sarkisov, O. M. *Phys. Lett. A* 2008, *372*, 5193.
[46] Parlett, C. M. A.; Durndell, L. J.; Machado, A. *Catal. Today* 2014, *229*, 46.
[47] Fröba, M.; Köhn, R.; Bouffaud, G. *Chem. Mater.* 1999, *11*, 2858.
[48] Bouvy, C.; Chelnokov, E.; Zhao, R.; Marine, W.; Sporken, R.; Su, B.-L. *Nanotechnology* 2008, *19*, 105710.
[49] Zhang, S.; Jiang, D.; Tang, T.; Li, J.; Xu, Y.; Shen, W.; Xu, J.; Deng, F. *Catal. Today* 2010, *158*, 329.
[50] Zhao, S.; Su, D.; Che, J.; Jiang, B.; Orlov, A. *Mater. Lett.* 2011, *65*, 3354.
[51] Pillewan, P.; Mukherjee, S.; Roychowdhury, T.; Das, S.; Bansiwal, A.; Ravalu, S. *J. Hazard. Mater.* 2011, *186*, 367.
[52] Yu, H.; Jiang, D.-M.; Zhai, Q.-Z.; Hu, W.-H.; Fan, L. *J. Lumin.* 2012, *132*, 474.
[53] Srinivasan, N. R.; Bandyopadhyaya, R. *Micropor. Mesopor. Mater.* 2012, *149*, 166.
[54] Du, C.; Yang, H. *RSC Adv.* 2013, *3*, 13990.
[55] Sellam, D.; Bonne, M.; Arrii-Clacens, S.; Lafaye, G.; Bion, N.; Tezkratt, S.; Royer, S.; Marécot, P.; Duprez, D. *Catal. Today* 2010, *157*, 131.
[56] Caponetti, E.; Pedone, L.; Saladino, M. L.; Martino, D. C.; Nasillo, G. *Micropor. Mesopor. Mater.* 2010, *128*, *101*.
[57] Wang, S.; Choi, D.-G.; Yang, S.-M. *Adv. Mater.* 2002, *14*, 1311.
[58] Dimos, K.; Koutselas, I. B.; Karakassides, M. A. *J. Phys. Chem. B* 2006, *110*, 22339.
[59] Gao, F.; Lu, Q.; Liu, X.; Yan, Y.; Zhao, D. *Nano Lett.* 2001, *1*, 743.
[60] Barry, L.; Copley, M.; Holmes, J. D.; Otway, D. *J. Solid State Chem.* 2007, *180*, 3443.
[61] Brühwiler, D.; Seifert, R.; Calzaferri, G. *J. Phys. Chem. B* 1999, *103*, 6397.
[62] Parala, H.; Winkler, H.; Kolbe, M.; Wohlfart, A.; Fischer, R. A.; Schmechel, R.; Seggern, H.; *Adv. Mater.* 2000, *12*, 1050.
[63] Winkler, H.; Brikner, A.; Hagen, V.; Wolf, I.; Schmechel, R.; Seggen, H.; Fischer, R. A.; *Adv. Mater.* 1999, *11*, 1444.
[64] Zhang, J.; Ma, Z.; Jiao, J.; Yin, H.; Yan, W.; Hagaman, E. W.; Yu, J.; Dai, S. *Langmuir* 2009, *25*, 12541.
[65] Du, C.; Yi, G.; Su, Y.; Liu, Z. *J. Mater. Sci.* 2012, *47*, 6305.
[66] Huang, S.; Cheng, Z.; Ma, P. a.; Kang, X.; Dai, Y.; Lin, J. *Dalton Trans.* 2013, *42*, 6523.
[67] Angelatos, A. S.; Wang, Y.; Caruso, F. *Langmuir* 2008, *24*, 4224.
[68] Yuu, M.-H.; Yeou, J.-W.; Kim, J. H.; Lee, H. I.; Kim, J. M.; Kim, S.; Jung, Y. *Macromol. Res.* 2011, *19*, 421.

[69] Liu, Y.; Zhao, G.; Liu, G.; Wu, S.; Chen, G.; Zhang, W.; Sun, H.; Jia, M. *Catal. Commun.* 2008, *9*, 2022.
[70] Mal, N. K.; Fujiwara, M.; Tanaka, Y.; Taguchi, T.; Matsukata, M. *Chem. Mater.* 2003, *15*, 3385.
[71] Yoshitake, H.; Yokoi, T.; Tatsumi, T. *Chem. Mater.* 2002, *14*, 4603.
[72] Fu, Q.; Rao, G. V. R. ; Ista, L. K.; Wu, Y.; Andrzejewski, B. P.; Sklar, L. A.; Ward, T. L.; López, G. P. *Adv. Mater.* 2003, *15*, 1262.
[73] Li, Y.; Yan, B.; Liu, J.-L. *Nanoscale Res. Lett.* 2010, *5*, 797.
[74] Hsu, Y.-T.; Chen, W.-L.; Yang, C.-M. *J. Phys. Chem. C* 2009, *113*, 2777.
[75] Hoffmann, F.; Cornelius, M.; Morell, J.; Fröba, M. *Angew. Chem. Int. Ed.* 2006, *45*, 3216.
[76] Melero, J. A.; Grieken, R. van; Morales, G. *Chem. Rev.* 2006, *106*, 3790.
[77] Brunelli, N. A.; Venkatasubbaiah, K.; Jones, C. W. *Chem. Mater.* 2012, *24*, 2433.
[78] Mukhopadhyay, K.; Sarkar, B. R.; Chaudhari, R. V. *J. Am. Chem. Soc.* 2002, *124*, 9692.
[79] Jana, S.; Dutta, B.; Bera, R.; Koner, S. *Langmuir* 2007, *23*, 2492.
[80] Li, S.; Song, H.; Li, W.; Ren, X.; Lu, S.; Pan, G.; Fan, L.; Yu, H.; Zhang, H.; Qin, R.; Dai; Wang, T. *J. Phys. Chem. B* 2006, *110*, 23164.
[81] Li, Y.; Yan, B.; Li, Y. *J. Solid State Chem.* 2010, *183*, 871.
[82] Zuo, Q.; Li, B.; Zhang, L.; Wang, Y.; Liu, Y.; Zhang, J.; Chen, Y.; Guo, L. *J. Solid State Chem.* 2010, *183*, 1715.
[83] Xia, Y.; Mokaya, R. *J. Phys. Chem. B* 2003, *107*, 6954.
[84] Wu, Z. Y.; Wang, H. J.; Zhuang, T. T.; Sun, L. B.; Wang, Y. M.; Zhu, J. H. *Adv. Funct. Mater.* 2008, *18*, 82.
[85] Bérubé, F. O.; Khadhraoui, A.; Janicke, M. T.; Kleitz, F.; Kaliaguine, S. *Ind. Eng. Chem. Res.* 2010, *49*, 6977.
[86] Zhao, L.; Dong, Y.; Zhan, X. *Catal. Lett.* 2012, *142*, 619.
[87] Vinu, A.; Sawant, D. P.; Ariga, K.; Hossain, K. Z.; Halligudi, S. B.; Hartmann, M.; Nomura, M. *Chem. Mater.* 2005, *17*, 5339.
[88] Guo, Y.; Zhengwang, L.; Guangjian, W.; Huang, Y.; Kang, F. *Appl. Surf. Sci.* 2011, *258*, 1082.
[89] Selvaraj, M.; Pandurangan, A.; Sinha, P. K. *Ind. Eng. Chem. Res.* 2004, *43*, 2399.
[90] Seredych, M.; Bandosz, T. *Langmuir* 2007, *23*, 6033.
[91] Gu, Z.; Deng, B.; Yang, J. *Micropor. Mesopor. Mater.* 2007, *102*, 265.
[92] Tiseanu, C.; Kumke, M. U.; Parvulescu, V. I.; Martens, J. *J. Appl. Phys.* 2009, *105*, 063521-1.
[93] Gavallaro, G.; Pierro, P.; Palumbo, F. S.; Testa, F.; Pasqua, L.; Aiello, R. *Drug Deliv.* 2004, *11*, 41.
[94] Park, J.-Y.; Suh, M.; Kwon, Y.-U. *Micropor. Mesopor. Mater.* 2010, *127*, 147.
[95] Sutra, P.; Brunel, D. *Chem. Commun.* 1996, *21*, 2485.
[96] Cabrero-Antonino, J. R.; García, T.; Rubio-Marqués, P.; Vidal-Moya, J. A.; Leyva-Pérez, A.; Al-Deyab, S. S.; Al-Resayes, S. I.; Díaz, U.; Corma, A. *ACS Catal.* 2011, *1*, 147.
[97] Desá, G. F.; Malta, O. L.; Donegá, C. D.; Simas, A. M.; Longo, R. L.; Santa-Cruz, P. A. *Chem. Rev.* 2000, *196*, 165.
[98] Li, Y.; Yan, B. *J. Mater. Chem.* 2011, *21*, 8129.

[99] Wang, H.; Ma, Y.; Tian, H.; Tang, N.; Liu, W.; Wang, Q.; Tang, Y. *Dalton Trans.* 2010, *39*, 7485.
[100] Xu, M.; Wang, W.; Seiler, M.; Buchholz, A.; Hunger, M. *J. Phys. Chem. B* 2002, *106*, 3202.
[101] Martínez, M. L.; Gómez Costa, M. B.; Monti, G. A.; Anunziata, O. A. *Micropor. Mesopor. Mater.* 2011, *144*, 183.
[102] Huang, L.; Huang, Q.; Xiao, H.; Eic, M. *Micropor. Mesopor. Mater.* 2008, *111*, 404.
[103] Gallo, J. M.; Bisio, C.; Gatti, G.; Marchese, L.; Pastore, H. O. *Langmuir* 2010, *26*, 5791.
[104] Wang, S.; Shi, Y.; Ma, X.; Gong, J. *ACS Appl. Mater. Interfaces* 2011, *3*, 2154.
[105] Peng, R.; Zhao, D.; Dimitrijevic, N. M.; Rajh, T.; Koodali, R. T. *J. Phys. Chem. C* 2012, *116*, 1605.
[106] Ringenbach, C. R.; Livingston, S. R.; Kumar, D.; Landry, C. C. *Chem. Mater.* 2005, *17*, 5580.
[107] Aktas, O.; Yasyerli, S.; Dogu, G.; Dogu, T. *Ind. Eng. Chem. Res.* 2010, *49*, 6790.
[108] Kilicarslan, S.; Dogan, M.; Dogu, T. *Ind. Eng. Chem. Res.* 2013, *52*, 3674.
[109] Li, F.; Yu, F.; Li, Y.; Li, R.; Xie, K. *Micropor. Mesopor. Mater.* 2007, *101*, 250.
[110] Szczodrowski, K.; Prélot, B.; Lantenois, S.; Douillard, J.-M.; Zajac, J. *Micropor. Mesopor. Mater.* 2009, *124*, 84.
[111] Gokulakrishnan, N.; Pandurangan, A.; Sinha, P. K. *Ind. Eng. Chem. Res.* 2009, *48*, 1556.
[112] Kumar, A.; Srinivas, D. *Catal. Today* 2012, *198*, 59.
[113] Wang, N.; Shen, K.; Huang, L.; Yu, X.; Qian, W.; Chu, W. *ACS Catal.* 2013, *3*, 1638.
[114] Jana, S. K.; Takahashi, H.; Nakamura, M.; Kaneko, M.; Nishida, R.; Shimizu, H.; Kugita, T.; Namba, S. *Appl. Catal. A: Gen.* 2003, *245*, 33.
[115] Goswami, P.; Ganguli, J. N. *Dalton Trans.* 2013, *42*, 14480.
[116] Castro, Y.; Julián-López, B.; Boissière, C.; Viana, B.; Grosso, D.; Sanchez, C. *Micropor. Mesopor. Mater.* 2007, *103*, 273.
[117] Zhang, Z.; Dai, S.; Fan, X.; Blom, D. A.; Pennycook, S. J.; Wei, Y. *J. Phys. Chem. B* 2001, *105*, 6755.
[118] Shin, H. J.; Ryoo, R.; Liu, Z.; Terasaki, O. *J. Am. Chem. Soc.* 2001, *123*, 1246.
[119] Prashar, A. K.; Hodgkins, R. P.; Chandran, J. N.; Rajamohanan, P. R.; Devi, R. N. *Chem. Mater.* 2010, *22*, 1633.
[120] Juan, H.; Chong, Y.; Li, S.; Yang, H. *J. Phys. Chem. C* 2012, *116*, 6512.
[121] Yen, C.-W.; Lin, M.-L.; Wang, A.; Chen, S.-A.; Chen, J.-M.; Mou, C.-Y. *J. Phys. Chem. C* 2009, *113*, 17831.
[122] Wang, Y.; Yao, J.; Li, H.; Su, D.; Antonietti, M. *J. Am. Chem. Soc.* 2011, *133*, 2362.
[123] Gu, J.-L.; Shi, J.-L.; You, G.-J.; Xiong, L.-M.; Qian, S.-M.; Hua, Z.-L.; Chen, H.-R. *Adv. Mater.* 2005, 17, 557.
[124] Lu, A.-H.; Nitz, J.-J.; Comotti, M.; Weidenthaler, C.; Schlichte, K.; Lehmann, C. W.; Terasaki, O.; Schüth, F. *J. Am. Chem. Soc.* 2010, *132*, 14152.
[125] Schröder, F.; Hermes, S.; Parala, H.; Hikov, T.; Muhler, M.; Fischer, R. A. *J. Mater. Chem.* 2006, *16*, 3565.
[126] Wu, Z.; Hao, N.; Xiao, G.; Liu, L.; Webley, P.; Zhao, D. *Phys. Chem. Chem. Phys.* 2011, *13*, 2495.

[127] Liu, B. S.; Chang, R. Z.; Jiang, L.; Liu, W.; Au, C. T. *J. Phys. Chem. C* 2008, *112*, 15490.
[128] Garcia, F. A. C.; Silva, J. C. M.; de Macedo, J. L.; Dias, J. A.; Dias, S. C. L.; Filho, G. N. R. *Micropor. Mesopor. Mater.* 2008, *113*, 562.
[129] Ryu, S. Y.; Balcerski, W.; Lee, T. K.; Hoffmann, M. R. *J. Phys. Chem. C* 2007, *111*, 18195.
[130] Fox, M. A.; Pettit, T. L. *Langmuir* 1989, *5*, 1056.
[131] Peng, R.; Zhao, D.; Baltrusaitis, J.; Wu, C.-M.; Koodali, R. T. *RSC Adv.* 2012, *2*, 5754.
[132] Zhao, A.; Zhang, X.; Chen, X.; Guan, J.; Liang, C. *J. Phys. Chem. C* 2010, *114*, 3962.
[133] Chen, S.-F.; Liu, C.-P.; Eliseev, A. A.; Petukhov, D. I.; Dhara, S. *Appl. Phys. Lett.* 2010, *96*, 111907.
[134] Dai, Q.; Song, H.; Bai, X.; Pan, G.; Lu, S.; Wang, T.; Ren, X.; Zhao, H. *J. Phys. Chem. C* 2007, *111*, 7586.
[135] Huang, S.; Li, C.; Yang, P.; Zhang, C.; Cheng, Z.; Fan, Y.; Lin, J. *Eur. J. Inorg. Chem.* 2010, *2010*, 2655.
[136] Tu, J.; Li, N.; Chi, Y.; Qu, S.; Wang, C.; Yuan, Q.; Li, X.; Qiu, S. *Mater. Chem. Phys.* 2009, *118*, 273.

In: Comprehensive Guide for Mesoporous Materials. Volume 3 ISBN: 978-1-63463-318-5
Editor: Mahmood Aliofkhazraei © 2015 Nova Science Publishers, Inc.

Chapter 13

ON THE CONCEPTION AND ASSESSMENT OF MESOPORE NETWORKS: DEVELOPMENT OF COMPUTER ALGORITHMS

Fernando Rojas-González[1],[*] *Graciela Román-Alonso*[2],
Salomón Cordero-Sánchez[1], *Miguel Alfonso Castro-García*[2],
Manuel Aguilar-Cornejo[2] *and Jorge Matadamas Hernández*[2]

[1]Universidad Autonóma Metropolitana, Unidad Iztapalapa,
Departamento de Química, México D. F., Mexico
[2]Universidad Autonóma Metropolitana, Unidad Iztapalapa,
Departamento de Ingeniería Eléctrica, México D. F., Mexico

PACS 61.43.Gt, 07.05.Tp, 61.43.Bn
Keywords: Computer modeling and simulation, disordered solids, Monte Carlo Methods, parallel computing, parallel greedy algorithms, structure of porous materials

1. Introduction

There exist a great variety of phenomena taking place in porous solids that are strongly affected by the morphological and topological characteristics of these media, among these processes we can mention: (a) the immiscible displacement of a given fluid by another, (b) imbibition and drying processes, (c) separation of fluid mixtures, (d) heterogeneous catalysis, and (e) catalytic deactivation, etc. [1, 2, 3, 4, 5, 6]. The characterization of mesoporous and macroporous materials, especially the issue regarding the determination of the pore size distribution of these substrates from experimental data, is a subject of great practical importance that involves the development of both theoretical and experimental methods. Some of the experimental techniques, as for instance NMR, SAXS, and SANS, require sophisticated instruments while some others such as Hg porosimetry and sorption of vapors require simple devices that are available to many laboratories. In order to understand the textural results provided by these methods, a crucial issue consists in the development of

---
[*] E-mail: fernando@xanum.uam.mx

a theory that can appraise the topological properties of these media. Among the already available theories the Dual Site-Bond Model (DSBM) [7] results outstanding since this approach represents practical advantages due to its simplicity and capability to recognize topological correlations among pore entities. The DSBM [8] is particularly suitable for: (a) conceiving [7, 9] representing [10], and classifying [11] pore networks; (b) understanding and predicting the mechanisms of capillary processes [11, 12, 13, 14, 15, 16] (such as capillary condensation and evaporation, mercury intrusion, imbibition, immiscible displacement, etc.); (c) determining the texture of porous materials [17, 18]; and (d) accounting for the peculiarities of heterogeneous surfaces of adsorbents (adsorption equilibrium and surface diffusion, chemisorption, reaction rate in the adsorbed phase, and surface characterization) [19]. It also has been applied to the description of other kinds of disordered systems such as arboreous [20] and dense [21] aggregates resulting from the sol-gel transition, as well as to the assessment of the morphology of polymers [22]. One way for conveniently modeling porous media through the DSBM consists in the representation of the pore assemblage in terms of a regular pore network. In this case, the porous space is represented via two pore entities: the sites and the bonds. A site is a cavity that is surrounded and interconnected to C fellow entities through bonds, which are necks or channels that communicate two neighboring sites. The C parameter is usually labeled as the connectivity of the pore network. From the very nature of sites and bonds, it surges the theoretical background on which the DSBM rests and that will be described in detail in the next section.

The applications of pore networks are extensive, going from the interpretation of adsorption-desorption of vapors to Hg porosimetry experiments. Besides, refinements of the DSBM have been developed as to incorporate geometric restrictions between neighboring pores [23], as well as in the description of dimorphic pore networks [24]. Furthermore, pore networks simulated from the DSBM have also been employed for describing disordered substrates different from porous media, such as heterogeneous energetic surfaces [25] and pharmaceutical drug release matrices [26]. Additionally, the DSBM has also been applied to study the properties of correlated percolating clusters [27]. Pore networks, therefore, represent an important physical model for describing the structure of disordered media.

A methodology for the in silico construction of pore networks is, indeed, relatively simple to implement [28]. Briefly, it consists in departing from an arbitrary initial state stablished after performing a random assignment of sizes to the sites and bonds of the pore network sampled from the defined distribution of pore sizes. A *Markov* chain of successive states is then obtained by swapping the sizes of two sites or two bonds chosen at random, until a valid pore network configuration is attained; i.e., pore elements fulfill the expected statistical properties stated by the DSBM [8, 34]. However, as the size correlation among pore entities increases, the involved computational time increases exponentially, something that really represents a problem given that the overall computational time turns out to be of the order of several months, for the extreme cases of highly correlated networks [29]. This problem has been recently surmounted by proposing an efficient greedy computational algorithm, called *the NoMISS algorithm* (No Mistake Initial Seeding Situation)[29].

The distinctive feature of *the NoMISS algorithm* is based on achieving, from the start, an adequate linking among sites that are suitable of being interconnected together and then continues with the assemblage of the remaining network elements. The adequate linking

Table 1. Approaches employed for detailing the basic structural features of porous media and energetically heterogeneous surfaces

Approach	Ref. No.
Dual Site Bond Model Fundamentals	[7, 8, 9, 10, 11, 12]
Precluding Geometric Interferences among Pore Entities	[23]
Design of Dimorphic Pore Networks	[24]
Description of Energetically Heterogeneous Surfaces	[18, 19, 25]
Description of Diverse Disordered Substrates	[20, 21, 22]
Pharmaceutical Drug Release Systems	[26]
Percolation Clusters	[27]
Pure Monte Carlo Methods	[28]
Greedy Algorithms	[29]
Parallel Computing Algorithms	[30]

Table 2. Description and treatment of diverse capillary and physicochemical phenomena occurring in porous media and heterogeneous surfaces

Approach	Ref. No.
Sorption mechanisms	[2, 11, 13, 14, 15]
Determination of the Texture of Porous Materials	[17, 19]
Transport in porous media (Immiscible displacement)	[1, 3]
Imbibition and drainage	[5]
Molecular diffusion	[16]
Catalytic deactivation	[4]

of pore elements is developed from an initial ordering of sites according to their sizes by the use of linked lists storing the sizes of the sites and their C-bonds. A seeding process continues with the instauration of smaller pore network subunits, chosen adequately from the linked lists and allocated randomly throughout the pore network, which are conveniently joined together at a later and final stage. *The NoMISS algorithm* has also been recently implemented with parallel simulation techniques that have distributed the memory consumption and speeded up computing time, issues intrinsic to large pore networks [30].

Table 1 summarizes the different approaches to describe heterogeneous structures and their corresponding references. Likewise, Table 2 summarizes the physicochemical phenomena studied by means of the latter approaches and their corresponding references.

The present chapter thoroughly describes the computational techniques developed in the two last references as well as the preliminary results that have been obtained for the simulation of porous networks subjected to geometrical restrictions. Also, the structure characterization of the constructed pore networks is updated within the framework of the fractal and percolation theory, something missing in previous publications. The chapter is organized as follows. Section 2, presents the theoretical background of the DSBM approach. Section 3, describes the incorporation of geometrical restrictions into the DSBM. Section 4, covers the topological characterization of pore networks. Section 5, accounts for the algorithms for the *in silico* construction of pore networks. Section 6, describes the parallel computational techniques for implementing the previous algorithms. Section 7, displays the simulated pore networks and the results obtained therefrom. Finally, in Section 8, the perspectives and the conclusion of this chapter are stated.

2. The Dual Site-Bond Model (DSBM)

As mentioned previously, porous media can be described adroitly by considering sites (antrae, cavities) and bonds (capillaries, throats, passages), which inevitably alternate to form an interconnected network [8]. The connectivity C, is the mean number of bonds meeting at a site. For simplicity, the size of each entity is expressed by using only one quantity, R, defined as follows: for sites, considered as hollow spheres, R is the radius of the sphere, while for bonds, idealized as hollow cylinders open at both ends (owing to their function of passages), R is the radius of the cylinder. A twofold (sites and bonds) size distribution is established by means of the normalized size probability density functions, $F_S(R_S)$ and $F_B(R_B)$, of sites and bonds respectively. The fractions of sites, $S(R)$, or bonds, $B(R)$, of sizes smaller than a particular value R are:

$$S(R) = \int_0^R F_S(R_S) \, dR; \quad B(R) = \int_0^R F_B(R_B) \, dR \tag{1}$$

Pore networks possess a very special property [7, 9]: the size of a site must be always larger than (or at least equal to) the size of any one of its delimiting bonds. This Construction Principle (CP) is of the upmost importance for the case of highly overlapped $F_S(R_S)$, $F_B(R_B)$ twofold structures, so that the elements are not free to distribute fully at random. Two self-consistency laws guarantee the fulfillment of the CP. The first law states that bonds must be enough and sufficiently small as to link the sites corresponding to a given size distribution.

A second law is nevertheless still necessary since when there exists an overlap between the site and bond size distributions topological correlations arise. Thus, the events of finding a site of size $R_S \in (R_S, R_S + dR_S)$ and a size $R_B \in (R_B, R_B + dR_B)$ for a given one of its C bonds are not independent. In this case, the joint probability for such an event is:

$$F(R_S, R_B) = F_S(R_S) F_B(R_B) \Phi(R_S, R_B) \, dR_S dR_B \tag{2}$$

An expression of the second law can be deduced from the last equation as follows:

$$\text{SecondLaw}: \Phi(R_S, R_B) = 0, \forall R_S < R_B \tag{3}$$

Here, the correlation function $\Phi(R_S, R_B)$ incorporates all the information about the site-bond assignment that will arise after constructing the pore network. For the simplest of cases, called the Self Consistent situation, sites and bonds are assigned to each other in the most random way as allowed by the CP, then, $\Phi(R_S, R_B)$ attains the following form:

$$\Phi(R_S, R_B) = \frac{\exp\left(-\int_{S(R_B)}^{S(R_S)} \frac{dS}{B-S}\right)}{B(R_S) - S(R_S)} = \frac{\exp\left(-\int_{B(R_B)}^{B(R_S)} \frac{dB}{B-S}\right)}{B(R_B) - S(R_B)} \tag{4}$$

If we denote by Ω the overlapping area between the site and bond probability density functions, as shown in Figure 1 for the case of Gaussian distributions, Φ involves the following properties: (i) $\Phi_{\Omega \to 0}(R_S, R_B) = 1$, $\forall R_S, R_B$; this means that sites and bonds are distributed completely at random, and (ii) $\Phi_{\Omega \to 1}(R_S, R_B) \propto \delta(R_S - R_B)$; this indicates that sites and bonds group together in the form of macroscopic patches, each one of these regions having their own value of R. Thus, Ω is a fundamental parameter for describing the topology of the pore network according to this model.

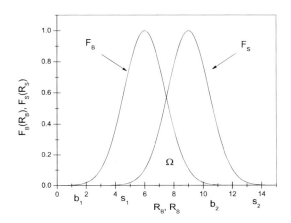

Figure 1. Overlapping (Ω) between the bond $F_B(R_B)$ and the site $F_S(R_S)$ size distributions. The s_1 and s_2 labels represent the smallest and the largest sites, respectively; similarly, the b_1 and b_2 labels symbolize the smallest and the largest bond entities, respectively.

3. Geometrically Restrained Pore Network Modeling

Beyond the Self-Consistent situation, physical restraints arise when perceiving or simulating pore networks. On the basis of the DSBM approach, site-bond interconnections extend all over the available space in order to conform a pore network. A physical constraint immediately surges when implementing the previous representation: two adjacent bonds (of sizes R_{B_1} and R_{B_2}) could interfere each other before arriving to the converging site if their sizes fail to comply with the next condition:

$$R_{B_1}^2 + R_{B_2}^2 \leq R_S^2 \qquad (5)$$

The above physical restraint generates a strong correlation among the sizes of the incumbent site and those of its incoming bonds, not to mention limitations imposed on the sizes of the different pairs of neighboring converging bonds.

The considerations above presented lead us to formulate the following alternative *Construction Principle* (CP) concerning the simulation of pore networks endowed with geometrical restrictions among their constituting void elements: "While every bond that converges into a site must be of a size smaller than this last cavity, two adjacent bonds converging to a site have still to assume the right combination of sizes in order to avoid any physical interference between them before meeting together into the site".

The overlap Ω between the site- and bond-size distributions is a crucial parameter for the attainment of a given pore topology (i.e., the exact sequence in which sites and bonds are linked together throughout the pore network) is. Ω controls the extent of the correlations existing among the sizes of pore elements and determines the arousal of phenomena such as pore-size segregation, pore connectivity segregation, or the particular aspect of the diverse pore zones.

4. Structural Characterization of Pore Networks

The characterization of constructed pore networks may be carried out through the equations next described. From here on out, we designate as T1-networks those porous structures that comply with the construction principle, and as T2-networks those porous structures that comply with both the construction principle and geometrical restrictions. Note that xms and xmb correspond to the average sizes of sites and bonds respectively, and σ to the standard deviation of the previous distributions.

4.1. Conditional Probabilities

Two conditional probabilities can be formulated as follows:

$$P(R_S/R_B) = \frac{F_S(R_S)\Phi(R_S, R_B)}{F_B(R_B)} \qquad (6)$$

$$P(R_B/R_S) = \frac{F_B(R_B)\Phi(R_S, R_B)}{F_S(R_S)} \qquad (7)$$

Equation 6 refers to the conditional probability density for a given bond of the pore network of size equal to R_B to be connected to a site of size R_S. Conversely, equation 7 defines the conditional probability density for a given site of the pore network of size equal to R_S to be connected to a bond of size R_B. These two last probabilities can be numerically evaluated by examining the sizes of the pore elements in the constructed pore networks. Hence, by using numerical values of the probability densities in Equations 6 or 7, numerical values of the correlation function $\Phi(R_S, R_B)$ can be evaluated as well. Obviously, the degree of size correlation among connected pore elements produces different graphs for the quantities $P(R_B/R_S)$ and $P(R_S/R_B)$ vs. R_B and R_S, respectively. To exemplify this point, we present in Figure 2 some values of these quantities for two networks with very different degrees of correlation. As can be appreciated in Figure 2(a), $P_{\Omega \to 0}(R_B/R_S) \approx F_B(R_B)$. On the other hand, in Figure 2(b), $P_{\Omega \to 1}(R_B/R_S) \approx \delta(R_S - R_B)$.

Since the function $\Phi(R_S, R_B)$ is bivariate, several different pairs of values for R_S and R_B, representative of the mathematical domain of both $F_S(R_S)$ and $F_B(R_B)$, have to be examined in order to characterize the pore network thoroughly. So, it is better to define an average value of $\Phi(R_S, R_B)$ accounting for the whole specific network:

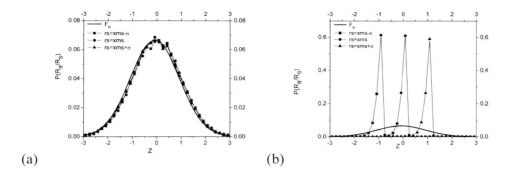

Figure 2. Conditional probability densities, $P(R_B/R_S)$, for three fixed sizes of sites: $R_S = \{xms - \sigma, xms, xms + \sigma\}$ evaluated in T1-networks. (a) $\Omega = 0$ (b) $\Omega = 0.92$.

$$\overline{\Phi(R_S, R_B)} = \left\langle \sum_{R_S=s_1}^{R_S=s_2} \sum_{R_B=b_1}^{R_B=b_2} \Phi(R_S, R_B) \right\rangle \tag{8}$$

Where b_1, b_2, s_1 and s_2 account for the sizes of the smallest bond, biggest bond, smallest site and biggest site, respectively.

Combination of equations 6, 7 and 8 results in the definition of two average conditional probability densities:

$$\overline{P(R_S/R_B)} = \frac{F_S(R_S)\,\overline{\Phi(R_S, R_B)}}{F_B(R_B)} \tag{9}$$

$$\overline{P(R_B/R_S)} = \frac{F_B(R_B)\,\overline{\Phi(R_S, R_B)}}{F_S(R_S)} \tag{10}$$

Figures 3 and 4 present results for T1 and T2 networks, accounting for the average probability densities $\overline{P(R_B/R_S)}$ and $\overline{P(R_S/R_B)}$ as functions of Z, a normalizing variable associated to the standard normal distribution:

$$Z = \begin{cases} \frac{R_B - xmb}{\sigma}; & for\ bonds \\ \frac{R_S - xms}{\sigma}; & for\ sites \end{cases} \tag{11}$$

Also, Figure 5 shows results for T1 and T2 networks, accounting for $\overline{\Phi(R_S, R_B)}$.

The average probability density and the average value of the correlation function, both calculated in T1-networks, reveal a mathematical evolution consistent with the value of the overlaping. For the case of lowly correlated networks it is observed that $\overline{P_{\Omega \to 0}(R_B/R_S)} \approx F_B(R_B)$ and $\overline{P_{\Omega \to 0}(R_S/R_B)} \approx F_S(R_S)$(Figure 3); which is understandable taking in mind that $\Phi_{\Omega \to 0}(R_S, R_B) = 1$ for any combination of both R_S and R_B (hence $\overline{\Phi(R_S, R_B)} = 1$, as shown in Figure 5(a)). On the other hand, for the opposite case, it is observed that $\overline{P_{\Omega \to 1}(R_B/R_S)}$ and $\overline{P_{\Omega \to 1}(R_S/R_B)}$ are $\approx 1/(a-b)$, i.e., the uniform distribution (Figure 3). This last finding can be understood if it is realized that each value

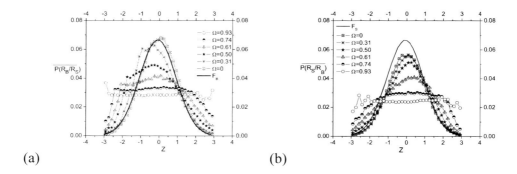

Figure 3. Average conditional probability densities, (a) $\overline{P(R_B/R_S)}$ and (b) $\overline{P(R_S/R_B)}$, evaluated in T1-networks.

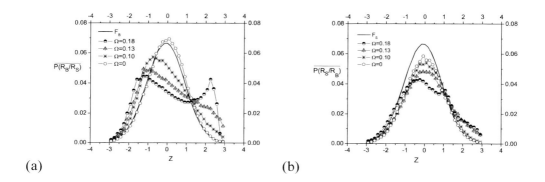

Figure 4. Average conditional probability densities, (a) $\overline{P(R_B/R_S)}$ and (b) $\overline{P(R_S/R_B)}$, evaluated in T2-networks.

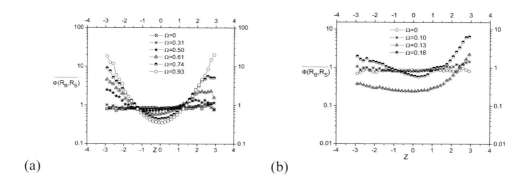

Figure 5. Average value of $\Phi(R_S, R_B)$ calculated by means of Equations 8 and 9. (a) T1-networks. (b) T2-networks.

of $P(R_B/R_S)$ and $P(R_S/R_B)$ making up the respective average value, is equal to the delta function when $\Omega \to 1$; hence, the set of $\overline{\delta(R_S - R_B)}$, for any combination of R_S, R_B, produces the uniform distribution. Lastly, the $\overline{\Phi(R_S, R_B)}$ function takes the form of $k_1 cosh(Z^2)$, k_1 being a proportionality constant, when $\Omega \to 1$; this is a result complementary to the uniform distribution verified via the average probability densities, which ensures the $\Phi(R_S, R_B)$ value of being reciprocal to the corresponding values of $F_S(R_S)$ and $F_B(R_B)$. The networks arising between these two extremes cases represent intermediate situations. The previous discussion can be summed up by means of the next three equations, which are graphed in Figure 6.

$$\overline{P(R_B/R_S)} = \begin{cases} F_B(R_B); \Omega \to 0 \\ \frac{1}{b-a}; \Omega \to 1 \end{cases} \quad (12)$$

$$\overline{P(R_S/R_B)} = \begin{cases} F_S(R_S); \Omega \to 0 \\ \frac{1}{b-a}; \Omega \to 1 \end{cases} \quad (13)$$

$$\overline{\Phi(R_S, R_B)} = \begin{cases} 1; \Omega \to 0 \\ k_1 cosh(Z^2); \Omega \to 1 \end{cases} \quad (14)$$

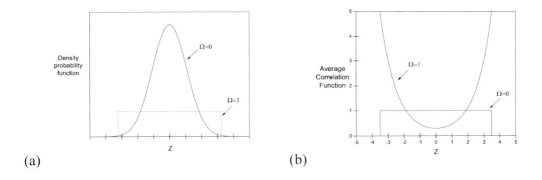

Figure 6. Adjusted functions representing the average conditional probability densities and the average correlation function for the two extremely correlated cases as verified for T1-networks. (a) Equation 11 or 12. (b) Equation 13.

An equivalent analysis can be formulated for the T2-networks. In fact, the case where $\Omega \to 0$ is equivalent to the corresponding case in the T1-networks (Figures 4 and 5(b)). However, the extremely correlated case differs. It is observed that both $\overline{P_{\Omega \to 1}(R_B/R_S)}$ and $\overline{P_{\Omega \to 1}(R_S/R_B)}$ present two maxima, instead of the one obtained for the T1 case. These two peaks correspond to small and big sizes of pore elements. The maximum located at the smaller size is related to sites of small size, which are the type of sites most difficult to accommodate with C-bonds without geometrical interferences. On the other hand, the maximum located at the bigger size corresponds to bonds of large size, which are the type of bonds that have to be connected to the largest sites in order to comply with the geometrical restrictions. It is important to note that the interconnected bonds and sites differ considerably in size, since these entities have to comply with geometrical restrictions: the most correlated T2-network is equivalent to $\Omega = 0.18$ as this relatively low overlapping value ensures that the sizes of connected sites and bonds differ enough as to comply the geometrical restrictions. The networks remaining between these two extreme cases represent intermediate situations. The previous discussion can be summarized by means of the next three equations (k_2, k_3, k_4 and k_5 are proportionality constants), which are plotted in Figure 7.

$$\overline{P(R_B/R_S)} = \begin{cases} F_B(R_B); \Omega \to 0 \\ k_2 cosh(Z); \Omega \to 1 \end{cases} \quad (15)$$

$$\overline{P(R_S/R_B)} = \begin{cases} F_S(R_S); \Omega \to 0 \\ k_3 cosh(Z); \Omega \to 1 \end{cases} \quad (16)$$

$$\overline{\Phi(R_S, R_B)} = \begin{cases} 1; \Omega \to 0 \\ cosh(k_4 ln(Z + k_5)); \Omega \to 1 \end{cases} \quad (17)$$

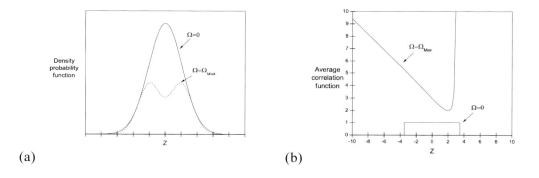

(a) (b)

Figure 7. Adjusted functions representing the average conditional probability densities and the average correlation function for the two extremely correlated cases as verified for T2-networks. (a) Equation 14 or 15. (b) Equation 16.

4.2. Correlation Length

The information supplied by the average quantities given above is important, but still renders no account for the topological domains or regions where the sizes of pore elements have similar sizes, a characteristic which becomes a very important issue when choosing the appropriate network size to statistically represent heterogenous media. For example, below the percolation threshold, the average size of clusters has to be at least ten times smaller than the size of the pore network, in order to assess the correct percolating cluster. This condition is ensured if the average size of the regions where pore elements have similar sizes is calculated. To this end, we define a correlation length via the following equation:

$$\xi = -\frac{r}{LogC(r)} \quad (18)$$

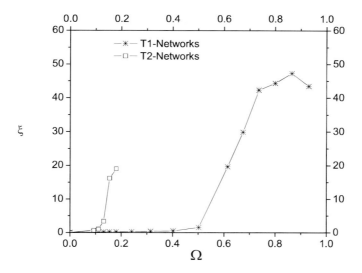

Figure 8. Correlation length, ξ, as a function of the overlapping, Ω, evaluated for T1 and T2 networks.

Where ξ is the correlation length and $C(r)$ is the correlation coefficient between the sizes of two pore elements separated by a distance r, given by:

$$C(r) = \frac{\left\langle (R_{S_i}(r_i) - xms)\left(R_{S_j}(r_j) - xms\right)\right\rangle}{\sigma^2} \quad (19)$$

Being R_{S_i} and R_{S_j} the sizes of two sites with positions given by the vectors r_i and r_j, respectively, and separated by a distance $r = |r_i - r_j|$.

The parameter ξ evaluates the average size of the patches in the network, where the sizes of pore elements have similar sizes. In order to ensure the construction of a statistically representative pore network the correlation length has to be at least five times smaller than the length of the porous network. Figure 8 shows the correlation length evaluated for T1 and T2 networks. From this figure, it is appreciated that the parameter ξ reaches higher values for the T2 networks (the highest size ~ 40) than those reached for the T1 networks (the highest size ~ 20). The domains with similar site sizes in the case of the most correlated T1-networks, almost double the corresponding values found in the most correlated T2-networks; this great difference can be explained if we consider that T2-networks are subjected to geometrical restrictions, in contrast to the T1-networks, which are subjected to a more relaxed restricting construction principle, thus allowing for the existence of domains of larger sizes inside them.

4.3. Percolation Cluster

One important application of pore networks consists in the calculation of percolation properties. Percolation may be studied either via bonds (bond percolation) or sites (site per-

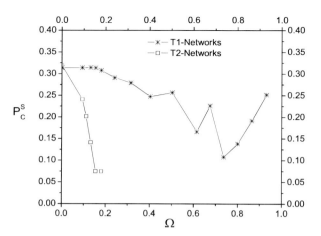

Figure 9. Site percolation threshold, P_C^S, as function of the overlapping, Ω, evaluated for T1 and T2 networks.

colation). Besides, percolation may be assessed either in correlated or uncorrelated pore networks. One percolation property that varies greatly with correlation is the percolation threshold. To date it has been found that this quantity decreases considerably with the overlapping (Ω); although it has only been evaluated for square networks [27]. Figure 9 presents the site percolation threshold, P_C^S, evaluated in T1 and T2-networks [35]. From this figure, it can be appreciated that the value of the percolation threshold attains the value of 0.31117 when $\Omega = 0$, for both T1 and T2 networks; this is the expected value reported for simple cubic lattices [32]. In accordance with [27] the percolation threshold decreases as Ω increases. However, it decreases sharply for T2-networks ($P_C^S = 0.074$ when $\Omega = 0.92$), whilst it decreases smoothly up to $\Omega \sim 0.4$ and then increases and decreases alternatively up to $\Omega \sim 0.9$, for T1-networks.

The previous results can be complemented with the fractal dimension, D_F, of the spanning clusters, calculated at P_C^S (Figure 10). The information presented in the latter figure shows the following behaviours. T2-networks: the value of D_F always decreases as Ω increases (from 2.5 for $\Omega = 0$ to 1.9 for $\Omega = 0.19$). T1-networks: the value of D_F fluctuates between 2.5 for $\Omega = 0$ to 2.25 for $\Omega = 0.92$. At points located between the extreme values of Ω, D_F attains local minima and maxima. The decrease of the D_F value in T2-networks, indicates that the level of detail of the spanning cluster is lost as Ω increases. In other words, the spanning cluster densifies as Ω increases. On the other hand, the values of D_F obtained for the T1-networks indicate that the texture of the spanning cluster changes drastically as Ω varies, densifying and spreading alternatively. This constitutes an unusual result that has to be studied by further research. However, these results also indicates that correlated T1-networks provide interesting structures with unusual percolating properties. Some spanning clusters are presented in Figure 11. Table 3 lists the values of the data corresponding to Figures 8-10.

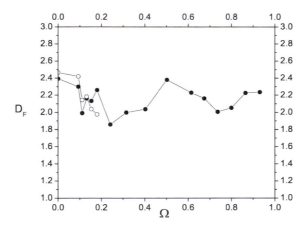

Figure 10. Fractal dimension, D_F, as function of the overlapping, Ω, evaluated for T1 and T2 networks.

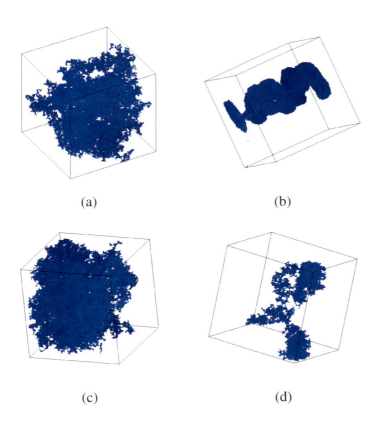

Figure 11. Spanning clustsers. (a) T1-network, $\Omega = 0$; (b) T1-network, $\Omega = 0.93$; (c) T2-network, $\Omega = 0$; (d) T2-network, $\Omega = 0.18$.

Table 3. Values of the data presented in Figures 8–10

Ω	$\xi(T1)$	$\xi(T2)$	$P_C^S(T1)$	$P_C^S(T2)$	$D_F(T1)$	$D_F(T2)$
0	0.02092	0.01557	0.31409	0.31963	2.39232	2.466
0.09289	0.1685	0.6809	0.31423	0.24089	2.29908	2.42
0.11066	0.1789	1.02182	0.31430	0.20167	1.99018	2.144
0.13093	0.19561	3.38213	0.31433	0.14139	2.16254	2.182
0.1539	0.22234	16.19462	0.31374	0.07503	2.13316	2.038
0.17972	0.24448	19.00756	0.30777	0.0748	2.25954	1.976
0.24065	0.29915	-	0.29097	-	1.85842	-
0.31461	0.38179	-	0.27926	-	1.99764	-
0.40196	0.53481	-	0.24721	-	2.03647	-
0.50229	1.5261	-	0.25645	-	2.37928	-
0.61437	19.657	-	0.16623	-	2.22896	-
0.67422	29.883	-	0.22597	-	2.16147	-
0.73618	42.325	-	0.10799	-	2.00565	-
0.79989	44.385	-	0.13817	-	2.05232	-
0.86493	47.27	-	0.19166	-	2.22551	-
0.93089	43.478	-	0.25148	-	2.23528	-

5. Pore Network Construction

The construction of virtual pore networks can proceed through different alternative routes. In this section, we describe iterative and greedy algorithms, for constructing pore networks under the DSBM. We also propose a hybrid greedy-iterative algorithm for constructing pore networks subjected to geometrical restraints.

5.1. Pure Monte Carlo (PMC) Procedure

In this iterative method, sites are randomly chosen from their precursory distribution and seeded accordingly throughout all nodes of the lattice; in turn, bonds are chosen haphazardly from their own starting distribution and seeded arbitrarily in-between sites. This careless seeding of sites and bonds leads, most of times, to the construction of pore networks that violate the CP, especially if Ω is considerable. The CP contraventions existing between sites and bonds and between adjacent bonds are settled by means of an intense attempting exchange between pairs of pore elements chosen at random positions throughout the whole structure. For this exchange, Monte Carlo steps (MCS) are required; a MCS consists in a number of exchanging attempts equal to the total number of hollows that constitute the pore network. A Metropolis [31] swapping scheme is used to perform a successful exchange: if after interchanging two pore elements, the number of CP violations is smaller than, or at most equal to, the number of initial violations, the swapping is accepted otherwise is rejected. The drawback of this method arises when the computing execution time that is required, not only to eliminate CP violations but to also provide topologically credible networks, becomes too long, especially when Ω is considerable.

5.2. NoMISS Procedure

The execution time shortcoming evidenced by the PMC method can be overcome through a seeding approach labeled as the NOMISS greedy method; under this scheme, a valid network (i.e., a CP errorless network) is set just from the very start of the simulation. To do this, a list containing the sizes of sites in ascending order is set. Bonds are next linked sequentially to the smaller, intermediate, and larger site sizes; the smaller cavities complete their whole pore contour (i.e., C bonds are surrounding each site), while respecting the CP.

5.2.1. Multiplex Interconnection for the Creation of Pore Networks

The way by which a pore network is being simulated determines the overall fulfillment of the CP, as well as the adequate topology of the resultant arrangement. Since every bond is shared by two sites, the unit cell of the whole pore arrangement is a multiplex entity, which consists of a central site surrounded by C half bonds; the term half bond surges as a consequence of what was mentioned above, i.e., that any bond is shared by two sites. In a cubic network for every site there exist 6 half bonds while in a 2-D square network there only exist 4 half bonds per site. In total, in the cubic network, there are $N \times N \times N$ sites and $3 \times N \times N \times N$ complete bonds or, alternatively, $6 \times N \times N \times N$ half bonds, each of these shared by two neighboring sites.

Now, imagine the following interconnection procedure for building a pore network containing $N \times N \times N$ sites. Initially, C bonds are going to be assigned to each site; this should be done while respecting the CP. The next step consists in bringing together the collection of site-bond complete contour multiplexes.

When two of these arrangements are bonded to each other; the largest of the facing bonds is used to interconnect the two sites while the remaining bond is set free, and should be allocated to sites having incomplete bond contours. A major problem with this kind of interconnection pattern resides in the way by which the network is to be constructed; for instance, one could firstly prefer to grow a multiplex to construct a 2-D plane of interconnected multiplexes along a given direction and then to grow this plane into a 3-D array.

5.2.2. Multiple Seeding (MS) Procedure

It has been found [29] that the growth of a pore network from a single multiplex (a site surrounded by C bonds) seed through the attachment of further multiplex arrangements, leads to anisotropic pore arrangements. This undesired anisotropy quality consists in that the statistical properties of the voids, together with their interconnection properties show no homogeneity along all directions. To overcome this problem, several multiplex seeds are randomly assigned to some pore network locations. Then, the seeds are grown in order to fill the network space.

5.3. Construction of Pore Networks Subjected to Geometrical Restraints from the DSBM: The Case of a Cubic Lattice

This method is a hybrid approach that consists of two main steps: the execution of a greedy procedure to construct an initial pore network where some locations may not accomplish the Geometrical Restraints, and the execution of an iterative algorithm that is applied on the initial network in order to eliminate the possible violations to the Geometrical Restraints; here, several Monte Carlo exchanges are applied until no mistakes are found in the pore network.

In order to simulate cubic DSBM pore networks, subjected to geometric restrictions between pore entities, such as prohibiting any interpenetration between cylindrical capillaries (bonds) previous to their meeting at a spherical cavity (site), it is first required to select a cubic lattice (i.e., of a connectivity equal to six) with a total size equal to L^3, together with two precursory size distributions: one for sites (cavities, chambers) and another for bonds (throats, necks). Specifically, in the greedy algorithm step an L_S list is created when the sites proceeding from the precursor site size distribution are ordered according to a descending size fashion. A similar procedure is made to generate a L_B list of bonds, proceeding from the given precursory bond size distribution, these void elements are also ordered in a descending size manner. Each of these collections of hollow elements is then accommodated throughout a given geometrical lattice (e.g., triangular, cubic) in the following manner: sites are located at the intersecting nodes while bonds are inserted in-between nodes. The detailed outline of the greedy algorithm action is itemized in the following paragraphs.

5.3.1. Seeding of Sites to Start the Generation of Network Clusters Throughout a Cubic Lattice

- Sites from the L_S hierarchical list are seeded one by one, starting from the first one (i.e., the largest site) to the last one (i.e., the smallest site), at diverse node positions all over the cubic lattice.

- To start an i, j, k node position is selected at random.

- The largest site in L_S is then assigned to this place. This site size is stored, together with its lattice position, in a different Assigned-Site L_{AS} list, while keeping the same site size descending sequence.

- More sites are successively chosen (one after another) from L_S along each coordinate axis in order to generate a cubic cell; this process is continued until eventually building, around the first seeded site, a cubic cluster of K^3 total size (this K size is related to the number of nodes composing the cluster along a given coordinate direction). This cluster continues growing along all directions of the lattice while respecting the same desired (i.e., cubic) geometry. When a site is positioned close to the outer face of the lattice, the cluster is now grown at the opposite face of the lattice, in a similar same way as if boundary conditions were imposed.

- The already assigned sites are continuously transferred to the L_{AS} list according to the same pre-established (i.e., descending) size order, together with their accompanying lattice positions.

- It should be noted that since clusters are being built one by one, according to the size descending ordering of the L_S list, the first cluster that is set is composed of the largest sites while the last cluster is formed by the smallest sites. Every time that clusters overlap, the already assigned positions are kept untouched, whilst empty positions are seeded with the still available site sizes.

- Additional clusters of the same K^3 size are grown similarly as the first cluster at different random positions of the cubic lattice. The number of clusters (seeds) is selected beforehand. Every time that clusters overlap each other, the already assigned sites are kept untouched while empty positions are occupied by available sites.

- Once all foresighted site clusters have been seeded and grown to their final sizes, and since some node positions could still be empty, a recovering layer is started from the first cluster to all vacant node positions until reaching a final L^3 total site seeding extent.

5.3.2. Bond Assignation to Sites

The policy of trying to link firstly the largest bond sizes to sites, according to the descending size order in which these last elements appear in the L_S list, is followed. In this case, however, there can still surge geometric restraints when two adjacent bonds are going to be connected to the same site and preventive action is required. The steps below are carried out for achieving a rightful linking of bonds to sites.

- Bonds are assigned to each site according to the decreasing sequence prescribed by the L_B list.

- Bond assignation starts with the site having the largest size, and the process continues with the ensuing sites according to the order established by the L_{AS} list.

- Every site requires C bonds for completing its corresponding bond contour; it is thus expected that some violations of the CP would occur when a pair of bonds interfere with each other. Thus, every additional bond that is linked to a site requires to be appropriately chosen in order to not contravene the CP.

- Nonetheless, any bond that is going to be attached to a site already connected to some other bonds should be of the largest possible size, according to the hierarchic order dictated by the L_B list. This action precludes any eventual lack of sufficiently small bonds required to complete the pore network, especially when dealing with highly overlapped site- and bond-size distributions.

- The above process is continued orderly with every one of the remaining site elements of L_S until a completion of bond contours around all sites is fulfilled.

- It may happen that one given bond could not be linked to a site if, already, there are not enough available bonds of sufficiently small sizes; in this case, one bond of the largest size is attached to the site regardless of introducing a CP violation.

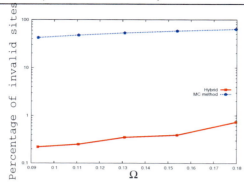

Figure 12. Percentage of invalid sites generated via a random initialization and with the hybrid algorithm running under different mean site size values (Ω overlap).

5.3.3. Pore Swapping to Eliminate Geometrical Errors and Improve Pore Network Isotropy

The previous strategy allows the initialization of a pore network which tolerates a small number of CP violations. The site violations percentage of the hybrid algorithm is considerably smaller than that obtained from a random initialization; the random initialization of a $100x100x100$-size network generates 63% errors when having a maximum Ω overlap, whilst the hybrid algorithm renders only 0.8% errors, as shown in Figure 12. In order to eliminate the existing violations generated through the greedy algorithm step, we execute the iterative algorithm that succesively applies the pure Monte Carlo exchanges, as explained before. Finally, additional Monte Carlo iterations are performed to improve the pore network isotropy.

6. Parallel Simulation of Pore Networks

The three methods described above for the construction of pore networks require the processing of millions of data items. This type of applications can definitely take advantage of parallel computing technologies. In this section, we present two basic parallel approaches to construct pore networks considering T1 constraints. The proposed algorithms are based on the distributed memory programming model, using the Message Pasing Interface (MPI) library and a set of collaborative processing cluster-nodes. The first approach is the parallel Monte Carlo version which follows the main construction behavior described in 5.1. The second approach is a parallel version of the NoMISS algorithm specified in 5.2.. In both approaches a set of cluster-processors is organized as a 3D NxNxN logical torus topology; each processor has 6 neighbouring-processors which are directly connected to it, as in the example shown in Figure 13 where a 3x3x3 torus topology is represented. This organization is very suitable for the partitioning of the 3D pore network structure, then obtaining NxNxN smaller subnetworks; thus, the processors work simultaneously for the construction of a subnetwork. A restriction is considered when working with the outer subnetwork faces. Since the pore network was partitioned into NxNxN subnetworks some of the outer face pores in a subnetwork have not a complete bond contour (as shown in Figure 14), thus

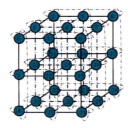

Figure 13. A 3x3x3 processor torus example.

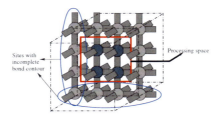

Figure 14. An example of a pore slide in a subnetwork.

the processing is restricted to be applied only to the interior pore structures of a subnetwork.

In order for the outer face pores to be initialized, some pore transfers among processors should be done [33]. Half of each subnetwork is transferred on the processor torus towards the x, y, and z axis directions, thus allowing the initialization of the whole pore network space. Figure 15 shows how a 4x4 pore slide is initialized after the three transfers. With a 3D subnetwork, 7 transfers should be applied throughout the x, y, x, z, x, y, and x axis to obtain a complete pore initialization. In the following subsections, we respectively present the particular behaviours of the Monte Carlo and greedy parallel algorithms employed for the pore network simulation.

6.1. Parallel Monte Carlo Method for Pore Network Construction

The processors start creating their own subnetwork space. The parallel Monte Carlo method can be divided into three steps:

1. Pore-size initialization
   The torus processors simultaneously perform the same pore size initialization algorithm, as in the sequential version, thus working on their local space. So, the initialization time T_{ini} is reduced to T_{ini}/N, where N is the number of used processors.

2. Pore exchanges
   Here, a succesive number of pore-size exchanges is applied to eliminate the CP violations, as in the sequential version. However, after a number of local exchanges, each processor transfers half of its subnetwork to its neighbouring processors on the torus, as previously explained. This process allows the whole set of pores to be exchanged and it is repeated until all of them arrive to a valid configuration.

3. Isotropy improvement

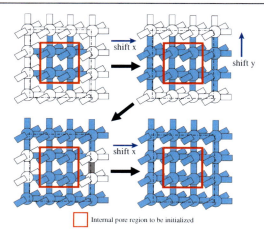

Figure 15. Representation of a subnetwork slide initialization through subnetwork transfers.

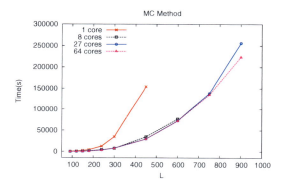

Figure 16. Parallel Monte Carlo performance.

When there are no more CP violations in a pore network, an additional number of pore-size exchanges is still required in order to improve the network isotropy. In this step the processors remain working in parallel by alternately applying a certain number of local pore exchanges via processor transfers along the three axis of the torus architecture.

This method and the next one were tested on a cluster in which each node had two Quad-Core Xeon Intel processors at 3 GHz and 16 GB RAM equally distributed among eight cores. Therefore, in each node there were eight processing units available. In turn, the sequential version was tested by using a single core with 2GB RAM; the parallel versions were tested by means of 8, 27 and 64 cores.

Figure 16 shows how the parallel Monte Carlo version clearly outperforms the sequential version. Besides, the availability of a higher volume of memory distributed on the torus processors, allows the generation of pore networks which can reach larger sizes.

Table 4. Average simulation response time of NoMISS and MC methods when constructing pore networks of sizes $L = 90, 120, 150, 180, 240$, and 300, using $1, 8, 27$, and 64 cores

Cores	MC(s)	NoMISS(s)
1	9346.2	7586.8
8	2581.8	831.7
27	2433.7	121.8
64	2074.5	36.3

6.2. Parallel NoMISS Algorithm for Pore Network Construction

In this version each torus processors starts creating its own subnetwork space and generating its local ordered site list that contains L^3/P sites, with P =number of processors. The parallel greedy algorithm can be divided into the following three steps:

1. Parallel pore-cluster seeding
   The cluster seeding procedure is performed by each processor, working on its internal current subnetwork space. This procedure is applied alternately with several torus processor transfers towards the x, y, and z axis. As a result, the allocated pore-clusters present a non-uniform distribution throughout the global pore network space.

2. Filling the empty space
   After the seeding procedure execution, the processors should initialize the remaining non-initialized subnetwork pores by constructing a spanning cluster. However, they do not still have the same number of items in their remaining pore lists. It is possible that one processor is not able to complete the spanning cluster because its list is empty whereas another processor has non used list items after completing its subnetwork initialization. To handle this situation a data distribution algorithm can be applied, such as the linear balancing algorithm shown in [30].

3. Isotropy improvement
   As in the Monte Carlo parallel version, once the pore network has no CP violations, an additional number of pore-size exchanges should be applied to improve the network isotropy. In this step, the processors also work in parallel by alternately applying a certain number of local pore exchanges with transfers along the three axis of the torus architecture.

Figure 17 shows how the parallel greedy algorithm version outperforms the sequential NoMISS version. Comparing Figure 16 and 17 we observed that, since the NoMISS parallel method requires additional memory for sorting pore sizes, the obtained pore networks were shorter than those obtained by the parallel Monte Carlo version, when using the same number of processors. However, we observed that the parallel greedy NoMISS version, in general, achieved a better performance than the Monte Carlo version, as shown in Table 4.

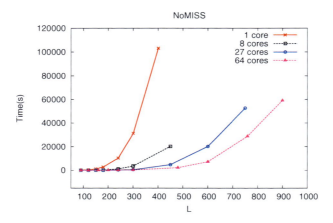

Figure 17. Parallel NoMISS performance.

7. Results

As part of the results, in Figures 18 and 19 two examples of porous networks are shown. Both types of networks comply with the CP, with the difference that in Figure 18 these pore networks involve no geometric constraints; whereas Figure 19 shows pore networks subjected to geometric restrictions. Both networks are $100x100x100$ in size.

Figure 18(a) shows the initial graph constructed by using 800 clusters of size 10. This network (without geometrical restrictions) is constructed with a high overlap between the bond and site distributions (xmb=26 and xms=29). In order to improve the isotropy of the network 50,000 Monte Carlo steps are applied, the result is shown in Figure 18(b).

Similarly, Figure 19(a) shows the initial pore network under geometric constraints, occupying 800 clusters of size 10. Unlike the previous graph (without geometric constraints) this network (with geometric constraints) can only be built from a lower overlap (xmb=26 and xms=42). Again, this initial network is subjected to Monte Carlo steps of swappings to improve its isotropy, the result is shown in Figure 19(b).

In Figure 20 we show several pore networks subjected to geometrical restrictions. These networks are 100x100x100 in size (1 million of sites), with xmb=26 and xms extending from 42 to 46. The Figure shows different pore networks for different overlaps (between sites and bonds), starting from high overlap with xms=42 to low overlap with xms=46.

In terms of numerical results, Figure 21(a) plots the initial errors (y-axis) when a pore network including geometric constraints is constructed. Remember that the Hybrid algorithm initially builds a network depicting construction errors, ie. sites or bonds that do not meet the geometric constraints; then these errors are corrected by exchanging pore sizes. The graph shows these initial errors regarding the overlap (x-axis) among sites and bonds, for different initial configurations. In these configurations we have varied the number of clusters and their sizes. As expected, we observe that with a higher overlap (xms = 42) the number of initial errors is greater, while at lower overlap (xms = 46) the number of errors decreases. Also observe that when the number of clusters increases (e.g., the number of clusters in cs10nc1500 is 1500), the number of errors increases. This is because it is more difficult to build a complete network based on many clusters that grow independently but

On the Conception and Assessment of Mesopore Networks 365

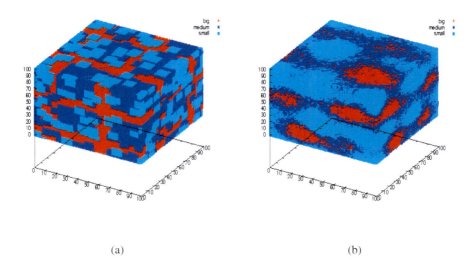

(a) (b)

Figure 18. Porous media under the DSBM (no geometrical restrictions allowed); xmb=26,xms=29,sigma= 6, cluster size=10, number of clusters= 800; without pore exchanges 18(a), and after 50000 pore exchanges 18(b). Large pore sizes are represented in gray color, medium pore sizes in dark gray color, and small pore sizes in light gray color.

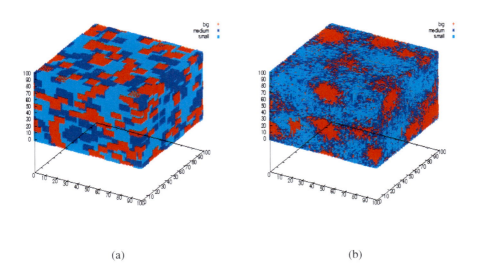

(a) (b)

Figure 19. Porous media under the DSBM, xmb=26,xms=42,sigma=6, cluster size=10, number of clusters= 800; without pore exchanges 19(a), and after 50000 pore exchanges 19(b).

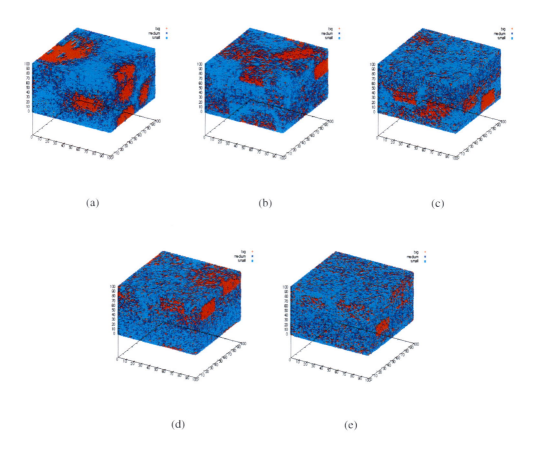

(a) (b) (c)

(d) (e)

Figure 20. Porous media under the DSBM, xmb=26, xms from 42 to 46, sigma=6, cluster size=30, number of clusters=50, with 40000 pore exchanges. xms=42 20(a), xms=43 20(b) ... xms=46 20(e).

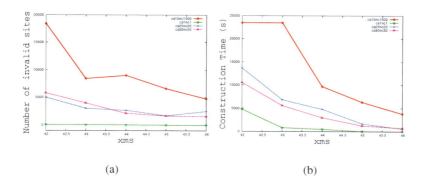

(a) (b)

Figure 21. Porous media under the DSBM, L=100, xmb=26,sigma=6, varing xms, the cluster size and number of clusters; number of sites with T2-CP violations 21(a), and construction time 21(b).

which at sometime overlap arises between them.

Finally, Figure 21(b) displays the execution time in seconds (y-axis), needed to remove these initial errors, as varying the overlap (x-axis) for different network configurations.

Conclusion

Mesoporous materials are physical networks in which geometrical restrictions surge among their interconnected pore elements as a result of the physical nature of these substrates. Therefore, the in-silico simulation of mesoporous networks should take into account not only a myriad (i.e., billions) of voids of nanometric dimensions but also the physical restraints that can arise because of their mutual interconnection. The geometrical restrictions endow the networks with interesting and different textural properties. This is verified by the continuos decrease of both the site percolation threshold and fractal dimension, as function of the overlap. Networks involving no geometrical restrictions do not attain clear fractal and percolation properties; as it is appreciated from the discontinuos decrease of both quantities: the site percolation threshold and the fractal dimension as function of the overlap. Several computer algorithms were implemented for the simulation of pore networks: the basic Monte Carlo method, the greedy NoMISS algorithm, and a hybrid approach. The two first methods were intended for the construction of networks that involve not geometrical restrictions among pore elements, whilst the third method is related to the case of geometrical restrictions arising in pore networks. All approaches were conceived from the DSBM model. We explored the simulation performance by proposing two parallel computing algorithms for the Monte Carlo version and greedy NoMISS approach, which accelerated the running execution time and allowed the construction of large pore networks with billions of voids. Our results indicated that the parallel greedy algorithm performed the best.

Acknowledgments

FRG and SCS thank to CONACYT for the support provided under the project No. 83659 "Estudio Fisicoquímico de la obtención y de las propiedades de los sólidos mesoporosos". Thanks are also due to the SEP-PROMEP Network "Fisicoquímica de Sistemas Complejos Nanoestructurados", Project "Estudio de Propiedades Fisicoquímicas de Sistemas Complejos Nanoestructurados. (UAM-I-CA-31 Fisicoquímica de Superficies)". JMH thanks the CONACyT for the scholarship support. Thanks to the "LSVP, UAM-I" for allowing the authors to use the Aitzaloa cluster.

References

[1] M. Sahimi, Flow phenomena in rocks: from continuum models to fractals, percolation, cellular automata, and simulated annealing, *Rev. Mod. Phys.* 1993, 65(4), 1393-1534.

[2] N. A. Seaton, Determination of the Connectivity of Porous Solids from Nitrogen Sorption Measurements, *Chem. Eng. Sci.* 1991, 46(8), 1895-1909.

[3] M. Sahimi, Flow and Transport in Porous Media and Fractured Rock, VCH, New York, 1995.

[4] I. Kornhauser, F. Rojas, R. J. Faccio, J. L. Riccardo, A. M. Vidales, G. Zgrablich, Structure characterization of disordered porous media: A memorial Review dedicated to Vicente Mayagoitia, Fractals 1997, 5(3), 355-377.

[5] Y. C. Yortsos, B. Xu, D. Salin, Phase Diagram of Fully Developed Drainage in Porous Media, *Phys. Rev. Lett.* 1997, 79(23), 4581-4584.

[6] G. Drazer, R. Chertcoff, L. Bruno, M. Rosen, Tracer dispersion in double porosity porous media with nonlinear adsorption, *Physica A* 1998, 257, 371-375.

[7] V. Mayagoitia, M. J. Cruz and F. Rojas, Mechanistic studies of capillary processes in porous media. Part 1: Probabilistic description of porous media, *J. Chem. Soc.* Faraday Trans. 1, 1989, 85(8), 2071-2078.

[8] V. Mayagoitia, , F. Rojas and I. Kornhauser, Twofold description of porous media and surface structures: a unified approach to understand heterogeneity effects in adsorption and catalysis, Langmuir 1993, 9(10), 2748-2754.

[9] V. Mayagoitia, , I. Kornhauser, Principles and Applications of Pore Structural Characterization; J. M. Haynes, P. Rossi-Doria, Eds.; Arrowsmith: Bristol, 1985; p. 15.

[10] M. J. Cruz, V. Mayagoitia, and F. Rojas, Mechanistic studies of capillary processes in porous media. Part 2: Construction of porous networks by Monte-Carlo methods, *J. Chem. Soc.* Faraday Trans. 1, 1989, 85(8), 2079-2086.

[11] V. Mayagoitia, F. Rojas and I. Kornhauser, Domain complexions in capillary condensation. Part 1: The ascending boundary curve, *J. Chem. Soc.* Faraday Trans. 1, 1988, 84(3), 785-799.

[12] V. Mayagoitia, B. Gilot, F. Rojas and I. Kornhauser, Domain complexions in capillary condensation. Part 2: Descending boundary curve and scanning, *J. Chem. Soc.* Faraday Trans. 1, 1988, 84(3), 801-813.

[13] V. Mayagoitia. In Characterization of Porous Solids II; F. Rodriguez-Reinoso, J. Rouquerol, K. S. Sing and K. K. Unger, Eds.; Elsevier: Amsterdam, 1991; p. 51.

[14] V. Mayagoitia, F. Rojas, I. Kornhauser, G. Zgrablich, and J. L. Riccardo. In Characterization of Porous Solids III; F. Rodriguez-Reinoso, J. Rouquerol, K. S. Sing and K. K. Unger, Eds.; Elsevier: Amsterdam, 1994; p. 141.

[15] V. Mayagoitia and F. Rojas. In Fundamentals of Adsorption III; A. B. Mersman and , S. E. Scholl, Eds.; The Engineering Foundation: New York, 1991; p. 563.

[16] V. Mayagoitia, Fundamentals of the Textural Characterization of Catalysts, *Catal. Lett.* 1, 1993, 22(1-2), 93-105.

[17] V. Mayagoitia, F. Rojas, V. D. Pereyra and G. Zgrablich, Mechanistic study of surface processes on adsorbents: I. Statistical description of adsorptive surfaces, *Surf. Sci.*, 1989, 221(1-2), 394-408.

[18] V. Mayagoitia, F. Rojas, J. L. Riccardo, V. D. Pereyra and G. Zgrablich, Dual site-bond description of heterogeneous surfaces, *Phys. Rev. B.*, 1990, 41(10), 7150-7155.

[19] V. Mayagoitia and I. Kornhauser. In Fundamentals of Adsorption IV; M. Suzuki, Ed.; Kodansha: Tokyo, 1993; p. 421.

[20] V. Mayagoitia, A. Domínguez and F. Rojas, Twofold description of the morphology of colloid aggregates and other complex structures, *J. Non-Cryst. Solids*, 1992, 147-148,183-188.

[21] V. Mayagoitia, F. Rojas and I. Kornhauser, Twofold description of the silica gelling process, *J. Sol-Gel Sci. Technol.*, 1994, 2(1-3), 259-262.

[22] G. B. Kuznetsova, V. Mayagoitia and I. Kornhauser, Twofold Description of the Morphology of Polymers, *Int. J. Polym. Mater.*, 1993, 19(1-2), 19-28.

[23] V. Mayagoitia, F. Rojas, I. Kornhauser, G. Zgrablich, R.J. Faccio, B. Gilot, C. Guiglion, Refinements of the twofold description of porous media, Langmuir, 1996, 12(1), 211-216.

[24] U. Gil-Cruz, M. A. Balderas-Altamirano, S. Cordero-Sánchez, Textural study of simulated dimorphic porous substrates, *Colloids Surf. A*, 2010, 357(1-3), 8492UTF201390.

[25] M. Quintana, M. Pasinetti, A.J. Ramirez-Pastor and G. Zgrablich, Adsorption thermodynamics of monomers on diluted-bonds triangular lattices, *Surface Sci.*, 2006, 600(1), 33-42.

[26] R. Villalobos, A. Ganem, S. Cordero, A. M. Vidales, A. Domínguez. Effect of the drug-excipient ratio in matrix type controlled release systems: Computer simulation study. *Drug Dev. Ind. Pharm.*, 31 (6), 535-543, 2005.

[27] O. Cruz, R. Hidalgo, S. Alas, S. Cordero, L. Meraz, R. López and A. Domínguez, Is the Alexander-Orbach Conjecture Suitable for Treating Diffusion in Correlated Percolation Clusters?, *Adsorpt. Sci. Techno.*, 2011, 29(7), 663-676.

[28] J.L. Riccardo, W.A. Steele, A.J. Ramírez-Cuesta, G. Zgrablich, Pure Monte Carlo Simulation of Model Heterogeneous Substrates: From Random Surfaces to Many-Site Correlations, Langmuir, 1997, 13(5), 1064-1072.

[29] G. Román-Alonso, F. Rojas-González, M. Aguilar-Cornejo, S. Cordero-Sánchez, M.A. Castro-García, In-silico simulation of porous media: Conception and development of a greedy algorithm, *Micropor. Mesopor. Mat.*, 2011, 137(1-3) 18-31.

[30] J. Matadamas-Hernández, G. Román-Alonso, F. Rojas-González, M.A. Castro-García, Azzedine Boukerche, M. Aguilar-Cornejo, and S. Cordero-Sánchez, Parallel Simulation of Pore Networks Using Multicore CPUs, *IEEE T. Comput.*, 2014, 63(6), 1513-1525, 10.1109/TC.2012.197.

[31] N. Metropolis, A. W. Rosenbluth, M. N. Rosenbluth, A. H. Teller, and E. Teller, Equation of State Calculations by Fast Computing Machines *J. Chem. Phys.*, 1953, 21(6), 1087-1092.

[32] D. Stauffer, and A. Aharony, Introduction to Percolation Theory. Taylor and Francis. Second Edition. 2007.

[33] M.D. Kalugin, and A.V. Teplukhin, Parallel Monte Carlo study on caffeine-DNA interaction in aqueous solution, IPDPS, pp.1-8, 2009 IEEE International Symposium on Parallel & Distributed *IP Processing*, 2009.

[34] J.M. Hammersley, and D.C. Handscomb, Monte Carlo Methods, Methuen's Monographs on Statistics and Applied Probability, 1964.

[35] J Martín-Herrero and J Peón-Fernández, Alternative techniques for cluster labelling on percolation theory, *J. Phys. A: Math. Gen.* 2000,33(9) 1827-1840.

ABOUT THE EDITOR

Dr. Mahmood Aliofkhazraei
Assistant Professor
Faculty of Engineering
Department of Materials Engineering
Tarbiat Modares University
P.O. Box 14115-143
maliofkh@yahoo.com

INDEX

A

α-terpineol, 104
adsorbents, 44, 78, 148, 149, 320, 344, 369
adsorption, 344, 368
algorithm, 344, 345, 356, 358, 360, 363, 367, 369
alumina, 58, 62, 74, 154, 189, 317
anisotropy, 357
annealing, 367
automata, 367

B

base, 234, 322
batteries, 286
behaviors, 237
bioactive glasses, 19, 20, 21, 22, 36, 37, 38, 39, 40, 41, 272
bioceramics, 19, 20, 21, 28, 38
biomass conversion, 67, 68
bonds, 344, 345, 346, 347, 348, 349, 351, 352, 353, 356, 357, 358, 359, 364, 369
Bronsted acid, 43, 51, 56, 74, 110

C

caffeine, 370
cancer cells, 333
capacitances, 231, 236, 238
capillary, 344, 345, 368
catalysis, 309, 368
catalysts, 43, 44, 51, 52, 53, 54, 58, 60, 61, 62, 67, 68, 69, 70, 71, 72, 73, 74, 75, 77, 78, 79, 80, 81, 85, 96, 97, 102, 104, 148, 149, 150, 153, 159, 163, 164, 165, 180, 181, 185, 187, 188, 189, 190, 194, 196, 198, 199, 200, 202, 203, 205, 207, 208, 209, 286, 288, 289, 291, 295, 303, 304, 318, 320, 323, 324, 325, 326, 327, 329, 330
chemisorption, 344
clinoptilolite, 149, 151, 160, 166, 195, 196, 197, 201
clusters, 290, 292, 344, 352, 354, 358, 359, 363, 364, 365, 366
computing, 343, 345, 356, 360, 367
condensation, 344, 368
configuration, 344, 361
connectivity, 344, 346, 348, 358
construction, v, 344, 345, 348, 353, 356, 360, 364, 367
consumption, 345
contour, 357, 359, 360, 361
correlation, 344, 346, 347, 348, 349, 351, 352, 353, 354
correlation coefficient, 353
correlation function, 346, 348, 349, 351, 352
correlations, 344, 346, 348
crystals, 287
cysteine, 271

D

data distribution, 363
dealumination, 75, 147, 151, 152, 153, 154, 155, 156, 157, 158, 159, 160, 161, 162, 163, 164, 165, 166, 167, 168, 169, 175, 176, 177, 179, 191, 192, 193, 196, 197, 198, 199, 200, 201, 203, 210
dehydrogenation, 99, 303, 309, 330
deposition, 291, 297, 331
desorption, 344
diffraction, 271, 273
diffusion, 344, 345
diodes, 331
disordered systems, 344
dispersion, 368
displacement, 343, 344, 345

distributed memory, 360
distribution, v, 343, 344, 346, 349, 351, 356, 358, 363
DNA, 370
drainage, 345
drug controlled release, 1, 3, 273
drug delivery, 271
drug dissolution, 1, 2, 4, 5, 7, 8, 11, 12, 14, 15
drug loading capacity, 9
drug release, 344
drying, 343

E

energy, 294
energy conversion, 116
energy storage, 213, 215, 217, 238, 302, 303
epoxidation, 57, 62, 104, 209, 297, 303, 304, 309, 323, 324, 325, 326
equilibrium, 344
etching, 287
evaporation, 344
evolution, 349
execution, 356, 357, 358, 363, 367

F

fibers, 333
films, 306, 317
fluid, 271, 272, 273, 343
fossil fuel sources, 67
fossil fuels, 294
fractal dimension, 354, 367

G

gel, 273, 286, 301, 327, 333, 344
geometry, 358
growth, 357
growth factor, 31

H

heterogeneity, 368
homogeneity, 357
hybrid, 356, 357, 360, 367
hydro dehalogenation, 44

I

imbibition, 343, 344
initial state, 344
interference, 348
ion adsorption, 332
ionic surfactants, 23
irradiation, 323, 324, 328, 333
isotherms, 277

L

lattices, 354, 369
laws, 346
lead, 347
light, 365
limonene, 71, 72, 75, 76
Liquid Crystal Templating (LCT), 43, 44, 46, 51
lithium ion battery, 213, 214, 215, 218
local order, 363

M

macroporous materials, v, 343
Markov chain, 344
materials, 237, 273, 280, 292, 303, 309, 316, 317, 318, 319, 321, 323, 326, 329, 330, 331, 334, 336, 367
matrix, 369
MBG, 19, 24, 25, 26, 27, 28, 29, 30, 32, 34, 35, 39
media, 343, 344, 345, 352, 365, 366, 368
memory, 345, 362, 363
mercury, 344
mesopores, 1, 3, 5, 8, 9, 10, 11, 13, 27, 32, 34, 41, 44, 48, 53, 54, 58, 69, 77, 89, 92, 95, 108, 114, 147, 150, 158, 159, 160, 165, 169, 171, 172, 173, 178, 181, 184, 185, 186, 187, 188, 189, 192, 199, 200, 202, 205, 206, 208, 210, 218, 221, 224, 226, 227, 232, 271, 277, 290, 302, 305, 309, 316, 319, 331, 336
mesoporous molecular sieve, 16, 38, 60, 75, 104
mesoporous silica film, 308
mesoporous silicas, 73, 104, 322, 328
mesoporous silicates, 1, 2, 3, 4, 7, 9, 10, 12, 15, 16, 309
metal oxides, 285, 286
metals, 309
methane, 188, 326
methodology, 344
Mexico, 343
micelles, 14, 18, 23, 25, 26, 31, 44, 46, 108, 112, 186, 286

microscopy, 246, 247, 261, 262
models, 367
monomers, 369
Monte Carlo method, 361, 367
montmorillonite, 272
morphology, 344, 369
MPI, 360

N

nanoparticles, 318, 322, 328
network elements, 344
NMR, v, 343
nodes, 356, 358, 360
normal distribution, 349

O

OMC, 301
ores, 277
organic compounds, 294
overlap, 346, 348, 359, 360, 364, 367

P

parallel, v, 343, 345, 360, 361, 362, 363, 367
parallel algorithm, 361
parallel simulation, 345
PCH, 190
percolation, v, 345, 352, 353, 354, 367, 370
percolation theory, v, 345, 370
pharmaceutical, 344
photoactivity, 87, 97, 100, 106
photocatalyst, 86, 87, 92, 93, 96, 98, 100, 101, 103, 104, 107, 115
photocatalytic applications, 85, 87
photovoltaics, 85
PMOs, 67
policy, 359
polyelectrolytes, 185, 317
polymers, 344
pore size, v, 1, 3, 9, 11, 13, 31, 37, 43, 48, 49, 50, 62, 67, 69, 71, 75, 77, 81, 87, 92, 113, 116, 149, 171, 175, 176, 184, 185, 189, 194, 200, 218, 220, 221, 226, 229, 238, 243, 244, 254, 255, 277, 282, 286, 296, 297, 298, 299, 303, 306, 308, 309, 316, 325, 326, 328, 344, 361, 363, 364, 365
porosity, 368
porous materials, 343, 344
porous media, 344, 345, 346, 368, 369
porous space, 344
probability, 346, 347, 348, 349, 350, 351, 352
probability density function, 346
programming, 360
project, 367
propene, 156, 179
proportionality, 351, 352

R

radionuclides, 149
radius, 346
random assignment, 344
reaction rate, 344
reactions, 273
reparation, 204
response time, 363
restrictions, v, 344, 345, 347, 348, 351, 352, 353, 358, 364, 365, 367
routes, 356

S

SANS, v, 343
SAXS, v, 343
seeding, 345, 356, 357, 359, 363
segregation, 348
sensors, 286
silica, 369
silicon layers, 241, 242, 268
simulation, v, 343, 345, 347, 357, 361, 363, 367, 369
sol–gel process, 108
solution, 370
sorption, v, 343
spectroscopy, 246, 258, 263
stability, 286, 289, 307, 323, 324, 327
standard deviation, 348
states, 344, 346
structure, v, 285, 287, 292, 296, 343, 344, 345, 356, 360
subnetworks, 360
substrates, v, 343, 344, 367, 369
supercapacitors, 213, 214, 217, 237, 238, 286, 302
surface area, 238, 290, 299, 316
surface structure, 368
surfactant, 193
synthesis, 273, 304, 305, 309

T

techniques, v, 343, 345, 346, 370
temperature, 242, 271, 273, 274, 293, 297, 303, 309, 326

Index

textural properties, 21, 24, 25, 29, 31, 32, 33, 68, 79, 167, 175, 177, 223, 224, 238, 272, 282, 367
texture, 344, 354
topology, 347, 348, 357, 360
torus, 360, 361, 362, 363
transmission, 267
treatment, 330, 345
tungstosilicic, 73

V

vapor, 291, 297, 331

W

water, 274
water treatment, 93

Z

zeolite, 68, 72, 147, 148, 149, 150, 151, 152, 153, 154, 156, 157, 158, 159, 160, 161, 162, 163, 164, 165, 166, 167, 168, 169, 170, 171, 172, 173, 174, 175, 176, 177, 178, 179, 180, 181, 182, 183, 184, 185, 186, 187, 188, 189, 190, 191, 192, 193, 194, 195, 196, 197, 198, 199, 200, 201, 202, 203, 204, 205, 206, 207, 208, 209, 210, 211, 323